The Local Group is a small cluster of galaxies of which 35 members are currently known, including the Milky Way. It is believed that at least half of all galaxies in the Universe belong to similar groups. Galaxies of the Local Group can be used as "stepping stones" to determine the distance to more remote galaxies, and thus they help to measure the size and age of the Universe. Studying stars of differing ages in different members of the Local Group allows us to see how galaxies evolve over timescales in excess of 10 billion years. The oldest stars in the Local Group galaxies also provide critical information on the physical conditions of the early Universe. The Local Group thus provides many valuable clues to understanding the rest of the Universe.

This authoritative volume provides a comprehensive and up-to-date synthesis of what is currently known about the Local Group of galaxies. It includes a summary of our knowledge of each of the individual member galaxies, as well as those galaxies previously regarded as possible members. After examining each galaxy in detail, the book goes on to examine the mass, stability, and evolution of the Local Group as a whole. The book includes many important previously unpublished results and conclusions.

With characteristic clarity, Professor van den Bergh provides in this book a masterful summary of all that is known about the galaxies of the Local Group and their evolution, and he expertly places this knowledge in the wider context of ongoing studies of galaxy formation and evolution, the cosmic distance scale, and the conditions in the early Universe.

T0206337

THE GALAXIES OF THE LOCAL GROUP

Cambridge astrophysics series

Series editors

Andrew King, Douglas Lin, Stephen Maran, Jim Pringle, and Martin Ward

THE GALAXIES OF THE LOCAL GROUP

SIDNEY VAN DEN BERGH

Dominion Astrophysical Observatory,
National Research Council of Canada

CAMBRIDGE UNIVERSITY PRESS
Cambridge, New York, Melbourne, Madrid, Cape Town, Singapore, São Paulo

Cambridge University Press
The Edinburgh Building, Cambridge CB2 8RU, UK

Published in the United States of America by Cambridge University Press, New York

www.cambridge.org
Information on this title: www.cambridge.org/9780521651813

First published 2000
This digitally printed version 2007

A catalogue record for this publication is available from the British Library

Library of Congress Cataloguing in Publication data
Van den Bergh, Sidney, 1929–
The Galaxies of the Local Group / Sidney van den Bergh.
    p.   cm. – (Cambridge astrophysics series : 35)
ISBN 0-521-65181-6
1. Local Group (Astronomy)   I. Title.   II. Series.
QB858.8.L63V36   1999
523.1′12 – dc21   99-31357
                                                    CIP

ISBN 978-0-521-65181-3 hardback
ISBN 978-0-521-03743-3 paperback

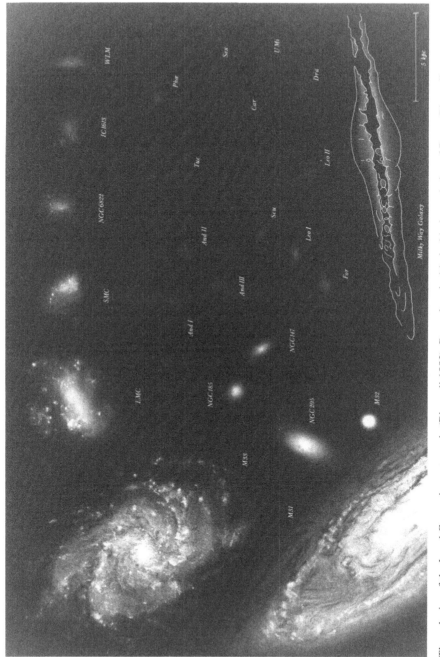

The galaxies of the Local Group drawn to scale (Binggeli 1993). Reproduced with the kind permission of Bruno Binggeli.

Contents

Preface

In April of 1968 I gave a series of lectures on the structure, evolution, and stellar content of nearby galaxies at the University of California in Berkeley. An outline of these talks was printed as a slender volume entitled "The Galaxies of the Local Group" (van den Bergh 1968a). Since the publication of this booklet the number of known members of the Local Group has doubled. Furthermore both the quantity, and the quality, of the data that are available on the previously known Local Group members have increased enormously.

Particularly exciting developments since 1968 have been (1) the discovery of the Sagittarius dwarf, which is the nearest external galaxy, (2) the discovery of six dwarf spheroidal companions to the Andromeda nebula, (3) the application of CCD detectors to studies of stellar populations in various Local Group systems, and (4) deep high-resolution observations of various objects in the Local Group with the *Hubble Space Telescope*. With the presently available enlarged sample, and the improved quality of data on individual objects, we are now in a much better position to start exploring the evolutionary history of the Local Group and its constituent galaxies. Finally (5) it has become clear during the past quarter century that the masses of dark matter halos are typically an order of magnitude greater than the masses of the baryonic galaxies that are embedded within them.

The distance scale within the Local Group remains somewhat controversial, even though the adopted distances to most individual galaxies have not changed by more than $\sim$10% over the past three decades. In the present volume the highest weight has been given to distance determinations based on observations of Cepheids and RR Lyrae variables. However, recent observations with the *HIPPARCOS* satellite have cast some doubt on the most widely accepted luminosity calibrations of classical Cepheids and RR Lyrae stars.

Literature citations in this book are complete for papers that arrived at the Dominion Astrophysical Observatory before February 1, 1999. All coordinates in this volume refer to equinox J2000.

The astronomical literature on the galaxies of the Local Group is so vast that it is quite impossible to do justice to all of it. The present volume has therefore been written in the spirit of Winston Churchill's *History of the English-Speaking Peoples*, of which Clement Attlee said that it should have been called "things in history that interested me."

It is a particular pleasure to acknowledge the help and encouragement by friends and colleagues too numerous to be thanked individually. I should also like to express my gratitude to Donald Lynden-Bell, the Institute of Astronomy, and Clare College, Cambridge, where the first outline of this book was written many years ago, and to the Dominion Astrophysical Observatory of the National Research Council of Canada, where it was finished. Thanks are also due to Janet Currie for typing many drafts of this manuscript, to David Duncan for drawing the majority of the figures, to Eric LeBlanc for helping to find numerous obscure references, and to text editor Ellen Tirpak for many helpful suggestions. Thanks are also due to Stéphane Courteau and Chris Pritchet for help with the redetermination of the solar apex, and of the slope of the Local Group luminosity function, respectively. I am also indebted to the Observatories of the Carnegie Institution of Washington and to the Cerro Tololo Inter-American Observatory, where I obtained most of the plates that are reproduced in this volume. I am deeply grateful to Eva Grebel, Jim Hesser, and Mario Mateo for their careful reading of the manuscript. Finally, I thank my wife Paulette for her support, patience, and understanding.

1

Introduction

The galaxies of the Local Group are our closest neighbors in the Universe. Because most of them are nearer than one megaparsec (Mpc) they are easily resolved into stars. This enables one to study these objects in much more detail than is possible for more distant galaxies. The members of the Local Group are therefore the laboratories in which individual objects, such as star clusters, planetary nebulae, supernova remnants, etc., can be studied in detail. Furthermore, important empirical laws, such as the Cepheid period–luminosity relation (Leavitt 1907), the maximum magnitude versus rate-of-decline relation for novae, and the luminosity distribution of globular clusters, can be calibrated in Local Group galaxies. For an earlier review of the properties of some of these galaxies the reader is referred to the proceedings of the symposium on *The Local Group: Comparative and Global Properties* (Layden, Smith & Storm 1994). Reviews of more recent work are provided in *New Views of the Magellanic Clouds* = IAU Symposium No. 190 (Chu, Hesser & Suntzeff 1999), in *The Stellar Content of the Local Group of Galaxies* = IAU Symposium No. 192 (Whitelock & Cannon 1999), and in *Stellar Astrophysics for the Local Group* (Aparicio, Herrero & Sánchez 1998).

1.1   Is the Local Group typical?

Inspection of the *Palomar Sky Survey* (Minkowski & Abell 1963) shows (van den Bergh 1962) that only a small fraction of all galaxies are isolated objects or members of rich clusters. The majority of galaxies in nearby regions of the Universe are seen to be located in small groups and clusters resembling the Local Group. Our Milky Way system is therefore situated in a rather typical region of space. Hubble (1936, pp. 128–129) already pointed out that study of individual galaxies in the Local Group was important because they were (1) "the nearest and most accessible examples of their particular types" and (2) because they provided "a sample collection of nebulae, from which criteria can be derived for further exploration." In Chapter 2 it will be seen that the Local Group contains examples of most major types of galaxies, except giant ellipticals. In fact Shapley (1943) commented that if Fate or Chance had placed a giant elliptical (or spiral) at the distance of the Magellanic Clouds then "The astronomy of galaxies would probably have been ahead by a generation, perhaps even fifty years." In fact, the presence of a giant elliptical at a distance of only 50 kpc would have disrupted the Milky Way galaxy, so that human beings (and hence astronomers) probably would not have come into existence!

1

The Local Group contains an early-type spiral (M31) and a late-type spiral (M33), a luminous irregular (LMC) and a dim irregular (Leo A), a spheroidal[1] (NGC 205) and a dwarf spheroidal (Sculptor), and a single dwarf elliptical (M32). The Local Group does not contain a giant elliptical or cD galaxy, or a blue compact galaxy,[2] although IC 10 is, perhaps, a presently rather inactive example of this class. Finally, the Local Group is not known to contain an oversized dwarf spheroidal (dSph) such as F8D1 in the M81 Group (Caldwell et al. 1998). However, such a large low surface-brightness object would be difficult to find. If F8D1, which has a central visual surface brightness of only 25.4 mag arcsec^{-1}, were placed at the distance of the LMC it would have a scale-length of over a degree. Clearly such objects would be very difficult to discover in the Local Group.

1.2   Discovery of the Local Group

The term "The Local Group" was introduced by Edwin Hubble (1936, pp. 124–149) in his seminal book *The Realm of the Nebulae*. It refers to the small cluster of stellar systems surrounding our own galaxy. In addition to the Milky Way system (= the Galaxy[3]), Hubble assigned its two satellites, the Large Magellanic Cloud (LMC) and the Small Magellanic Cloud (SMC), the Andromeda galaxy (M31 = NGC 224) and its companions M32 (= NGC 221) and NGC 205, the Triangulum galaxy (M33 = NGC 598), NGC 6822, and IC 1613 to the Local Group. He also mentioned the heavily reddened objects NGC 6946, IC 342, and IC 10 as *possible* Local Group members. However, we now know that only IC 10 is a true member of the Local Group (Wilson et al. 1996, Sakai, Madore & Freedman 1998) and that NGC 6946 and IC 342 are more distant objects. Krismer, Tully & Gioia (1995) quote a distance $D = 3.6$ Mpc for IC 342, and Sharina, Karachentsev & Tikhonov (1997) give $5.5 \lesssim D(\text{Mpc}) \lesssim 7.0$ for NGC 6946. So the total number of true Local Group galaxies known to Hubble was nine. More recently Baade (1963, p. 24) listed 18 probable Local Group members, of which we now know that IC 5152 is too distant to be gravitationally bound to the Local Group (Zijlstra & Minniti 1999). The increase from nine to 17 Local Group members was due to the addition of the distant M31 satellites NGC 147 and NGC 185 and to the discovery of the intrinsically faint dwarf spheroidal galaxies Sculptor, Fornax, Leo I, Leo II, Draco, and Ursa Minor. Finally, the recent discovery of additional faint dwarf irregular (dIr) and dSph galaxies brings the total number of presently known probable members of the Local Group up to 35. Figure 1.1 shows that the number of known members of the Local Group has been increasing at the rate of about four per decade.

---

[1] Since NGC 205 is quite luminous ($M_V = -16.4$) it seems inappropriate to call it a dwarf spheroidal. In this book objects such as NGC 147, NGC 185, and NGC 205 will therefore be referred to as spheroidals, whereas objects fainter than $M_V = -15$ (such as Sculptor and Fornax) will be referred to as dwarf spheroidals. It should, however, be emphasized that the spheroidals and dwarf spheroidals are bright and faint members of the same physical class of galaxies.

[2] Blue compact galaxies were formerly known as intergalactic H II regions.

[3] Throughout this book "Galaxy" and "Galactic" refer to the Milky Way system, while galaxy and galactic refer to other extragalactic systems.

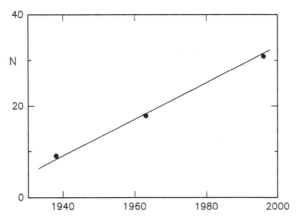

Fig. 1.1. Number of known Local Group members as a function of time. Over the past half century the number of known Local Group members is seen to have increased by about four per decade. Last point is for 1997.

2

Local Group membership

2.1   Introduction

Critical discussions of membership of individual galaxies in the Local Group have recently been given by van den Bergh (1994a,b, 1999), by Grebel (1997), and by Mateo (1998). A galaxy distance $\lesssim 1.5$ Mpc was used as a preliminary selection criterion for Local Group membership. A detailed discussion of additional selection criteria will be given in Chapter 16 of the present volume, which deals with the membership of individual galaxies located near the outer fringes of the Local Group. Table 2.1 presents a summary of the observational parameters for the 35 most probable members of the Local Group. This table lists the name and (where appropriate) one alias for each Local Group member, its DDO classification (mostly from van den Bergh 1966a), the J2000 coordinates of each galaxy, its observed heliocentric velocity, its integrated magnitude in the V band, and its reddening $E(B - V)$. The derived parameters for each of these galaxies, based on the discussion in the present volume, will be given in Table 19.1.

2.2   Incompleteness of the sample

The fact that IC 1613 ($M_V = -14.9$) has been known (Dyer 1895) for about a century indicates that our Local Group sample (at least outside the zone of avoidance at low Galactic latitudes) is almost certainly complete for objects brighter than $M_V = -15$. The fact that Irwin (1994) discovered only a single new Local Group member during a survey of $\sim 20,000$ square degrees at high Galactic latitudes suggests that the search for Galactic satellites at high latitudes is probably complete to at least $M_V = -10$. However, the fact that all known Local Group members fainter than $M_V = -9$ are satellites of the Galaxy suggests that the sample of more distant faint Local Group galaxies is still incomplete. Two lines of evidence support this conclusion: (1) A number of faint new members of the M31 subgroup of the Local Group were discovered by Armandroff, Davies & Jacoby (1998) and by Karachentsev & Karachentseva (1999), and (2) Bellazini, Fusi Pecci & Ferraro (1996) have noted that the dwarf spheroidal companions to the Milky Way system that have the *lowest* surface brightnesses are situated closest to the Galaxy. This is surprising because one would have expected only those satellites with the *highest* densities to survive tidal disruption close to the Galaxy.[1] This strongly suggests that

---

[1] Alternatively, low surface brightness might be the signature of an advanced state of disruption.

Table 2.1. *Observational data on Local Group members*

Name	Alias	Type	α (J2000) δ	V_r (kms^{-1})	V	E(B − V)
WLM	DDO 221	Ir IV–V	00^h 01^m 57$.^s$8 −15°27′51″	−120	10.42	0.02
IC 10	. . .	Ir IV:	00 20 24 +59 17 30	−344	10.4	≈0.85
NGC 147	. . .	Sph	00 33 11.6 +48 30 28	−193	9.52	0.17
And III	. . .	dSph	00 35 17 +36 30 30	. . .	14.21	0.05
NGC 185	. . .	Sph	00 38 58.0 +48 20 18	−202	9.13	0.19
NGC 205	. . .	Sph	00 40 22.5 +41 41 11	−244	8.06	0.04
M32	N 221	E2	00 42 41.9 +40 51 55	−205	8.06	0.06
M31	N 224	Sb I–II	00 42 44.2 +41 16 09	−301	3.38	0.06
And I	. . .	dSph	00 45 43 +38 00 24	. . .	12.75	0.04
SMC	. . .	Ir IV/IV–V	00 52 36 −72 48 00	+148	1.97	0.06
Sculptor	. . .	dSph	01 00 04.3 −33 42 51	+110	8.8	0.00
Pisces	LGS 3	dIr/dSph	01 03 56.5 +21 53 41	−286	14.26	≃0.03
IC 1613	. . .	Ir V	01 04 47.3 +02 08 14	−232	9.09	0.03
And V	. . .	dSph	01 10 17.1 +47 37 41	. . .	15	0.16
And II	. . .	dSph	01 16 27 +33 25 42	. . .	12.71	0.08
M33	N 598	Sc III	01 33 50.9 +30 29 37	−181	5.85	0.07
Phoenix	. . .	dIr/dSph	01 51 03.3 −44 27 11	. . .	. . .	0.02
Fornax	. . .	dSph	02 39 53.1 −34 30 16	+53	7.3	0.03
LMC	. . .	Ir III–IV	05 19 36 −69 27 06	+275	0.4	0.13
Carina	. . .	dSph	06 41 36.7 −50 57 58	+223	10.6	0.05
Leo A	DDO 69	Ir V	09 59 23.0 +30 44 44	+24	12.69	0.02
Leo I	Regulus	dSph	10 08 26.7 +12 18 29	+287	10.2	0.02
Sextans	. . .	dSph	10 13 02.9 −01 36 52	+226	10.3	0.04
Leo II	DDO 93	dSph	11 13 27.4 +22 09 40	+76	11.62	0.03
Ursa Min.	DDO 199	dSph	15 08 49.2 +67 06 38	−247	10.6	0.03
Draco	DDO 208	dSph	17 20 18.6 +57 55 06	−293	11.0	0.03
Milky Way	. . .	S(B)bc I–II:	17 45 39.9 −29 00 28	+16	. . .	. . .
Sagittarius	. . .	dSph(t)	18 55 04.3 −30 28 42	+142	. . .	0.15
SagDIG[a]	. . .	Ir V	19 29 58.9 −17 40 41	−79	14.2	0.07
N 6822	. . .	Ir IV–V	19 44 56.0 −14 48 06	−56	8.52	0.25
Aquarius[a]	DDO 210	V	20 46 53 −12 50 58	−131	13.88	0.04
Tucana[a]	. . .	dSph	22 41 48.9 −64 25 21	. . .	15.15	0.00
Cassiopeia	And VII	dSph	23 26 31 +50 41 31	. . .	15.2	0.17
Pegasus	DDO 216	Ir V	23 28 34 +14 44 48	−182	12.59	0.15
Pegasus II	And VI	dSph	23 51 39.0 +24 35 42	. . .	14.1	0.04

[a]Local Group membership needs to be confirmed.

observational selection effects may have biased the presently known sample of faint dwarf galaxies against distant objects of low surface brightness. The suspicion that the effect noted by Bellazini et al. is due to observational selection effects is strengthened by the observation (Grebel 1999) that the correlation between central surface brightness and distance from M31 is much less pronounced for the Andromeda dwarfs than it appears to be for the dwarf spheroidal companions to the Milky Way system.

M31 has fewer known satellites than the Galaxy, even though the Andromeda galaxy is more luminous than the Milky Way system.[2] This strongly suggests that more faint members of the Andromeda subgroup of the Local Group remain to be discovered. A deep CCD search for objects within 100 kpc (7°5) of M31 would probably prove to be particularly rewarding. On the basis of the work by Einasto et al. (1974) one might expect such a search of the inner corona of M31 to turn up mainly dSph galaxies, whereas dIr/dSph galaxies, which have managed to retain some hydrogen gas, might be more common at larger distances from M31. A recent deep search for H I galaxies with radial velocities <7,400 km s^{-1} by Zwaan et al. (1997) shows no evidence for large numbers of hydrogen-rich low-mass galaxies. A single object with a velocity of $\sim +350$ km s^{-1}, relative to the centroid of the Local Group, was found by Burton & Braun (1998). A more detailed discussion of this point will be given in Chapter 17, which deals with interstellar matter in the Local Group.

2.3   Substructure within the Local Group

It has been known for many years that the Local Group exhibits considerable substructure (see Table 2.2). The two main subgroups of the Local Group are centered on M31 and on the Galaxy (Ambartsumian 1962; van den Bergh 1968a; Gurzadyan, Kocharyan & Petrosian 1993). In the Andromeda subgroup M31, M32, and NGC 205 form a tight core. Within the Galactic subgroup the Large and Small Magellanic Clouds constitute a close interacting pair. The low-luminosity members of the Local Group, which are not listed in Table 2.2, are also subclustered. Fornax, Sculptor, Sagittarius, Sextans, Leo II, Ursa Minor, and Draco (and perhaps Leo I) are associated with the Galaxy. The positions of Andromeda I and Andromeda III on the sky (see Figure 2.1) strongly suggest that these

Table 2.2. *Substructure in the Local Group^a*

M31	
M32	
NGC 205	Andromeda subgroup
NGC 147	
NGC 185	
M33	
IC 10	
NGC 6822	
IC 1613	
WLM	
LMC	
SMC	Galaxy subgroup
Galaxy	

^{a}Only galaxies with $M_V < -14.0$ listed.

[2] It should, however, be emphasized that the M31 companions NGC 147, NGC 185, and NGC 205 are much more luminous than any of the dwarf spheroidal companions to the Galaxy.

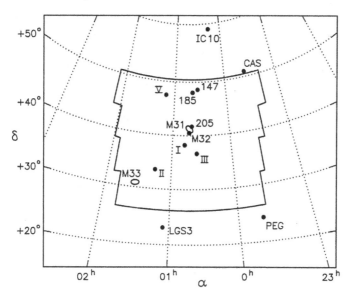

Fig. 2.1. The Andromeda subgroup of the Local Group.

objects are satellites of M31. Even though Andromeda II is located closer to M33 ($d = 11°$) than to M31 ($d = 20°$), it might, in fact, be gravitationally bound to the more massive Andromeda galaxy, rather than to the less massive Triangulum galaxy. Alternatively And II might be regarded as a free-floating member of the Andromeda subgroup of the Local Group. The Pisces dwarf (= LGS 3) also appears to be a member of the Andromeda subgroup of the Local Group, as are the recently discovered dwarf spheroidals And V, And VI, and And VII. IC 10 may also be an outlying member of the Andromeda subgroup.

Among the brighter Local Group members NGC 6822, IC 1613, and the Wolf-Lundmark-Melotte (WLM) system seem to be free-floating objects. The fainter Phoenix (dSph/dIr), DDO 210 (dIr), Tucana (dSph), and Pegasus (dIr) systems also appear to be unattached giant Local Group members. Available evidence does not appear to favor Ambartsumian's (1962) suggestion that dwarf galaxies might fill the Local Group uniformly.

From its observed radial velocity dispersion, van den Bergh (1981a) derived a mass of $(7.5 \pm 3.9) \times 10^{11}\ M_\odot$ for the Andromeda subgroup of the Local Group. This value is comparable to the mass of $(4.9 \pm 1.1) \times 10^{11}\ M_\odot$ that Kochanek (1996) obtained for the volume within 50 kpc of the Galactic center but perhaps smaller than the $(4.6–12.5) \times 10^{11}\ M_\odot$ mass that Zaritsky et al. (1989) and Zaritsky (1998) found from distant satellites of the Milky Way system. On the basis of the $7.5 \times 10^{11}\ M_\odot$ and $4.9 \times 10^{11}\ M_\odot$ masses of the Andromeda and Milky Way subgroups, respectively, it will be assumed that the center of mass of the Local Group is situated in the direction of the Andromeda galaxy, but that it is closer to M31 than to the Galaxy.

2.4   Summary

The number of known Local Group galaxies, which is presently 35, continues to increase at the rate of about four per decade. All three of the most luminous Local

Group members are spiral galaxies. However, the 22 faintest known members of the Local Group are all of types dSph, dIr, or dSph/dIr. The luminosity function of the Local Group contains fewer faint dwarfs than do the luminosity functions of many rich clusters of galaxies. This difference is too large to be entirely accounted for by observational bias against the discovery of very low surface-brightness Local Group members. Most of the galaxies in the Local Group belong to subclusterings that are centered on M31 and the Galaxy. Even though the Andromeda galaxy is more luminous than the Milky Way system, its known complement of satellites is smaller than that of the Galaxy. It might prove rewarding to search for additional faint companions to M31 with modern detectors.

3

The Andromeda galaxy (M31)

3.1 Introduction

The great spiral in Andromeda is the most luminous member of the Local Group. The first known record of this object is by al-Sufi (903–986). M31 was first viewed through a telescope by Marius, who described it as looking "like a candle seen through a horn," in 1612. The spiral nature of the Andromeda galaxy was first clearly shown in photographs obtained by Roberts (1887) with a 0.5-m reflector. Ritchey (1917) referred to these arms as "great streams of nebulous stars." It is listed as object number 31 in Messier's catalog of nebulae. An atlas of finding charts for clusters, associations, variables, etc. in M31 has been published by Hodge (1981). Van den Bergh (1991b) has written a long review paper on this object. The monograph *The Andromeda Galaxy* by Hodge (1992) provides a detailed historical discussion of research on this object and an annotated bibliography for the period 1885–1950. A monograph entitled *Tumannost Andromedy* (*The Andromeda Nebula*) has been published by Sharov (1982). On the sky M31 covers an area of $92' \times 197'$ (Holmberg 1958). This large angular size makes the Andromeda galaxy particularly suitable for detailed studies of its structure and stellar population content. De Vaucouleurs (1959b) has shown that the distribution of surface brightness I in spiral galaxies can often be decomposed into a spheroidal component in which $I \sim \exp(-R^{1/4})$ and an exponential disk in which $I \sim \exp(-\alpha R)$, where R is radial distance and α is a constant. From photoelectric UBV surface photometry de Vaucouleurs (1958) concluded that $\sim 70\%$ of the V (yellow) light of M31 is emitted by an exponential disk that has an inclination[1] $i = 77°$ and that most of the remaining $\sim 30\%$ of the light is radiated by the central bulge that exhibits an $R^{1/4}$ radial profile. Additional population components are a (double) semistellar nucleus and an extended halo. The bulge of M31 has been resolved into red stars of "Population II" by Baade (1944), whereas the light of the Disk is dominated by bright blue main sequence stars and red supergiants of Population I.

The first (unsuccessful) attempts to obtain a spectrum of the Andromeda galaxy were made by Sir William and Lady Huggins in 1885 (Rubin 1997). The true nature of M31 was established by Scheiner (1899), who obtained an 7.5^h spectroscopic exposure of the central region of the Andromeda galaxy. From a comparison of this spectrum with one obtained of the Sun, with the same instrument, he concluded that "*die bisherige Vermuthung, dass die Spiralnebel Sternhaufen seien, zur Sicherheit erhoben ist*" (the

---

[1] Inclination is defined as the angle between the fundamental plane of a galaxy and the plane of the sky.

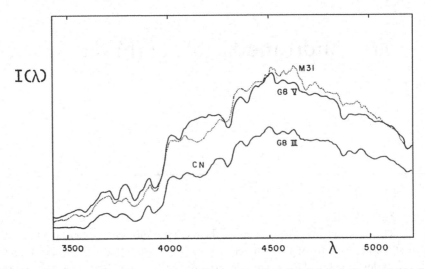

Fig. 3.1. Comparison of photoelectric spectrum scan (van den Bergh & Henry 1962) of the central region of M31 with those of a G8 III giant and a G8 V dwarf. Note the strong cyanogen absorption feature in the M31 and giant spectra, which is not present in the corresponding dwarf spectrum. These observations show that the light of the central region of M31 is dominated by strong-lined "cyanogen giant" stars (i.e., not by metal-poor stars of Baade's Population II).

previous suspicion that spiral nebulae are clusters of stars has now become a certainty). Almost two decades later Pease (1918) obtained a 79-hour exposure of the nuclear region of M31 with the slit aligned along the major axis of the galaxy. This spectrum exhibited inclined spectral lines, whereas the spectral lines on a second exposure with the slit aligned along the minor axis showed no inclination. These observations convincingly demonstrated that the Andromeda galaxy was a rotating system. It is of interest to note that this discovery was made *before* it was generally accepted that spiral nebulae were "island universes" resembling the Milky Way galaxy. The next great advance in spectroscopic studies of the central region of M31 was made in 1955, when Herbig and Morgan (Morgan & Mayall 1957) showed that the stars in the nuclear region of the Andromeda galaxy were strong-lined "cyanogen giants." This result was subsequently confirmed by the spectrophotometry of van den Bergh & Henry (1962), which is shown in Figure 3.1. The fact that the integrated light of the central region of M31 is dominated by strong-lined stars (see Morgan & Osterbrock 1969 for a review) shows that the central bulge of the Andromeda galaxy is populated by metal-rich stars, not as had previously been supposed, by the metal-poor stars of Baade's (1944) Population II. In fact, inspection of Figure 3.1 suggests that the giant stars in the bulge of M31 have λ 4216 CN-band absorption that is as strong as, or stronger than, that in normal giants in the solar neighborhood.

3.2   Reddening and distance

Using photoelectric photometry on the DDO system, McClure & Racine (1969) derived a reddening $E(B - V) = 0.11 \pm 0.02$ for the Andromeda galaxy. From a comparison between little reddened open clusters in M31 and the Galaxy, Schmidt-Kaler (1967) computed a foreground reddening $E(B - V) = 0.12 \pm 0.04$. Using the standard relation between neutral hydrogen column density and dust absorption, Burstein & Heiles obtained a value $A_B = 0.32$ mag, corresponding to $E(B - V) = 0.08$. Finally, Massey, Armandroff & Conti (1986) found a minimum foreground reddening of $E(B - V) = 0.08$ from color–color diagrams of OB associations in M31. More recently Schlegel, Finkbeiner & Davis (1998) have used 100-μm mapping to study the dust distribution over the entire sky. From these data Schlegel (1998, private communication) estimates that the foreground reddening in the direction of M31 is $E(B - V) = 0.062$, from which $A_V = 0.19$ mag. This value will be adopted in the subsequent discussion. It should, however, be emphasized that M31 occupies such a large area on the sky that point-to-point variations in foreground extinction are to be expected.

The first modern distance estimates for the Andromeda galaxy are by Oepik (1922) and by Lundmark (1923). By comparing novae in M31 with those in the nuclear bulge of the Galaxy in Sagittarius, Lundmark estimated that M31 was 63 times more distant than the Galactic novae. With a distance of ~8 kpc to the nucleus this yields a distance of ~500 kpc for M31. (Lundmark himself adopted an M31 distance of 4.3×10^6 light years, corresponding to 1.3 Mpc.) From what we would now refer to as the Tully–Fisher relation, in conjunction with an assumed mass-to-light ratio of 2.6, Oepik found a distance of 450 kpc for M31. This distance was more accurate than the value of 285 kpc that Hubble (1925a,b) derived from the Cepheids that he discovered in M31. The reason for this is now known to be (Baade 1954) that the luminosities of Cepheid variables had been underestimated by a factor of ~2. Prophetically Hubble (1925a) had written "The greatest uncertainty [in this distance] is probably in the zero point of Shapley's [period–luminosity] curve."

A representative sample of distance determinations to M31 is given in Table 3.1. Some caveats apply to the data listed in this Table. First, the Cepheid period–luminosity relation may be slightly dependent on metallicity (Nevalainen & Roos 1998). Also, the luminosities of red giant clump stars might depend on their metallicity and age (Cole 1998, Girardi et al. 1998). Furthermore, possible systematic problems with the *HIPPARCOS* calibrations of their luminosities have been discussed by Soderblom et al.

Table 3.1. *Recent distance determinations to M31*

Method	$(m - M)_0$	Reference
Cepheids	24.43 ± 0.12[a]	Freedman & Madore (1990)
Cluster red giants	24.47 ± 0.07	Holland (1998)
Red clump	24.47 ± 0.06	Stanek & Garnavich (1998)
Halo red giants	24.23 ± 0.15[b]	Pritchet & van den Bergh (1988)
Giant branch tips	24.62 ± 0.18	Salaris & Cassisi (1998)
Novae	24.27 ± 0.20	Capaccioli et al. (1989)
Field RR Lyrae	24.34 ± 0.15[b]	Pritchet & van den Bergh (1987a)
Carbon stars	24.50 ± 0.18[a]	Richer et al. (1990)
Planetary nebulae	24.26 ± 0.04	Ciardullo et al. (1989)
Globular cluster HB	24.60[b]	Fusi Pecci et al. (1996)
SN 1995A	24.7 ± 0.3	Fesen et al. (1998)
Faber–Jackson	24.12 ± 0.45	Di Nella-Courtois et al. (1998)

[a] LMC distance modulus $(m - M)_0 = 18.5$ assumed.
[b] M_V (RR) $= +0.6$ assumed.

(1998) and by Pinsonneault et al. (1998). Kaluzny et al. (1998) have found 12 eclipsing variables in M31, some of which may be semidetached. They estimate that a purely geometrical distance determination, with an accuracy of better than 5%, may be possible from detailed studies of these stars. Another geometrical distance determination for M31 can be obtained from observations of the remnant of SN 1885 by Fesen et al. (1998). From *Hubble Space Telescope* observations these authors find that the remnant of S Andromedae is circular with a diameter of $0\rlap{.}''70 \pm 0\rlap{.}''05$ and a maximum expansion velocity of $13,100 \pm 15,00$ km s^{-1}. Assuming that the remnant is spherical, and that its motion has not been decelerated, this gives a distance of (870 ± 120) kpc, corresponding to a distance modulus $(m - M)_0 = 24.7 \pm 0.3$. Table 3.1 lists a representative sample of a dozen recent distance determinations to M31. The formal unweighted mean of these values is $(m - M)_0 = 24.42 \pm 0.05$. In the subsequent discussion a distance modulus of $(m - M)_0 = 24.4 \pm 0.1$ will be adopted for the Andromeda galaxy.

3.3   The mass of the Andromeda galaxy

A comprehensive compilation of mass determinations for M31 has been given by Hodge (1992, p. 109). Most of these determinations fall in the range $(1.8–3.7) \times 10^{11}\, M_\odot$. The main reason these masses cover such a large range is that the fraction of the total mass that consists of dark matter increases dramatically with increasing distance from the nucleus of M31. The total H I mass of M31 is found to be $5.8 \times 10^9\, M_\odot$ (van de Hulst, Raimond & van Woerden 1957). Gottesman, Davies & Reddish (1966) estimate the total mass of H II in the Andromeda galaxy to be $\sim 2 \times 10^7\, M_\odot$, which is significantly smaller than the total H II mass in the Milky Way system, which they estimate to be $8.4 \times 10^7\, M_\odot$. The total H$_2$ mass of the Andromeda galaxy is found to be only $\sim 3 \times 10^8\, M_\odot$ (Dame et al. 1993). In other words only about 2% of the baryonic mass of M31 presently remains in the form of hydrogen gas. This shows that most of the gas that was originally present in the M31 protogalaxy has either been used up in star formation or was ejected.

Surprisingly, the radial velocity of M31 is not very well determined. A compilation of high-weight velocity determinations for the Andromeda galaxy has been published by Rubin & O'Dorico (1969). The weighted mean of their data yields $\langle V_r \rangle = -300 \pm 4$ km s^{-1}.

3.4   The nucleus of M31

High-resolution observations of the semistellar nucleus of M31 ($= BD + 40°148$) by Light, Danielson & Schwarzschild (1974) show that this object appears to be a distinct physical entity with $V = 12.6 \pm 0.3$, which corresponds to $M_V = -12.0$, that is embedded in the central bulge of the Andromeda galaxy. De Vaucouleurs & Corwin (1985) give the following astrometric position (equinox 1950) for this nucleus: $\alpha = 00^h 40^m 00\rlap{.}^s 13$, $\delta = +40°59'42\rlap{.}''7$. It is, however, not entirely clear (see below) to which component of the nucleus this value refers.

A puzzling feature of the high-resolution *Stratoscope II* balloon observations by Light et al. (1974) was that they found the nucleus of M31 to be offset relative to its outer parts (i.e., the nucleus was not located at the centroid of the outer isophotes of the nuclear bulge). This mystery was resolved by Lauer et al. (1993). Using the Planetary Camera of the *Hubble Space Telescope*, these authors were able to show that the nucleus of M31 is double. It consists of two components P$_1$ and P$_2$ that are separated by $0\rlap{.}''49$ (1.8 pc). The nuclear component P_2 is coincident with the bulge photocenter to $\sim 0\rlap{.}''05$. The brighter component P_1 is well resolved and separated from the photocenter of the bulge by $\sim 0\rlap{.}''5$. A beautiful yellow-light image of the off-center nucleus of M31, which was obtained with the *Hubble Space Telescope*, is shown in Figure 1 of King, Stanford & Crane (1995). A color image of the two nuclei is reproduced in Lauer et al. (1998). These authors find that P_2 is compact, but not pointlike. It has a half-light radius of ≈ 0.2 pc at 3,000 Å.

The hypothesis that P_1 is the nucleus of a recently captured companion to M31 is implausible because the timescale for merger, via dynamical friction, is so short ($<10^8$ years). Tremaine (1995) has made the attractive suggestion that the offcenter source P_1 is the apoapsis region of a thick eccentric disk composed of stars centered on a black hole and coincident with the lower surface brightness source P_2. On Tremaine's hypothesis P_1 is bright because stars linger in their orbits near apoapsis. The recent discovery (Lauer et al. 1996) that the nucleus of the Virgo galaxy NGC 4486B is also double suggests that eccentric nuclear disks may not be excessively rare. By combining photometric and spectroscopic observations, the mass of the black hole that coincides with P_2 is found to be $\sim 5 \times 10^7 \, M_\odot$ (Dressler & Richstone 1988, Kormendy 1988, Bacon et al. 1994). Tremaine, Ostriker & Spitzer (1975) suggested that this nucleus was built up from globular clusters that had been dragged into the center of the potential well of M31 by dynamical friction. To account for the observed luminosity of the nucleus of M31 ~ 70 globular clusters of average luminosity would have to have been dragged into the center of M31. An argument against the scenario suggested by Tremaine et al. is that the metallicity of the nucleus of M31 is higher (McClure 1969) than that of 97% of the present-day Andromeda globular clusters for which integrated spectra are available (van den Bergh 1969). However, the observation that the compact nucleus of M31 exhibits an ultraviolet excess relative to the inner $\sim 40''$ of the bulge (Sandage, Becklin & Neugebauer 1969) might be accounted for by the hypothesis that *a few* globulars were tidally captured by the nucleus. *Hubble Space Telescope* observations of the central region of M31 by King, Stanford & Crane (1995) show that the fainter optical peak (P_2), which is located at the centroid of the outer bulge isophotes, exhibits an ultraviolet (UV) upturn that is much greater than that of the brighter off-center source P_1 and than that of the main body of the M31 bulge. This observation supports the suggestion that the black hole at the geometrical center of M31 tidally captured *a small number* of globular clusters that contained a significant population of blue horizontal branch stars. Alternatively, the UV excess of the nucleus might be due to collisional stripping of the envelopes of some stars in this very dense region. A detailed discussion of this possibility is given in Lauer et al. (1998). Finally, Brown et al. (1998) have proposed that the observed UV spectrum of the nucleus of M31 might be accounted for if 2% of the stars in this region have progenitors of $\sim 0.5 \, M_\odot$ that have evolved through the extreme horizontal branch phase and post–extreme horizontal branch phase of evolution.

From observations of the strengths of magnesium and iron lines Sil'chenko, Burenkov & Vlasyok (1998) conclude that the nucleus of M31 is physically distinct from the bulge in which it is embedded. Using Hβ observations these authors find that the nuclear population is significantly younger than the bulge. Finally, Sil'chenko et al. have found evidence for a central disk that extends from $20''$ to $60''$ from the nucleus.

Radio observations by Crane, Dickel & Cowan (1992) yield a radio flux density of $28 \pm 5 \, \mu$Jy for the nucleus of M31. This value corresponds to only $\sim 1/30$ of the radio luminosity of Sgr A at the center of the Galaxy. Variability of the nuclear radio source in M31 has been suspected by Crane et al. (1993). Furthermore, X-ray observations using the *ROSAT* observatory by Primini, Forman & Jones (1993) yield a (possibly variable) luminosity of 2.1×10^{37} erg s^{-1} for the nucleus. These observations show that the black hole associated with the nucleus of the Andromeda galaxy must presently be in a very low state of activity. Observations by Sreekumar et al. (1994), with the EGRET gamma-ray telescope, have failed to detect M31.

3.5   The nuclear bulge of M31

The nuclear bulge of M31, which contributes $\sim$30% of the total visual luminosity of the Andromeda galaxy, has integrated colors $B - V = 1.02$ and $U - B = 0.48$ (de Vaucouleurs 1961). With an assumed foreground reddening of $E(B - V) \approx 0.06$ the corresponding intrinsic colors are $(B - V)_o \approx 0.96$ and $(U - B)_o \approx 0.44$. The nuclear bulge of M31 has an effective semi-major axis $a_e = 17\rlap{.}''5$ (3.8 pc)[2] and an apparent flattening $b_e/a_e \simeq 0.6$. The observed rotational velocity of the bulge is sufficient to account for this flattening (Richstone & Shectman 1980). Kent (1983, 1987) found that the outer isophotes of the nuclear bulge of M31 are slightly "boxy." Furthermore, the position angle of the major axes of the disk and bulge differ by $\sim$10°.

According to Lawrie (1983) the velocity dispersion in the bulge (excluding the nucleus) is $\langle\sigma\rangle = 146 \pm 6$ km s^{-1}. Kent (1983) finds that the bulge photometry and rotation curve give a mass-to-light ratio

$$M/L_B \simeq 3(\sigma/160)^2. \qquad (3.1)$$

Ultraviolet observations by Deharveng et al. (1982) limit the star formation rate in the bulge of M31 to $\sim$1 $\times$ 10^{-4} $M_\odot$ yr^{-1}, if a Salpeter luminosity function is assumed. This compares to 3×10^{-2} $M_\odot$ yr^{-1} of gas ejected by planetary nebulae in the bulge of M31 (Ford & Jacoby 1978a). From observations of emission lines (Rubin & Ford 1971) it is found that the total amount of ionized gas within 400 pc of the nucleus is $<10^5 M_\odot$. The high N/H and O/H values in the bulge gas (Rubin, Kumar & Ford 1972) show that this gas must have been ejected from evolving stars. To prevent a buildup of gas in the bulge one has to assume that the bulge of M31 is either (1) continuously being cleansed by stellar winds or (2) is cleared intermittently by explosive events. (A similar discrepancy between the rate of mass loss by stars and the rate of star formation in the Galactic nuclear bulge had been noted long ago by van den Bergh (1957).) It should, however, be emphasized that the central bulge of M31 is not entirely dust free. A special series of short exposures of the central region of M31 with the Hale 5-m telescope has been obtained by van den Bergh (unpublished) to study the distribution of dust clouds. These plates show that dust clouds occur up to a projected distance of $\sim$10" (37 pc) from the nucleus. The nucleus itself appears to be located in a dust-free area. The dust clouds within 3 arcmin (660 pc) of the center of M31 show evidence for shearing by differential rotation. There is, however, no unambiguous evidence for spiral structure in the innermost region of M31. A beautiful photograph published by Johnson & Hanna (1972) shows an intricate pattern of dust clouds superimposed on (or embedded in) the bulge of M31. Using a CCD

---

[2] The effective radius is defined as the radius that contains half of the light in projection.

camera and an interference filter that transmitted the light of Hα + [N II], Ciardullo et al. (1988) found a spiral-like emission feature with a diameter of $\sim$7$'$ that is centered on the nucleus. Surprisingly it appears that this spiral-like feature is being viewed almost pole-on. Since the gas in M31 exhibits a strong radial metallicity gradient (Blair, Kirshner & Chevalier 1981) the gas in the bulge might be metal rich (i.e., the presence of a few dust patches need not imply the presence of a large mass of gas inside the nuclear bulge).

Observations of the integrated light of the bulge of M31 (Morgan 1959, McClure & van den Bergh 1968) demonstrate that the dominant contribution to its light is produced by very metal-rich (μ Leonis-like) giant stars. Stellar photometry of the inner bulge of M31 by Rich & Mighell (1995) shows that (1) the red giant branch extends to $I = 19.5$ ($M_I = -5$) and that (2) the population in the inner bulge at $r < 40''$ does not differ substantially from that in an outer bulge field located $\sim$120$''$ south of the nucleus. The bulge of M31 has also been observed spectroscopically in the far ultraviolet with the *Hopkins Ultraviolet Telescope* (Ferguson & Davidsen 1993) and via UV imagery by Bertola et al. (1995) with the *Hubble Space Telescope.* Ferguson & Davidsen find that the spectral energy distribution of the nuclear bulge of M31 exhibits a smaller far-UV excess than does that of the giant E1 galaxy NGC 1399. The far-UV flux excess in (very metal-rich) supergiant ellipticals is thought to be produced by low-mass extreme horizontal branch stars and/or by "asymptotic branch *manqué*" stars (i.e, objects that have masses that are too low to allow them to ascend the classical post–asymptotic giant branch). Presumably such stars have low masses because high metallicity caused their stellar winds to lose more mass than would have been the case for metal-poor stars. Ferguson & Davidsen (1993) find that they can obtain an acceptable fit to the ultraviolet spectral energy of the M31 bulge if $\sim$65% of the flux (at $\sim$1,400 Å) comes from post–asymptotic branch stars and the remaining $\sim$35% from stars of the type(s) that produce the far-UV excess in NGC 1399. Such a picture is consistent with the conclusion by Peimbert (1990) that the specific frequency of planetary nebulae (which are only formed by classical post–asymptotic branch stars) decreases in galaxies as their metallicity increases.

From the distribution of "pixel luminosities" Renzini (1998) concludes that there is no solid evidence for the existence of an intermediate-age stellar population in the central region of M31.

3.6   The disk of the Andromeda galaxy

3.6.1   *Surface brightness of the disk*

The surface brightness of the disk component of a spiral galaxy can, as de Vaucouleurs (1959b) has shown, usually be represented by an exponential disk in which the surface brightness in linear units is of the form

$$I(R) = I_0 \exp(-R/h), \tag{3.2}$$

in which h is the radial scale-length of the disk. This result can also be conveniently represented in logarithmic form as

$$\sigma(R) = \sigma_0 + 1.086\,(R/h), \tag{3.3}$$

in which σ_0 is the central surface brightness expressed in mag arcsec^{-2}.

According to Walterbos & Kennicutt (1987) the disk of M31 has an integrated magnitude $V = 4.12$. The disk scale-length decreases with increasing wavelength from 34' (7.5 kpc) in U to 29' (6.4 kpc) in B and 26' (5.7 kpc) in V. This trend continues toward longer wavelengths (Hirimoto, Maihara & Oda 1983) with a scale-length of only 20' (4.4 kpc) in the K band (2.2 μm). These color observations show that the outer disk population of M31 is younger (bluer) than that of the inner disk. This result was already implicit in the early work of Babcock (1939), who "suspected that the relative strength of this line [Hδ] increased with increasing distance from the nucleus." A surprising result is that UBV photometry of the disk of M31 (Walterbos & Kennicutt 1988) shows uniform colors of $0.9 \lesssim B - V \lesssim 1.0$ and $0.45 \lesssim U - B \lesssim 0.55$. Possibly the blue light contributed by young blue stars in the outer disk of M31 is compensated for by the reddening produced in dusty gas, which has not yet been turned into stars in the outer regions of the Andromeda galaxy. Tinsley & Spinrad (1971) were able to show that the light of the disk of M31 is dominated by an old stellar population, which accounts for the red integrated colors of this disk. From color–magnitude diagrams in a number of disk fields Morris et al. (1994) were able to show that the red giant branch of disk stars exhibits considerable width. Contributions to this width are probably provided by both a range in metallicity of the old M31 disk stars and by the fact that the stars on the red giant branch of the M31 disk have funneled up from main sequence stars having a significant range in masses. From spectroscopic observations of H II regions and of supernova remnants, Blair, Kirshner & Chevalier (1981) were able to show that the disk of the Andromeda galaxy exhibits a significant radial abundance gradient. In this respect M31 resembles the Milky Way system (Smartt & Rolleston 1997) and many other giant and supergiant spiral galaxies. Venn et al. (1998) have found that A-type supergiants in the disk of M31 exhibit a radial abundance gradient similar to that derived from H II regions and supernovae.

From digital stacking of Palomar Schmidt plates Innanen et al. (1982) showed that the outermost parts of the optical disk of M31 are warped. This conclusion was subsequently confirmed by Walterbos & Kennicutt (1988). The observation that this warp shows up in both yellow *and* red light demonstrates that this warp is due to starlight and not to (red) emission nebulosity. This warping of the M31 disk was first seen by Baade (1963, p. 73), who noted that the outermost "Population II suddenly swirls off to one side." Radio observations by Newton & Emerson (1977) and by Cram, Roberts & Whitehurst (1980) show that the neutral hydrogen in M31 is warped in the same direction as the optical disk. The 21-cm isophotes do, however, appear to remain close to the fundamental plane of M31 out to somewhat greater distances from the nucleus than do the optical isophotes.

3.6.2   *Spiral structure*

The spiral structure embedded in the disk of the Andromeda galaxy has been described by Baade (1963, p. 60). He writes that N_1 and S_1 (see Table 3.2) are pure dust arms that contain no supergiants or H II regions. The first supergiants appear in N_2 and S_2, and H II regions are first seen in N_3 and in S_3. The maximum display of Population I stars is seen in N_4, N_5 and S_4, S_5. Finally Baade notes that arms N_6, N_7 and S_6, S_7 are defined only by scattered groupings of supergiants with no obvious sign of dust. The apparent lack of dust in the outermost region of the M31 disk is probably due to both a radial decrease in the dust-to-gas ratio and to the fact that dust can only be seen when it absorbs or scatters light from background luminous material. The high-resolution 21-cm

Table 3.2. *Main spiral arms of M31 according to Baade (1963)*

Arm	Distance[a]		Arm	Distance[a]	
N_1	3.́4	0.8 kpc	S_1	1.́7	0.4 kpc
N_2	8.0	1.8	S_2	10.5	2.3
N_3	25	5.5	S_3	30	6.6
N_4	50	11.1	S_4	47	10.4
N_5	70	15.5	S_5	66	14.6
N_6	91	21.1	S_6	95	21.0
N_7	110	24.3	S_7	116	25.6

[a]Distances are measured along the north-following and south-preceding parts of the major axis of M31.

observations by Emerson (1974) show that the dust lanes in M31 correspond to major concentrations of neutral hydrogen gas.

A 408 MHz continuum map of M31 by Pooley (1969) is shown in Figure 3.2. This map shows that the continuum radiation emitted by the Andromeda galaxy has two main components: (1) a centrally peaked source and (2) an annulus of emission that coincides with the broad ring of OB associations and H II regions at $R \sim 10$ kpc. The existence of the second component shows that particle acceleration in M31 must be closely linked with the formation (or death!) of massive stars.

3.6.3   OB associations

A comprehensive study of the distribution of young stars in the disk of M31 was undertaken with the Tautenburg 134-cm Schmidt telescope by van den Bergh (1964). By "blinking" plates in five colors he was able to isolate a total of 188 OB associations (see Figure 3.3). Using CCD photometry Magnier et al. (1993) were able to recover 174 associations. From far-UV photometry of a number of these OB associations Hill et al. (1993) have estimated the masses and ages of individual stars in some of these associations.

Figures 3.4 and 3.5 show plots of the distribution of van den Bergh's associations projected onto the fundamental plane of M31. A major axis position angle of $37°7$ and an inclination of $77°7$ were assumed for these deprojections. The figures suggest that the entire spiral structure of M31 has been distorted, possibly by tidal interactions with M32 (Schwarzschild 1954, Arp 1964). Due to the tidal interaction between M32 and the fundamental plane of M31, the bottom part of the rectified spiral pattern shown in Figure 3.5 may be systematically distorted. For $i = 77°7$ a deviation of 0.2 kpc from the fundamental plane will transform into a displacement of almost one kiloparsec. Baade (1963) proposed a schematic model (see Table 3.2) of the M31 spiral structure. In this model the spiral pattern of the Andromeda galaxy is interpreted in terms of two tightly coiled spiral arms, which can each be followed over three or four complete revolutions. However, van den Bergh's (1964) observations of the OB associations in M31, which are shown in Figure 3.5, suggest a much less ordered pattern in which the spiral structure of the Andromeda galaxy consists of short disjointed spiral-arm segments of differing age. Detailed study of the spiral pattern in M31 suggests that the relation among dust lanes, H II regions, and OB associations is often quite complex. In some (but not all!)

Fig. 3.2. Map of 408-MHz continuum radiation from M31. Note the source associated with the nucleus and the ring-shaped distribution of emission that coincides with the ring of OB associations and H II regions at $R \sim 10$ kpc.

cases dust lanes are located along the inner edges of the spiral arms that are outlined by bright young stars. Occasionally strings of H II regions occur in regions that appear to be devoid of stars. As Baade (1963, p. 63) pointed out, "the spiral arm is like a chameleon – it can change its appearance." Van den Bergh (1964) used the clumpiness of the distribution of stars within associations to estimate their relative ages. Adopting an expansion velocity of 10 km s^{-1} he obtained expansion ages of 15×10^6 yr, 30×10^6 yr, and 40×10^6 yr for young, intermediate-age, and old associations, respectively. Young associations were found to contain the largest number of bright stars and H II regions. Furthermore, Figure 3.5 suggests that young associations (shown as filled circles) may be more strongly associated with well-defined spiral arm segments than old associations

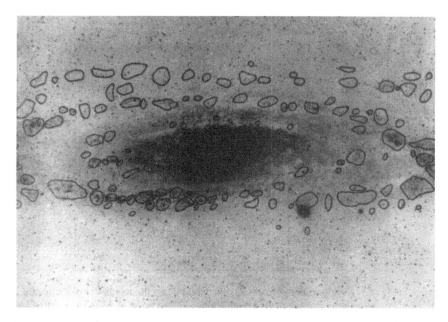

Fig. 3.3. The positions of OB associations in M31 according to van den Bergh (1964).

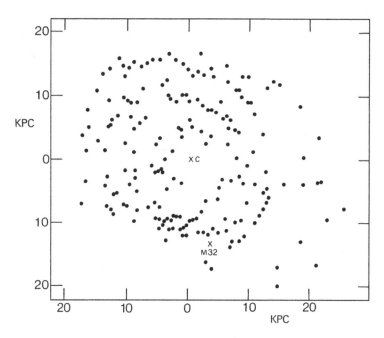

Fig. 3.4. Positions of OB associations projected onto the fundamental plane of M31. The locations of the nucleus of M31, and of M32, are marked by crosses. The apparent concentration of associations NE of M32 is probably due to a projection effect resulting from warping of the M31 Disk by M32.

N

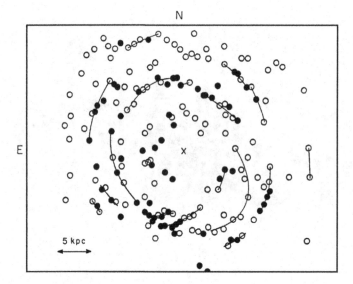

Fig. 3.5. Distribution of OB associations in M31 for an assumed disk inclination $i = 77°.7$. Possible spiral arm segments are indicated. The location of M32 is indicated by E. In the figure the major axis is vertical. The youngest associations are shown as filled circles.

(plotted as open circles). Inspection of this figure also shows that the most active region of star formation occurs between 9 kpc and 15 kpc from the nucleus of M31. Magnier et al. (1997a) have used the positions of individual stars in V versus $U - V$ diagrams, in conjunction with theoretical evolutionary tracks, to derive the age distributions of stars in different associations. In addition Magnier et al. (1997b) have speculated that NGC 206, which is the largest association in M31, was formed by the intersection of two spiral arms.

Because the density of young stars in the disk of M31 is rather low, it is possible to trace individual associations in this galaxy out to rather large radii. This contrasts with the situation in M33 (Humphreys & Sandage 1980), in which associations appear to be small because they cannot be traced far into the rich stellar background in the disk of the Triangulum galaxy.

3.6.4   *Young star clusters*

Hodge (1992, pp. 163–182) gives a detailed discussion of open clusters in the disk of the Andromeda galaxy. He provides a catalog of ~400 probable open clusters in M31. A plot of the distribution of these clusters shows them to be mostly located in spiral arms, with the main concentration in a broad annulus that extends from $r = 45'$ (10 kpc) to $r = 75'$ (17 kpc). The brightest known young cluster in M31 is C^3 107 (= vdB0), for which $V = 14.94 (M_V = -9.6)$ and $B - V = 0.20$ (van den Bergh 1969). Cluster C 107 contains Hubble's (1929) Cepheid V 40, which has a period of 33 days. According to Hodge (1992, p. 174) the most remote open cluster in M31 is C 1, which is situated at $r = 139'$ (29 kpc). This places it about 4 kpc beyond the most distant spiral arm recognized by Baade (1963).

---

[3] C numbers are from Table 11.2 of Hodge (1992).

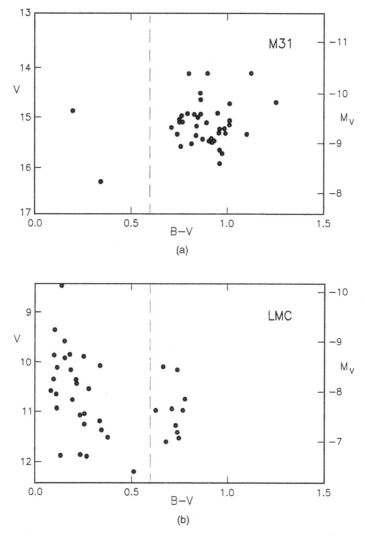

Fig. 3.6. (a) Color–magnitude diagram for the integrated colors of the brightest star clusters in M31. The plot shows that bright old red clusters are much more common than young blue ones in the Andromeda galaxy. This indicates that the rate of star and cluster formation must have been much higher in M31 in the distant past than it is at present. (b) Color–magnitude diagram for the brightest clusters in the Large Cloud. Young blue clusters are seen to be much more common in the LMC than are old red clusters.

Figure 3.6(a) shows a color–magnitude diagram for integrated colors of the brightest clusters in M31 (van den Bergh 1969). For comparison Figure 3.6(b) shows a similar plot (van den Bergh & Hagen 1968) for the integrated colors and magnitudes of the brightest clusters in the Large Magellanic Cloud. Intercomparison of these plots clearly shows that a much larger fraction of the M31 clusters are red than is the case for the LMC clusters. This difference implies that the star and cluster formation histories must have been very different in these two galaxies. In the Andromeda galaxy the star and cluster formation rate was initially quite high and then declined steeply, so that young blue clusters are now relatively rare. In contrast, young blue clusters greatly outnumber old red ones in the Large Cloud. This suggests that the star and cluster formation rate in the LMC has remained more nearly constant (or may actually have increased) with time.

3.6.5   H II regions

A catalog of 688 emission nebulae in M31 has been published by Baade & Arp (1964). Plots of the distribution of these H II regions by Arp (1964) show a large-scale distribution that is quite similar to that of the M31 associations plotted in Figures 3.4 and 3.5. Baade pointed out that the H II regions outline spiral arms like "beads on a string." A beautiful image of the Andromeda galaxy in the light of Hα has been published by Devereux et al. (1994). This picture shows that the Hα emission of M31 is mainly associated with three distinct components: a large star forming annulus with a diameter of ~1°65, which contributes 66% of the total flux, a bright nuclear component that contributes 6% of the flux, and an extensive system of filaments interior to the great star forming ring, which contributes the remaining 28% of the total M31 Hα flux of ~$7 \times 10^6 L_\odot$.

Hα photographs of H II regions in M31 have also been published by Walterbos & Braun (1994). These authors show that the integrated Hα emission in the disk of M31 implies a rather low star formation rate of 0.35 $M_\odot$ per year. Their data show that the diameter distribution of the M31 H II regions obeys van den Bergh's law (Petit, Sivan & Karachentsev 1988), that is, the diameter distribution is of the form

$$N(D) = N(0) \exp - (D/D_o), \tag{3.4}$$

in which $D_o \simeq 50$ pc.

Arp & Brueckel (1973) have obtained small-scale Hα images of M31 and M33 in order to study H II regions in a way similar to that in which galaxies at larger redshifts are investigated. In this work the Andromeda and Triangulum galaxies were photographed using exactly the same plate, filter, and exposure combinations. Arp & Brueckel found that the M33 emission regions are frequently single large loops and rings, whereas the M31 emission regions tend to be more stellar and occur in associated groups. If the larger H II complexes in M31 are classed as single regions, then there is a contradiction with the generally quoted result that H II regions are larger in Sc galaxies than they are in objects of Hubble type Sb. The three largest H II regions in M31 are found to have diameters of 180″, 153″, and 64″, corresponding to linear dimensions of 660, 560, and 240 pc, respectively. For comparison the largest H II regions in M33 have diameters of 111″, 89″, and 81″, which at an assumed distance of 795 kpc, corresponds to 430, 340, and 310 pc.

3.6.6   Large-scale distribution of gas and dust

The large-scale distribution of neutral hydrogen gas in M31 has been studied by Emerson (1974). He finds that the distribution of H I in the Andromeda galaxy is quite asymmetric and that this asymmetry is particularly pronounced in the 10 kpc ring. He also finds that there is generally excellent agreement between the location of strings of OB associations and ridges of H I gas. Emerson notes that 186 of van den Bergh's 188 OB associations are located in regions having surface density in the plane of M31 greater than $\sigma_{\mathrm{H_I}} = 8 \times 10^{19}$ cm^{-2}. Emerson finds that

$$\sigma_{\mathrm{H_I}} \propto \sigma_{\mathrm{OB}}^{m}, \tag{3.5}$$

in which m is greater in associations that van den Bergh (1964) classifies as young than it is in those to which he assigns older ages. On scales of more than one kiloparsec there is also good agreement between $\sigma_{\mathrm{H_I}}$ and the distribution of H II regions (Baade & Arp 1964). Agreement on smaller scales is not expected because H II regions are, in fact, holes in the general distribution of neutral and molecular hydrogen.

Devereux et al. (1994) note that the Hα and far-infrared emission of the Andromeda galaxy have a very similar distribution. This suggests that these two emission components have a common origin (i.e., hot stars embedded in dusty gas clouds). Dame et al. (1993) have used a survey at 115 GHz to show that the molecular gas in M31 is strongly concentrated in the same 1°65 diameter annulus that also contains most of the neutral hydrogen, OB associations, young star clusters, and H II regions in the Andromeda galaxy. This region also contains a magnetic belt that represents the most conspicuous feature exhibited by the magnetic field of M31 (Moss et al. 1998). Rotation measure observations of background sources by Han, Beck & Berkhuijsen (1998) appear to show that the regular magnetic field of the M31 disk extends from $\sim$5 kpc, through the great belt at $\sim$10 kpc, out to distances as great as $\sim$25 kpc. Dame et al. have noted that the total H$_2$ mass in M31, which is found to be $\sim$3 $\times$ 10^8 $M_\odot$, is an order of magnitude smaller than the total H I mass and only about a quarter of the H$_2$ mass of the Milky Way system computed in the same way. Wilson & Rudolph (1993) have measured a mass of 7×10^5 $M_\odot$ for a single massive molecular cloud in M31. This shows that the interstellar medium in the Andromeda galaxy is able to organize itself into quite massive interstellar clouds, even though its mean surface density is relatively low. From observations of ^{12}CO lines Loinard & Allen (1998) have been able to confirm that the inner disk of M31 contains large, massive, and cold clouds. They find that these clouds have kinetic temperatures close to that of the cosmic microwave background. Such extremely low temperatures might, if real, be due to the low heating rate that is a consequence of the fact that very little massive star formation is presently taking place in the Andromeda galaxy. However, Israel, Tilanus & Baas (1998) find that they can explain their neutral carbon observations by invoking clouds that have temperatures of $\sim$10 K and are (except, perhaps, for a higher metallicity) similar to those observed in the solar neighborhood. Haas et al. (1998) have used the *ISO* satellite observations at 175 μm to produce a beautiful map of the distribution of dust in M31. These authors find numerous small clumps superposed on the main dust ring at $R \sim$ 10 kpc. Some inner spiral structure and parts of a faint ring at $\sim$14 kpc are also visible. From a comparison of 100-μm and 175-μm maps Haas et al. conclude that the observations may be represented by a two-component model with large grains at $\sim$16 K and small grains heated to $\sim$45 K. They find that about half of all the dust in M31 is located within the broad 10-kpc ring. Haas et al. find a total dust mass of 3.7 $\times$ 10^7 $M_\odot$ and a dust-to-H I gas ratio of 1/130, which is close to the Galactic value of 1/170. Cesarsky et al. (1998) believe that the strange midinfrared spectrum of M31 can be explained by assuming that the observed emission features are due to hydrogenated amorphous carbon particles that have been synthesized in the atmospheres of carbon stars.

Neininger et al. (1998) have used carbon monoxide emission to trace the spiral arms in the Andromeda galaxy. They find a tight correlation between CO emission and dust extinction, from which they conclude that CO traces the molecular gas. Because so much gas is in molecular form, they find a poor correlation between I(CO) and I(HI). However, a tight correlation is found between I(CO) and I(HI) + 2N (H$_2$).

From color diagrams obtained with the *IRAS* satellite Xu & Helou (1994) conclude that very small grains (but not polycyclic aromatic hydrocarbons) are only half as abundant in M31 as they are in high-latitude Galactic "cirrus" clouds. Very small grains are the leading candidate for the carrier of the 2,175 Å bump in the interstellar absorption curve. These *IRAS* observations are consistent with recent *Hubble Space Telescope* observations (Hutchings et al. 1992, Bianchi et al. 1996), which show that the 2,175 Å bump is weaker and narrower in M31 than it is in the Galaxy. Since M31 and the Galaxy have similar metallicities, Xu & Helou speculate that this difference in the 2,175 Å bump is, in some way, related to the fact that the star formation rate in M31 is significantly lower than it is in the Galaxy.

3.6.7   *Reddening of starlight by dust*

From 21-cm line observations of nine globular clusters of known reddening Kerr & Knapp (1972) find that

$$n_H = \alpha \times 10^{21} E(B - V), \tag{3.6}$$

in which $\alpha = 4.7$ cm^{-2} mag^{-1}, and n_H is the number of neutral hydrogen atoms per cm^2 along the line of sight. From an unpublished study of the relation between the reddenings of 173 globular clusters in M31 and the surface density of neutral hydrogen at the positions of these clusters, van den Bergh found $\alpha = 10.5$ cm^{-2} mag^{-1}. Assuming that half of all of the globulars in M31 are located in front of the M31 gas layer, one then has $\alpha = 5.25$ cm^{-2} mag^{-1}, which agrees fortuitously too well with the Galactic value. Taken at face value this result suggests that the ratio of total-to-selective absorption in M31 is similar to that in the Galaxy. Within the accuracy of the present data there is no obvious radial variation of α over the disk of the Andromeda galaxy. There is, however, some evidence (van den Bergh 1968b) that the ratio of total-to-selective reddening may be lower for B 327 (which is the most heavily reddened cluster known in M31) than it is for other clusters. In this connection one is reminded of a story about Walter Baade (Schmidt-Kaler, private communication). When asked shortly before his death if he would want to become an astronomer again if he were to be given a chance to relive his life, he answered "only if I were first given a guarantee that the ratio of total-to-selective absorption is the same everywhere in the Universe."

3.7   The halo of M31

3.7.1   *Star counts in the halo*

The halo of the Andromeda nebula may be studied (1) by measurements of its integrated light, (2) from deep color–magnitude diagrams of individual halo fields, (3) by study of the radial distribution of star counts, and (4) by investigation of individual halo objects such as globular clusters, planetary nebulae, etc. The first modern photometry of the halo of the Andromeda galaxy is by de Vaucouleurs (1958), who used photoelectric

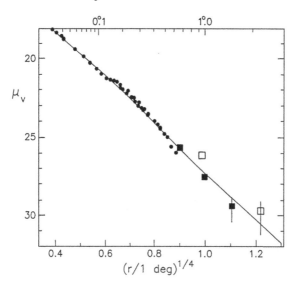

Fig. 3.7 Surface brightness in the halo of M31. Filled squares are observations by Pritchet & van den Bergh (1994) along the minor axis of the Andromeda galaxy. The two open squares are their observations of fields along the major axis of M31, which have a disk contribution. Dots are observations by Kent (1983), Walterbos & Kennicutt (1987), and de Vaucouleurs (1958).

UBV photometry to trace the surface brightness of the M31 halo out to $B = 26.8$ mag arcsec^{-2}, which was reached at 37' (8 kpc) from the nucleus along the minor axis. More recently Pritchet & van den Bergh (1994) were able to use counts of individual giant stars in the halo of M31, obtained in very good seeing with the Canada–France–Hawaii Telescope, in conjunction with assumptions about the halo luminosity function, to extend "surface brightness" measurements in the halo of the Andromeda galaxy down to $V \approx 30$ mag arcsec^{-2}. This surface brightness was reached at a distance of 5°0 (66 kpc) from the nucleus measured along the minor axis. Pritchet & van den Bergh believe that the technique of counting stars near the tip of the red giant branch could, in the not too distant future, be used to extend measurements of the halo surface brightness distribution to $B \sim 32$ mag arcsec^{-2}. (The great power of the star counting technique was first pointed out by Baum & Schwarzschild 1955.) By combining the data of Pritchet & van den Bergh (1994) with earlier photometry closer to the nucleus of M31 it is found (see Figure 3.7) that the surface brightness over the range $0.2 \lesssim R(\text{kpc}) \lesssim 20$ is reasonably well represented by a single de Vaucouleurs $R^{1/4}$ law. Alternatively the star counts in the outer halo of the Andromeda galaxy may also be modeled by a density law of the form $\rho(R) \propto R^{-5}$. This is much steeper than that for the M31 globular clusters, which appear to follow an R^{-3} radial distribution (Racine 1991). Racine's result suggests that the globular cluster component of the halo is more extended than the stellar component of this galaxy. At $V \approx 28$ mag arcsec^{-2} the axial ratio of the halo of M31 is found to be $b/a = 0.55 \pm 0.05$.

3.7.2   Color–magnitude diagrams

Color–magnitude diagrams for individual fields in the halo of the Andromeda galaxy have been published by Mould & Kristian (1986) and by Pritchet & van den Bergh

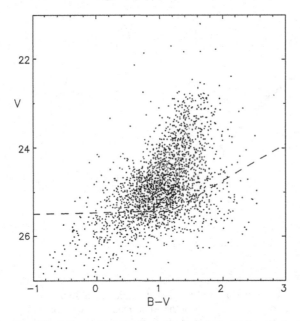

Fig. 3.8 Color–magnitude diagram of a field in the inner halo of M31. The large width of the red giant branch indicates that these stars have a significant range in metallicity.

(1988). In their study of a field situated at 40′ (8.8 kpc) from the nucleus along the minor axis, Pritchet & van den Bergh found that the red giant branch (see Figure 3.8) of the inner halo of M31 exhibits a significant width, which indicates that the stars that inhabit the halo have a considerable range in metallicity. Figure 3.9 (Pritchet & van den Bergh 1988) shows that the majority of the red giants in the inner halo of the Andromeda galaxy appear to have metallicities between those of M5 ([Fe/H] $= -1.4$) and 47 Tucanae ([Fe/H] $= -0.7$).[4] The color–magnitude diagram exhibits a horizontal branch that is only weakly populated on its blue side. Overall the color–magnitude diagram of the inner halo of M31 resembles that of the Galactic globular cluster NGC 6171, which has [Fe/H] $= -1.0$. A more recent study of the inner halo field 40′ SE of the nucleus of M31 by Durrell, Harris & Pritchet (1994) confirms the fact that the blue end of the horizontal branch is only weakly populated in this field. However, Durrell et al. derive a slightly higher mean metallicity of [Fe/H] ~ -0.6. These authors also draw attention to the fact that the metallicity of stars in the inner halo of M31 appears to be significantly higher than that of the M31 globular clusters, suggesting that these clusters formed *before* the majority of halo stars.

Pritchet & van den Bergh (1987a) have discovered RR Lyrae variables in the field 40′ SE of the nucleus of M31. They find that these objects have $\langle V \rangle = 25.68 \pm 0.06$, from which they derived $(m - M)_0 = 24.34 \pm 0.15$. The mean period of the RR Lyrae variables is found to be $\langle P_{ab} \rangle = 0.548$ days, which suggests that the cluster-type variables in the inner halo of M31 belong to Oosterhoff's (1939) Type I. The existence of RR Lyrae variables attests to the presence of a (small) metal-poor component among the stars of the M31 halo population. Observations of the metal-poor globular cluster Mayall IV

---

[4] [Fe/H] $= \log (\text{Fe/H})_* - \log (\text{Fe/H})_\odot$ with $12 + \log (\text{Fe/H})_\odot = 7.67$ (Anders & Grevesse 1989).

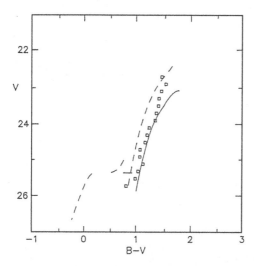

Fig. 3.9 The ridge line of the M31 halo color–magnitude diagram is denoted by small boxes. The colors of each of these boxes are uncertain by ± 0.03 mag. Also shown are the ridge lines of the color–magnitude diagrams of M5 (dashed line) and of 47 Tuc (continuous line). The figure shows that the mean metallicity of the inner halo of M31 is near [Fe/H] $= -1.0$.

(= G 219), which is located at a distance of $87'$ (19 kpc) from the nucleus, shows that this cluster is projected on a more metal-rich halo field with [Fe/H] ~ -0.7 (Christian & Heasley 1991). It is of some interest to note that the cluster Mayall II (= G1), which is located at a distance of $152'$ (34 kpc) from the center of M31, is projected on a halo population with a color–magnitude diagram (Rich et al. 1996) that is similar to that of the metal-rich Galactic globular cluster 47 Tuc (= NGC 104), for which [Fe/H] $= -0.7$.

Guhathakurta & Reitzel (1998) have pointed out that presently available data on the stellar density in the outer halo of M31 may be represented by a power-law density profile with a deprojected slope of -4.6. This is substantially steeper than that of the Galactic halo for which a power-law slope of -3.5 is observed (see Section 4.6). Furthermore the halo density (scaled for galaxy size) is at least an order of magnitude lower in the Galaxy than it is in M31. These results suggest (Freeman 1999) that the halo of M31 (which appears to merge smoothly with its bulge) might have been produced by violent relaxation following a major merger event.

Van den Bergh & Pritchet (1992) have obtained color–magnitude diagrams in halo fields located at $40'$, $60'$ and $90'$ along the minor axis of M31. They conclude that the median $B - V$ color of the halo giant branch in these three halo fields does not vary by more than 0.1 mag. Adopting $d(B - V)/d$[Fe/H] $\simeq 0.5$ at [Fe/H] $= -1$, this range of $B - V$ corresponds to ± 0.2 in [Fe/H]. These results suggest that the mean metallicity of stars in the halo of the Andromeda galaxy does not vary by more than a factor of ~ 1.6 between $R = 8.8$ kpc and $R = 19.9$ kpc. Searches for young stars in the halo of M31 (McCausland et al. 1993, Hambly et al. 1995) have so far remained inconclusive.

3.7.3   Dark matter in M31

The first convincing evidence for the existence of large amounts of invisible dark matter in the outer regions of spiral galaxies (see Figure 3.10) was obtained by Roberts

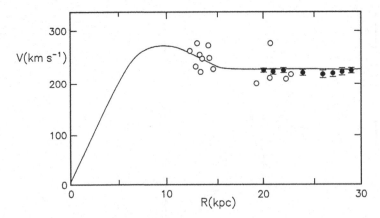

Fig. 3.10 Radial velocity observations along the south-preceding half of the major axis of M31, where there is less confusion with foreground Galactic H I gas than along the north-following side of the major axis. Optical observations (Rubin & Ford 1970) beyond 12 kpc are shown as open circles. Error bars show 21-cm observations from Roberts & Whitehurst (1975). Note that the rotation curve appears to remain flat beyond $R \approx 15$ kpc.

& Rots (1973) and Roberts & Whitehurst (1975).[5] Contrary to expectation, Roberts & Whitehurst found that the rotational velocity of the Andromeda galaxy remains constant beyond $R \sim 15$ kpc. They concluded that this "implies a mass that increases linearly with R over this range [10–30 kpc] and a mass-to-luminosity ratio $\gtrsim 200$ for this outer region." Roberts & Whitehurst suggested that the existence of so much unseen mass at large radii might be accounted for by the presence of vast numbers of dim dM5 stars. Ostriker, Peebles & Yahil (1974) linked the detection of dark matter in the halos of spiral galaxies to the missing mass problem in rich clusters of galaxies, which had first been noted by Zwicky (1933). The true nature of the missing mass that is observed in galactic halos and in clusters of galaxies remains a mystery. From a microlensing survey Crotts (1997) was able to exclude objects with masses in the range $\sim 10^{-7} M_\odot$ to $0.08\,M_\odot$ as major contributors to the "missing" mass in the inner disk and outer bulge of M31. In retrospect it is now clear that the "missing mass" problem in individual galaxies was first noted by Oort (1940). In his study of the edge-on spiral NGC 3115 he noted that "the distribution of mass in the system appears to bear almost no relation to that of light . . . one finds the ratio of mass to light in the outer parts of NGC 3115 to be about 250."

3.8   Globular clusters

Hubble (1932) was the first to observe "Nebulous objects . . . provisionally identified as globular clusters" in the Andromeda galaxy. He noted that these M31 globulars, of which he found 140, appeared to be significantly fainter than their Galactic counterparts. Apparently, Hubble never suspected that this luminosity difference was due to the fact that the distance he had adopted for M31 was too small. Additional globular

---

[5] This result was, to some extent, anticipated by Babcock (1939), who wrote: "The difference [in M/L] can be attributed mainly to the very great mass calculated in the preceding section for the outer parts of the spiral on the basis of the unexpectedly large circular velocities of these parts."

clusters in M31 have been discovered and catalogued by Mayall & Eggen (1953), Vetešnik (1962), Sargent et al. (1977), Battistini et al. (1980, 1987), Buonanno et al. (1982), Crampton et al. (1985), Racine (1991), Battistini et al. (1993), and by Mochejska et al. (1998).

3.8.1 *Distribution of globulars*

From the projected radial density profile of the globular clusters in M31 Racine (1991) concluded that M31 contains 270 ± 50 globular clusters. He estimated this value to be 1.8 ± 0.3 times larger than the corresponding number of Galactic globulars. This is consistent with the work of Battistini et al. (1993), who found the M31 cluster system to be 2.5 ± 1.0 times richer than that of the Galaxy. Fusi Pecci et al. (1993) estimate the total M31 globular cluster population to be $\sim400^{+70}_{-40}$, with $r_e = 24\rlap{.}'2$ (535 kpc), in which r_e is the radius containing half of all M31 globulars in projection.

According to Racine the projected surface density of clusters in M31 is proportional to R^{-2} over the range $6 < R(\text{kpc}) < 22$. Beyond ~22 kpc the cluster density in M31 appears to drop more steeply and possibly has a cutoff at a projected distance of ~40 kpc. Such a cutoff might have been imposed on the M31 cluster system by the tidal forces exerted by its close companions M32 and NGC 205. The reality of this cutoff is being checked by Frederici et al. (1994), who have so far found only six cluster candidates in the range $20 < R(\text{kpc}) < 70$.

Battistini et al. (1993) find that the projected radial density distribution of globular clusters in the inner 3 kpc of the Andromeda galaxy falls *below* the best-fitting $R^{1/4}$ law for the inner region of M31. Such a de Vaucouleurs law would predict the presence of 261 clusters with $R < 3$ kpc, which is ~2.5 times larger than the actual number observed; that is, about 150 clusters appear to be "missing" near the nucleus of M31. Possible explanations for this apparent deficit are (1) the census of clusters in the inner region of the Andromeda galaxy is incomplete, (2) clusters that once existed near the nucleus have been sucked into it by dynamical friction (Tremaine, Ostriker & Spitzer 1975), or (3) physical conditions in the central region of M31 were never conducive to the formation of globular clusters. Incompleteness of the cluster sample might be due to both the high surface brightness of the central region of M31 and to the fact that the radii of clusters become ever smaller as the nucleus is approached, thus making it more difficult to distinguish globulars from stars. Aurière, Coupinot & Hecquet (1992) observed a $7\rlap{.}'7 \times 7\rlap{.}'7 (1.7 \times 1.7$ kpc) field centered on the nucleus of M31 and found a dozen new globulars, in addition to 23 clusters that were already known. Most of these newly discovered clusters were fainter than $R = 17.0$. Aurière et al. also note that all of the newly discovered clusters have very small radii. Clearly it would be very useful to use the *Hubble Space Telescope* to place limits on the number of small faint globular clusters that could have been overlooked near the nucleus of M31. As has already been noted in Section 3.5, the fact that the true nucleus (P_2) of M31 exhibits a far-ultraviolet excess (King, Stanford & Crane 1995) suggests that *some* globular clusters containing blue horizontal branch stars might, indeed, have been sucked into the nucleus of M31 by tidal friction. By comparing the present radial distribution of M31 globular clusters with an inward extrapolation from the radial distribution of clusters with galactocentric distances >7 kpc, Capuzzo-Dolcetta & Vignola (1997) estimate that 85 globular clusters that once existed have disappeared (or still remain to be discovered).

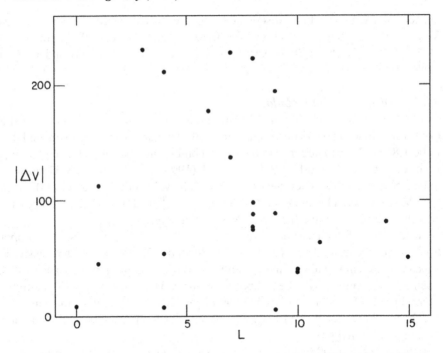

Fig. 3.11 Deviation from disk rotation velocity versus cluster metallicity index L (van den Bergh 1969). The plot suggests that only the most metal-rich M31 globular clusters exhibit disklike motions.

3.8.2   Kinematics of M31 clusters

The first study of the kinematics of the M31 globular cluster system was by van den Bergh (1969), who obtained radial velocities for 42 globular clusters associated with the Andromeda galaxy. He found that metal-poor globular clusters in M31 had a significantly higher velocity dispersion ($\sigma = 138$ km s^{-1}) relative to circular motion than did metal-rich clusters ($\sigma = 58$ km s^{-1}). Using the "projected mass method" of Bahcall & Tremaine (1981), van den Bergh (1981a) obtained $M = (9 \pm 2) \times 10^{10} M_\odot$ from the globular clusters with $r < 30'$ (66 kpc). For six globulars with $\langle R \rangle \sim 100'$ (22 kpc) he found a mass interior to this radius of $(2.4 \pm 1.2) \times 10^{11} M_\odot$. Using the much larger sample of globular cluster radial velocities that is presently available, Frederici et al. (1993) obtained a mass of $(6.0 \pm 0.7) \times 10^{11} M_\odot$ for the 18 globulars with $60' < r < 80'$ (13–18 kpc). The total number of M31 globular clusters for which radial velocity data are presently available is 181 (Battistini et al. 1993).

Recent work on the kinematics of the M31 globular cluster system has been reviewed by Huchra (1993). He finds that the globular cluster system of the Andromeda galaxy exhibits an overall rotational velocity of 80 ± 20 km s^{-1}, which is only marginally higher than the 60 ± 26 km s^{-1} systemic rotational velocity of the Milky Way globular cluster system. Figure 3.11 shows a plot of the cluster metallicity index L versus deviation of the cluster velocity from that of the M31 disk rotation curve. The figure shows that very metal-rich clusters with $L > 10$ have a relatively small velocity deviation from the disk rotation curve. The M31 cluster system appears to be slightly flattened with $b/a \sim 0.65$. Ashman & Bird (1993) have analyzed the positions and radial velocities of halo globulars in M31 and find marginal evidence for significant subclustering. If such subclustering is real, then each of the seven clusterings suspected by Ashman and Bird must be bound by a significant amount of dark matter. Possibly such subclustering is due to the survival of the remnants of the Searle & Zinn (1978) fragments that merged to form the M31 halo. [A hypothetical observer in the Andromeda galaxy, on looking back at the Milky Way system, might find that the four globular clusters associated with the Sagittarius dwarf galaxy (Ibata et al. 1995) constitute a similar clumping in phase space.] Holland, Fahlman & Richer (1995) have found that the M31 globular cluster G[6] 185 has a companion cluster at a projected separation of only 4″ (15 pc). It would be interesting to obtain a radial velocity of G 185b, to see if it is close to that of G 185a.

A color–magnitude diagram for the integrated colors of M31 globular clusters, which have been observed photoelectrically on the UBV system, is shown in Figure 3.12. This figure shows that most of the globulars in M31 are contained within a triangular area that is bounded on the blue side by $B - V = 0.66$ and on the red side by a reddening line of slope $R = A_V/E(B - V) = 2.5$. The fact that the slope of this line is smaller than the Galactic

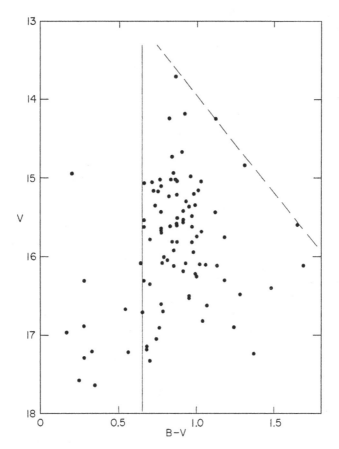

Fig. 3.12 Color–magnitude diagram for photoelectricity observed star clusters in M31. Open clusters have $B - V < 0.66$. The dashed line is a reddening line of slope $R = 2.5$.

---

[6] G numbers are from Sargent et al. (1977).

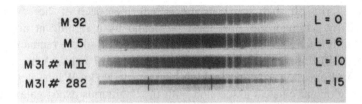

Fig. 3.13 Image tube spectra of some globular clusters in the Milky Way system and in the Andromeda galaxy. Note the strong lines in the M31 clusters Mayall II and 282, which have line-strength classes $L = 10$ and $L = 15$, respectively.

value $R \simeq 3.1$ was first noted by Kron & Mayall (1960). The blue objects with $B - V < 0.66$ are open clusters. The brightest of these, which is situated at $V = 14.94$, is vdB0.

3.8.3   *Spectroscopy of globular clusters*

The first spectroscopic observations of the globular clusters associated with the Andromeda galaxy (van den Bergh 1969) showed that these objects had approximately the same range in metallicities as their Galactic counterparts. The fact that some of the globular clusters in M31 have quite high metallicities is illustrated by the trailed Palomar 5-m image tube spectra (van den Bergh 1969) of the clusters M II (= G 1) and 282 (= G 280) shown in Figure 3.13. In that paper it was shown that the metallicity distribution of M31 clusters is skewed toward higher metallicities. This conclusion has been confirmed by more recent studies of the metallicities of globular clusters in M31 (Huchra, Brodie & Kent 1991) and in the Galaxy (Zinn 1985). Furthermore, it appears that the Andromeda galaxy does not exhibit the clear-cut dichotomy between metal-poor halo clusters with [Fe/H] < -1.0 and metal-rich "disk" clusters (most of which are actually associated with the bulge!) with [Fe/H] > -1.0 that is seen among Galactic globular clusters. Perhaps the most remarkable feature of the globular cluster system surrounding M31 is that quite metal-rich clusters, such as M II, occur far out in the halo. This observation shows that metal enrichment proceeded to a much higher level in the halo of M31 than it did in the Galactic halo, in which [Fe/H] remained below [Fe/H] ≈ -1.0 until star and cluster formation came to an end. This contrasts with the situation in the halo of M31, in which clusters with metallicities as high as [Fe/H] ~ -0.5 are found out to projected radial distances of $\sim$15 kpc. In fact, it is not clear if there is any metallicity gradient beyond $R = 5$ kpc in M31. The tentative conclusion that the globular clusters in the halo of M31 do not exhibit a radial metallicity gradient is confirmed by infrared observations (Cohen 1993). However, a metallicity gradient does appear to exist for $R < 5$ kpc. Alternatively one might describe the results of Huchra et al. (1991) by saying that clusters with [Fe/H] > -0.5 only occur within a projected distance of 5 kpc from the nucleus of the Andromeda galaxy. In other words, globular cluster formation may have continued in the bulge and inner halo of M31 until the metallicity of the interstellar medium had reached near solar values. In this respect cluster formation in the inner regions of M31 resembles that in many giant ellipticals (Forbes, Brodie & Grillmair 1997). Jablonka et al. (1998) have obtained spectra for some of the globular clusters in the bulge of M31. They find that these objects have metallicity $\langle [Z/Z_{\odot}] \rangle = -0.58$ with a dispersion $\sigma = 0.63$.

Figure 3.14 (Reed, Harris & Harris 1994) shows a comparison between the metallicity distributions in the Galaxy, M31, and the peculiar giant elliptical NGC 5128. The figure

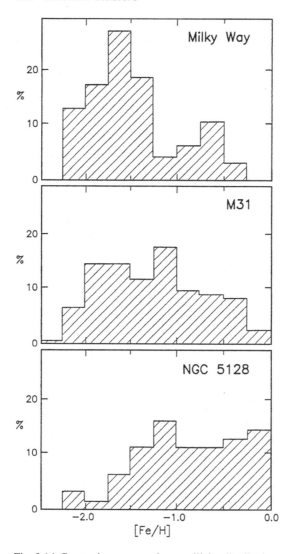

Fig. 3.14 Comparison among the metallicity distributions of globular clusters in the Galaxy, M31, and the giant elliptical NGC 5128 (= Centaurus A). The figure, which is adapted from Reed et al. (1994), shows that the metallicity distribution of globulars in the Andromeda galaxy is intermediate between that in the Milky Way and that in NGC 5128.

shows that the metallicity distribution of the globular clusters in M31 is intermediate between those in the Galaxy and in NGC 5128. Note in particular that the globulars in the Andromeda galaxy seem to differ from those in the Galaxy because they do not exhibit a metallicity distribution with two maxima, separated by a minimum at [Fe/H] ≈ -1.0. Furthermore, the metallicity distribution of M31 globulars does not exhibit the concentration of very metal-rich objects with [Fe/H] ~ 0.0 that is seen in the giant elliptical NGC 5128.

From inspection of his (little widened) image tube spectra, van den Bergh (1969) suspected possible subtle systematic differences between the integrated spectra of M31 and Galactic globular clusters. At a given metallicity the G-band appeared to be weaker in the M31 spectra, while the H- and K-lines seemed to be slightly too strong in the spectra of the globulars associated with the Andromeda galaxy. Subsequently Burstein et al. (1984) found that CN is enhanced significantly in several M31 globulars with respect to the Milky Way. Moreover, these authors found that M31 globulars also have significantly stronger Balmer lines than Galactic globulars. Van den Bergh's tentative conclusion that the Ca II H- and K-lines in M31 are enhanced, relative to Galactic globulars of similar [Fe/H], was subsequently confirmed by measurements of line-strength indices by Brodie & Huchra (1990, 1991). These authors also found the CN λ 4170 feature to be enhanced in the spectra of M31 globular clusters. Ponder et al. (1998) have used the *Hubble Space Telescope* to obtain spectra of the M31 globular clusters M II, M IV, K 58, and K 280. These authors find that the M31 clusters generally have stronger Mg_2 lines than do Galactic clusters with similar $\lambda\lambda$ 2600–3000 colors. Furthermore, the NH feature at λ 3360 in M31 globulars is greatly enhanced relative to its strength in Galactic globulars. This shows that the M31 globulars must be overabundant in nitrogen. (Or, perhaps, nitrogen is underabundant in Galactic globulars and in the Galactic halo and disk.)

A detailed discussion of the differences among the spectra of the M31 halo, disk, and bulge is given by Worthey (1998). He concludes that the nucleosynthetic history of the Andromeda galaxy must have been quite different from that of the Milky Way system: (1) M31 has substantially more N in its center, and (2) the metal-rich globulars in M31 have scaled-solar Fe, Mg, and Na but are significantly enhanced in N. According to Worthey this indicates that the Andromeda galaxy underwent a double nucleosynthesis process involving massive N production followed by a supernova-driven wind.

Detailed studies of the color–magnitude diagrams of M31 globular clusters, and eventually spectra of individual M31 globular cluster stars, will be required to understand these "second parameter" effects. An important first step in this direction has been taken by Fusi Pecci et al. (1996), who have used the *Hubble Space Telescope* to produce color–magnitude diagrams, which extend to one magnitude below the horizontal branch, for eight globular clusters in M31. Not unexpectedly, the three metal-poor clusters with $-1.8 <$ [Fe/H] < -1.5 are seen to have strong blue horizontal branches. In contrast, four metal-rich clusters with $-1.0 < 4$ [Fe/H] < -0.4 appear to have stubby red horizontal branches. Since the M31 globular clusters are all at approximately the same distance these data can be used to measure the (until recently quite controversial) relationship between metallicity and horizontal luminosity. If one adopts a distance modulus $(m - M)_0 = 24.40$ then Fusi Pecci et al. find that

$$M_V(\text{HB}) = (0.13 \pm 0.07)[\text{Fe/H}] + 0.98 \pm 0.09. \quad (3.7)$$

Taken at face value this result shows that the luminosity of horizontal branch stars, and of cluster-type variables, is *not* a strong function of metallicity. Weak support for the shallow slope of the M_V versus [Fe/H] relation, which was found above, is provided by the *HIPPARCOS* statistical parallaxes (Fernley et al. 1997a) of Galactic RR Lyrae stars. These authors find $M_V = +0.77 \pm 0.17$ for 84 halo RR Lyrae stars with $\langle[\text{Fe/H}]\rangle = -1.66$ and $M_V = +0.69 \pm 0.21$ for 60 metal-rich RR Lyrae variables with $\langle[\text{Fe/H}]\rangle = -0.85$. However, McNamara (1997a,b) has used the Baade–Wesselink technique to derive a steep M_V (RR) versus [Fe/H] relation, which differs at the $\approx 2\sigma$ level from that found by Fusi Pecci, using the horizontal branch luminosities of globular clusters in M31.

From a reasonably complete sample of 82 globular clusters in the halo of M31 Secker (1992) finds a mean cluster luminosity $\langle M_V \rangle = -7.51 \pm 0.15$, with a Gaussian dispersion of $\sigma \simeq 1.42$ mag. This value is 0.22 ± 0.20 magnitude brighter than the peak of the Galactic globular cluster luminosity distribution.[7] An important caveat is that the luminosity of halo globular clusters may differ systematically from that of globular clusters at smaller galactocentric distances. The conclusion that the M31 globular cluster luminosity function peaks at a slightly higher luminosity than does that for the Galaxy is confirmed by Reed, Harris & Harris (1994). These authors also show that the halo ($R > 4$ kpc) globular clusters in M31 are, on average, somewhat more metal rich (and hence redder) than those in the Galaxy. As a result the difference in the peak of the luminosity distributions of clusters should be slightly smaller in the B-band than it is in V.

3.8.4   Radii of globular clusters

Fusi Pecci et al. (1994) have used the *Hubble Space Telescope* to study the radial profiles of some of the globular clusters in M31. The radii of these clusters are found to be similar to those of Galactic globulars. Furthermore, the cluster G 105 is observed to have a collapsed core. From observations of the velocity dispersions in nine M31 globular clusters Dubath & Grillmair (1997) find M/L values that are similar to those of Galactic globulars. The ellipticities of the globular clusters associated with the Andromeda galaxy have been compared with those in the Milky Way system and in the Large Magellanic Cloud (Han & Ryden 1994, D'Onofrio et al. 1994). The clusters in both M31 and the Galaxy are found to be less flattened than those in the LMC. A notable exception to this rule is the metal-rich halo cluster Mayall II, which is the most luminous globular in M31. This cluster has $\epsilon = 0.22$. It is of interest to note that ω Centauri (= NGC 5139) is both the most highly flattened and the most luminous globular cluster associated with the Galaxy. Finally, NGC 1835 is both the most luminous and the most highly flattened cluster associated with the Large Cloud. These observations suggest (van den Bergh 1996a) that the most luminous globular cluster in any cluster system may also be the most highly flattened.

3.9   Variable stars

The modern era of the exploration of M31 began with the discovery of Cepheid variables by Hubble (1925a,b). An interesting sidelight on this discovery is provided by Christianson (1995), who writes that Humason told Sandage in 1956 (and stood by it until his death in 1972) that he had first discovered Cepheids in M31, but that Shapley "then calmly took out his handkerchief, turned the plates over, and wiped them clean of Humason's marks." The first systematic study of the variables in the Andromeda galaxy was carried out by Hubble (1929). Among the 50 variable stars that he discovered, 40 turned out to be Cepheids. Subsequently Baade and his colleagues used the Palomar 5-m reflector to study variable stars in four fields at different radial distances along the major axis of the Andromeda galaxy. Each of these "selected areas" was studied in detail with the Palomar 5-m telescope. Fields I, II, III, and IV are located at 15′ (3.3 kpc), 35′

---

[7] The mean absolute magnitude of the M31 globular clusters given above was based on $(m - M)_0 = 24.26 \pm 0.08$ and $A_V = 0.25 \pm 0.06$, resulting in $(m - M)_V = 24.51 \pm 0.10$. With our adopted value $(m - M)_V = 24.59$ (see Table 3.4) the mean cluster luminosity would become $\langle M_V \rangle = -7.6$.

Table 3.3. *Data on Cepheid variables in M31[a]*

Field	Distance	Period		$P < 4^{d}3$
		Longest	Median	
I	15'	$46^{d}9$	$13^{d}8$	16%
II	35'	42.3	7.0	36
III	50'	33.5	8.1	27
IV	96'	21.3	3.9	60

[a]From Baade & Swope (1965).

(7.7 kpc), 50′ (11 kpc), and 96′ (21 kpc) from the nucleus of M31. Each Field has an angular diameter of 15′. Field I takes in a region containing a spiral arm, which manifests itself by the presence of dust patches superposed on the outermost part of the nuclear bulge of the Andromeda galaxy. Field II encompasses a major dust arm and the super association NGC 206. Field III coincides with a broad spiral-arm feature that contains the largest concentration of resolved stars in M31. Many of the variable stars originally discovered by Hubble (1929) are located in this region. Finally, Field IV encompasses the outlying association OB 184. Some data on the Cepheid variable stars in Fields I–IV are collected in Table 3.3. The most striking feature of this table is that the mean period of Cepheid variables decreases dramatically as one moves outward from the center of M31. Part of this effect is, no doubt, due to the difficulty of finding faint short-period variables in Fields I and II (in which the unresolved stellar background is high). Nevertheless, the outward decrease in the period of the longest period Cepheids (which are bright and hence almost unaffected by selection effects) shows that some of this effect is almost certainly intrinsic to the Cepheid variable population in the Andromeda galaxy. A similar phenomenon was noticed for Cepheid variables in the Milky Way system by van den Bergh (1958), who interpreted this effect as the first evidence for a radial abundance gradient in the Galactic disk. A similar explanation probably accounts for the radial change in the periods of M31 Cepheids.

Baade and Swope (1963) were able to show that the period–luminosity relation for W Virginis stars (the Cepheids of Population II) parallels that for classical Cepheids of Population I but lies almost 2 mag below it. This observation provided a beautiful confirmation of Baade's (1954) hypothesis that the extragalactic distance scale was in error by a factor of ~2 because Shapley had adopted the zero-point of Population II variables to calibrate the period–luminosity relation for classical Cepheids of Population I. Freedman & Madore (1990) have published multicolor photometry of the Cepheids at ~3 kpc, ~10 kpc, and ~20 kpc from the nucleus of M31. After making corrections for reddening, Freedman & Madore found no significant difference between the Cepheid distance moduli derived from these three fields. Since metallicity in M31 is known to be a function of radius (Blair, Kirshner & Chevalier 1982) these results appeared to show that the Cepheid period–luminosity relation is independent of metallicity. However, this conclusion remains controversial because it is difficult to separate metallicity effects on (1) Cepheid luminosities and colors and (2) abundance-dependent changes in the reddening properties of interstellar grains. The dependence of Cepheid luminosities on their metallicities has been reviewed by Nevalainen & Roos (1998), who conclude that

$$\delta(m - M) = \gamma([\text{O}/\text{H}] - [\text{O}/\text{H}]_{\text{LMC}}), \tag{3.8}$$

with $\gamma = 0.31^{+0.15}_{-0.14}$ mag dex^{-1}. Note that an increase in γ results in an increase in the distance scale and hence in a decrease of the Hubble parameter H_0.

The most luminous variables in M31 (and in other late-type galaxies) belong to a class that was first described by Hubble & Sandage (1953). These are blue supergiants that exhibit cyclical variations with characteristic timescales of years. Four objects of this type (V19 = AF And, AE And, V15, and VA1) are known in M31. The light curve of AF And, which is the brightest of these objects, is shown in Figure 3.15. An additional candidate luminous blue variable in M31 has been discussed by King, Walterbos & Braun (1998). A decade of observations of V19 collected by Hubble (1929) already showed clearly that this object exhibited fluctuations in brightness with a timescale of about a year. Light curves for the other Hubble–Sandage variables in M31 have been illustrated by Sharov (1990). Spectroscopic observations by Humphreys (1975, 1978) show that these objects have a strong hot continuum that exhibits no Balmer discontinuity. P Cygni and η Carinae in the Galaxy, and S Doradus in the Large Cloud, are objects that belong to the same class. Such stars are typically found to have outflow velocities of 100–300 km s^{-1}, which results in stellar mass-loss rates of a few $\times 10^{-5}\, M_\odot$ yr^{-1}. In the Hertzsprung–Russell diagram the Hubble–Sandage variables occupy an inclined instability strip (Westerlund 1997, p. 83) bounded by $-11 \lesssim M_{\text{bol}} \lesssim -9$ and 14,000 K $\lesssim T_{\text{eff}} \lesssim$ 35,000 K.

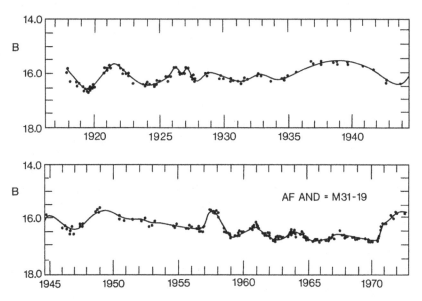

Fig. 3.15 Light curve of AF Andromedae, which is the most luminous Hubble–Sandage variable in M31.

3.10   Supernovae

When he composed his great paper on the variable stars in the Andromeda galaxy, Hubble (1929) was already clearly aware of the fundamental difference between novae and supernovae. He wrote "The occasional novae which have appeared in the smaller extragalactic nebulae and which have attained luminosities that are respectable fractions of the total luminosities of the nebulae [i.e., galaxies] themselves are clearly a different sort of object from the normal novae in M31 and 33. They are generally classed with S Andromedae, the nova of 1885 in M31, as a rare and peculiar type."

S And (= SN 1885A) is the only supernova that has been observed in M31 during historical times. A definitive discussion of this object was published at the time of the centenary of this event by de Vaucouleurs & Corwin (1985). S Andromedae occurred in the bulge of M31 at a projected distance of only $16''$ (59 pc) from the nucleus. If the probability of a supernova outburst were proportional to the surface brightness of background light, then the a priori probability of a supernova occurring so close to the nucleus of M31 would be only $\sim$0.5%. This suggests that one should at least keep the possibility in mind that the (super metal-rich) stars near the nucleus might be particularly prone to becoming supernovae. The hypothesis that S And was an unusual or peculiar object is supported by the observation that this object exhibited an unusually rapid decline in luminosity. Since the stellar population in the nuclear bulge of M31 resembles that in elliptical galaxies (in which only supernovae of Type Ia are observed to occur), it appears likely S And was also a supernova of this type. This hypothesis is supported by the discovery of the iron-rich remnant of this supernova by Fesen, Hamilton & Saken (1989). From the descriptions of the visual spectrum of S And, de Vaucouleurs & Corwin (1985) conclude that "nearly all of the features noted, however marginally, in the spectrum of S And by the spectroscopists of 1885 correspond closely to known intensity maxima or minima in the spectra of Type I supernovae." Fesen et al. (1999) have used the *Hubble Space Telescope* to study the absorption silhouette of the remnant of S And in the H&K lines of calcium. They find that this object is circular and has a diameter of $0\rlap{.}''70 \pm 0\rlap{.}''05$, corresponding to a linear radius of 1.3 pc. From the decrease of the intensity of M31 bulge light in the remnant Fesen et al. find that it lies $\sim$64 pc in front of the midpoint of the bulge. The spectrum of this remnant yields an average expansion velocity of $11,000 \pm 2,000$ km s^{-1}. Puzzlingly the center of the Ca I absorption feature at 4,227 Å appears to be redshifted by 1,100 km s^{-1}.

The remnants of supernovae exhibit larger [S II]/Hα line intensity ratios than do H II regions. Using this criterion, Blair, Kirshner & Chevalier (1982) were able to discover the remnants of 11 prehistoric supernovae in the disk of M31. More recently Braun & Walterbos (1993) have published an atlas of 52 "forbidden line" supernova remnants (SNRs) in the NE half of M31. The fact that the number of supernova remnants discovered in this galaxy is relatively small suggests that the low star formation rate in M31 has resulted in a low supernova rate. For galaxies of types Sab and Sb, van den Bergh & Tammann (1991) derived supernova rates of 0.49 h^2 SNu for Type Ia, 0.27 h^2 SNu for Type Ib, and 1.35 h^2 SNu for Type II, in which $h = H_0/100$, and one SNu corresponds to a rate of one supernova per century per $10^{10} L_\odot(B)$. Assuming $h = 0.7$ and a luminosity $L_B = 6 \times 10^{10} L_\odot(B)$ for M31, the corresponding predicted supernova rates are 1.4, 0.8, and 4.0 per century for Types Ia, Ib, and II, respectively. These predicted rates are clearly significantly higher than the observed rate of only one per century. Most of this discrepancy is probably due to the fact that M31 has a below-average rate of star formation. If the distance to M31 is approximately 2×10^4 light centuries, and the M31 supernova rate is about one per century, then the light of about twenty thousand supernovae that have already occurred is presently on its way to us from the Andromeda galaxy !

3.11   Novae

Two novae in M31 were discovered with the Mt. Wilson 1.5-m telescope in 1909. Subsequently a few more novae were observed by Duncan, Ritchey, and Shapley. A flood of more recent observations has brought the total number of nova discoveries in M31 to over 300. The first comprehensive study of these objects was made by Hubble (1929), who was able to establish the main features of the distribution, frequency, and light curves of novae in M31. In his paper Hubble appears so eager to establish the similarity between novae in M31, the LMC, and the Galaxy that he ignores the fact that objects such as nova Aquilae 1918, with $M(\max) = -9$, appeared to be significantly more luminous than any of the novae in the Andromeda galaxy. This, together with his (Hubble 1932) observation that Galactic globular clusters appeared to be less luminous than their counterparts in M31, should have alerted Hubble to the fact that there was a serious problem with the extragalactic distance scale.

Almost half a century after it was first used by Ritchey to discover novae in M31, the Mt. Wilson 1.5-m telescope was again dedicated to the study of novae in this galaxy. During the period June 1953 to January 1955, Arp (1956) discovered and obtained light curves for 30 novae in the Andromeda galaxy. These observations demonstrated that the novae in M31 followed a tight maximum magnitude versus rate-of-decline relationship similar to that which McLaughlin (1945) had found for novae in the Galaxy. This relation has subsequently been used to determine the distance to novae in the Virgo cluster (Pritchet & van den Bergh 1987b). More recent surveys of novae in M31 are by Rosino (1964, 1973), Capaccioli et al. (1989), and by Rosino et al. (1989). The most recent version of the maximum magnitude versus rate-of decline relation (Della Valle & Livio 1995) for novae in M31 is shown in Figure 3.16.

After correcting their data for various sources of systematic error, Capaccioli et al. found a rate of 29 ± 4 novae yr^{-1} for the Andromeda galaxy. From a comparison between the Galactic and M31 maximum magnitude versus rate-of-decline relations, Capaccioli et al. (1989) derive a distance modulus $(m - M)_B = 24.20 \pm 0.20$. Sharov (1993) found no statistically significant differences between the spatial distributions of fast and slow novae over the surface of the Andromeda galaxy. From the data by Arp (1956), Capaccioli et al. (1989), and Della Valle & Livio (1995) it is seen that novae in the nuclear bulge of M31 obey the same maximum magnitude versus rate-of-decline relation as do those located at larger galactocentric distances. However, fast and luminous novae appear to exhibit a higher relative frequency in the nuclear bulge than they do at larger galactocentric distances. It is not yet clear if this is an intrinsic effect, or whether it might be due to a selection effect resulting from the lower discovery probability of faint novae in the high surface brightness nuclear bulge. Figure 3.17 shows a plot of the rates of decline $\dot{m}$ of novae in the Andromeda nebula versus their distance $|X|$ from the minor axis of M31. This plot shows no evidence for a dependence of $\dot{m}$ on $|X|$ for $|X| > 2'$. About 5% of the novae in M31 appear to belong to a separate class that falls $\gtrsim 1$ mag above the standard maximum magnitude versus rate-of-decline relation. [One of seven novae in Virgo

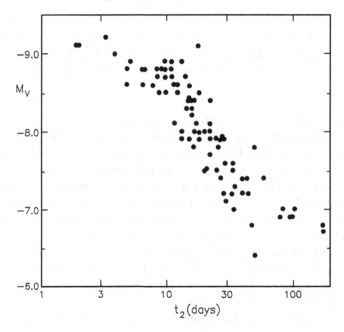

Fig. 3.16 Maximum magnitude versus rate-of-decline relation for novae in M31. Heavily absorbed novae and recurrent novae are not plotted. The figure shows that novae in M31, like those in the Galaxy, exhibit a clear-cut relationship between their magnitude at maximum light M_V and the time t_2 that it takes them to decline in brightness by two magnitudes. The data in the figure assume $(m - M)_o = 24.3$ for M31.

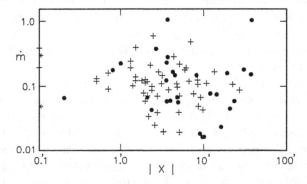

Fig. 3.17 Rate-of-decline $\dot{m}$ in magnitudes per day, versus distance $|X|$ from the minor axis of M31. Data from Arp (1956) are shown as dots and those by Capaccioli et al. (1989) as plus signs. The data show no evidence for a dependence of $\langle \dot{m} \rangle$ on $|X|$ for $|X| > 2'$. The deficiency of faint slow novae in the nuclear region of M31 might be due to a selection effect.

E galaxies (Pritchet and van den Bergh 1987b) also appears to belong to this bright nova class.] An unusually slow nova in the bulge of M31 has been discussed by Tomaney & Shafter (1993). They found this object to have physical characteristics very similar to those of slow novae in the Galaxy. A remarkable very slow nova (?) in the outer part of the M31 Disk has been discussed by Sharov et al. (1998). This object remained at $B \approx 20.5$ for >20 years and then flared up to $B = 17.25$ in 1992.

Ansari et al. (1999) report the discovery of a possible microlensing event that reached $R(\text{max}) = 18$ and $B - R(\text{max}) = 0.6$. This event may have been produced by lensing of a star with $R = 22$ that is located only $42''$ from the nucleus of M31. The possibility that this object was, in fact, a nova cannot be entirely ruled out. However, the faint value of $R(\text{max})$, and its fast rate of decline, would appear to militate against this possibility. Furthermore, the fact that this event occurred in the nuclear bulge of the Andromeda galaxy (in which absorption is low) makes it unlikely that is was a heavily obscured fast nova.

The Mt. Wilson 1.5-m telescope, with which most M31 novae have been discovered, has a field of $58' \times 81'$ at its Newtonian focus. This means that all of the bulge, but only part of the $92' \times 197'$ (Holmberg 1958) disk of the Andromeda galaxy, can be covered on a single exposure. This, together with significant dust absorption in the disk and incompleteness due to overexposure effects in the central region of the bulge, renders both the disk and the bulge nova rate in M31 rather uncertain. Capaccioli et al. (1989) concluded that $\sim$85% of the novae in M31 were associated with the bulge (and halo) population of the Andromeda galaxy. However, this result has recently been disputed by Hatano et al. (1997a,b) who conclude that most of the M31 novae are, in fact, associated with the disk rather than with the bulge. Ciardullo et al. (1990) find that NGC 5128 (= Centaurus A), the Large Magellanic Cloud, and Virgo ellipticals all have very similar nova rates normalized to their infrared K-band luminosities. This might be taken to suggest that the number of binary nova progenitors is similar in different environments. However, Ciardullo et al. find that the nova rate per $10^{10} L_\odot(K)$ in the disk of M31 is < 1.1 per year compared to a rate of $\sim$4 per year per $10^{10} L_\odot(K)$ for the other galaxies that they studied. These observations suggest the possibility that the nova rate in the disk of M31 may have been seriously underestimated. The total nova rate in M31 may therefore be significantly higher than the value of 29 ± 4 per year derived by Capaccioli et al. (1989). Alternatively, and perhaps more plausibly, Yungelson, Livio & Tutukov (1997) have argued that the currently observed nova rate in spiral galaxies is mainly determined by the recent disk star formation rate. Based on this hypothesis the low specific nova rate in the disk of M31 would be a consequence of the low rate of star formation in the disk of the Andromeda galaxy.

Because they emit strong Hα radiation it is very efficient to search for novae in the light of Hα in regions of high surface brightness (such as the inner bulge of M31 or in globular clusters) or in crowded star fields. Unfortunately it turns out that the Hα emission luminosity of novae does *not* exhibit a clear-cut correlation with the rate of decline of nova luminosity (Ciardullo et al. 1990). It is therefore not possible to use such Hα observations of novae to determine the distances to their parent galaxies.

3.12 Planetary nebulae

Over 300 bright planetary nebulae have been discovered in M31 by Ford & Jacoby (1978a,b). More recently Peimbert (1990) has estimated that the Andromeda galaxy contains $8,000 \pm 1,500$ planetaries, from which he finds that the total rate at which planetary nebulae are formed in M31 is 0.3 yr^{-1}. Nolthenius & Ford (1987) have used radial velocities of 37 planetaries with $R > 15$ kpc to study the dynamics of the outer disk of M31. They find that $\sim$1/3 of the objects in their sample are kinematically associated with the halo of the Andromeda galaxy. For the planetary nebulae that appear to be associated with the disk Nolthenius & Ford find a radial scale-length of 4.8 kpc, which is comparable to the value that Walterbos & Kennicutt (1988) obtained for the optical light of the M31 disk. From a velocity dispersion of 38 km s^{-1} at $\langle R \rangle = 19$ kpc Nolthenius & Ford (1987) derive a surprisingly large vertical scale-height of 1–3 kpc.

Jacoby & Ford (1986) find that the abundances derived at $\langle R \rangle = 19$ kpc in M31 are similar to those in the Orion nebula, which is located at a Galactocentric distance of only $R \simeq 9$ kpc. This result suggests that heavy element enhancement in M31 is more advanced than it is in the Galaxy. Furthermore observations of the halo planetaries M31-290 and M31-372, for which Henry (1990) finds logarithmic oxygen abundances [O/H][8] = 8.54 and 8.05, respectively (on a scale where the logarithmic hydrogen abundance is 12.0), provide direct evidence for chemical inhomogeneities in the halo of the Andromeda galaxy. Stasińska, Richer & McCall (1998) find that luminous planetaries in M31 and in the Galaxy have similar values of [O/H] = 8.67 and N/O ≈ 0.36. The chemical inhomogeneity of the planetaries in M31 is strongly confirmed by Jacoby & Ciardullo (1999), who found that $7.68 \leq 12 + \log(\text{O/H}) \leq 8.95$ for the planetaries in the bulge of the Andromeda galaxy. In other words the metallicities of planetary nebulae in M31 cover a decade in oxygen abundance. A similarly large abundance range is found for sulfur and neon. Unexpectedly none of the 12 bulge planetaries observed by Jacoby & Ciardullo approach the super metal-rich ([Fe/H] $\sim +0.25$) values that might have been expected. Their mean oxygen abundance is 8.40 ± 0.10. Surprisingly this is similar to the mean oxygen abundance of 8.44 found for the LMC, although it is larger than the mean SMC abundance of 8.24. It is not yet clear if the lack of luminous super oxygen-rich planetaries in the bulge of M31 is due to an unusual evolutionary history, or if the oxygen-rich planetaries never became luminous enough to be observed by Jacoby & Ciardullo. For the M31 planetaries, variations in N/O are found to be due almost entirely to variations in N, and independent of O. This shows that nitrogen is not being produced at the expense of oxygen.

Jacoby et al. (1992) have shown that the shape of the luminosity function of planetary nebulae (as measured in the λ 5007 line of [O III]) appears to differ little from galaxy to galaxy. This luminosity function can therefore be used as a distance indicator, which can be well calibrated in M31, and used out to distances comparable to that of the Virgo cluster. Ciardullo et al. (1989) find $(m - M)_0 = 24.26 \pm 0.04$ for M31 from the luminosity distribution of its planetary nebulae.

3.13   X-ray observations of M31

A number of recent studies of X-ray sources in the central region of M31 have been reviewed by Primini, Forman & Jones (1993). These authors used the *ROSAT* satellite to survey the innermost $\sim 34'$ (7.5 kpc) of the Andromeda galaxy. A total of 84 point sources associated with M31, having 0.2–4.0 keV luminosities greater than $\sim 2 \times 10^{37}$ erg s^{-1}, were found in the central region of M31. Of these sources 18 can be identified with globular clusters. The X-ray luminosity function of these cluster sources appears to have a rather well-defined upper limit near $L_X = 6 \times 10^{37}$ erg s^{-1}. In addition to the 18 sources identified with globular clusters, two can be identified with known supernova remnants, and one with the nucleus of M31. On the basis of *ROSAT* observations Kahabka (1998) concludes that M31 contains 16 "supersoft" sources produced by hot

---

[8] [O/H] $= \log(\text{O/H})_* - \log(\text{O/H})_\odot$, with $12 + \log(\text{O/H})_\odot = 8.93$ (Anders & Grevesse 1989).

$(3.0-4.5) \times 10^5$ K white dwarfs. Almost half of the X-ray sources in the central region of the Andromeda galaxy appear to be variable. A significant fraction of the X-ray luminosity of the inner part of the M31 bulge cannot be accounted for by known resolved sources. This diffuse radiation might be due to a hot component of the interstellar medium and/or to an unknown class of faint X-ray sources.

Primini et al. (1993) have pointed out that the projected density of X-ray sources per unit area that have $L_X > 10^{36}$ erg s^{-1} is much greater in the bulge of M31 than it is in the bulge of the Galaxy. This difference is particularly striking when the inner 0.8 kpc of the Andromeda galaxy and of the Milky Way system are intercompared. An additional difference between these two galaxies is that the central sources in the Milky Way have a distribution that is flattened toward the Galactic plane, whereas no such flattening is detected in M31. Ciardullo et al. (1987) have made the interesting suggestion that the central concentration of M31 X-ray sources is due to objects that were either ejected from globular clusters or had their origin in disrupted globulars.

Using the *Compton Gamma Ray Observatory*, Blom, Paglione & Carramiñana (1999) find an upper limit of 1.6×10^{-8} photons cm^{-2} s^{-1} for the γ-ray flux, above an energy of 100 MeV, from the Andromeda galaxy. If the Milky Way system were placed at the distance of M31, it would have over twice the lower limit quoted above. Presumably, this lower γ-ray emission of M31 results from the lower rate of star formation (and hence of supernovae) in this galaxy in the Milky Way.

3.14   Summary

The Andromeda galaxy is a spiral that resembles our own Milky Way system. M31 is, however, more luminous, and of earlier Hubble type, than the Galaxy. A summary of the fundamental characteristics of M31 is given in Table 3.4. The light of the nuclear bulge of M31, which Baade (1944) believed to consist of metal-poor stars of Population II, in fact turns out to be dominated by old metal-rich giant stars. The red integrated colors of M31 show that the present rate of star formation is low. Furthermore, the majority of star clusters associated with M31 are found to be old and red. In this respect the star and cluster forming history of the Andromeda galaxy differs dramatically from that of presently very active galaxies such as the Magellanic Clouds. The fact that the semistellar nucleus of M31 exhibits stronger metallic lines than the vast majority of M31 globular clusters militates against the suggestion that most of the starlight associated with the nucleus of the Andromeda galaxy was derived from globular clusters that had been captured by dynamical friction. However, the observation that the nucleus has bluer UV colors than the bulge of M31 suggests that blue horizontal branch stars from a *few* captured globulars may contribute to the present integrated light of the nucleus. Alternatively, collisions might have stripped the envelopes from large cool red giants. A striking difference between the halo populations of M31 and the Galaxy is that the stars and clusters in the halo of the Andromeda galaxy are, on average, significantly more metal rich than their Galactic counterparts. This suggests that metal enrichment proceeded more vigorously during the M31 halo evolutionary phase than it did during the formation of the stellar population in the Galactic halo. Alternatively it might be supposed that the Andromeda galaxy underwent a more rapid initial burst of star formation, which (temporarily) ejected most of the remaining metal-poor gas from the M31 halo. In this way the ratio of stellar ejecta to remaining gas would be relatively high, resulting in

Table 3.4. *Data on the Andromeda galaxy*

$\alpha(2000) = 00^h42^m44^s.2$ (14)		$\delta(2000) = +41°16'09''$ (14)
$\ell = 121°.17$		$b = -21°.57$
$V_r = -301\,\mathrm{km\,s^{-1}}$ (12,15)	$i = 77°.7$ (1)	p.a major axis $= 37°.7$ (1)
$V = 3.38$ (1,2)	$(B - V) = 0.91$ (1)	$(U - B) = 0.50$ (1)
$(m - M)_0 = 24.4 \pm 0.1$	$E(B - V) = 0.062$ (11)	$A_V = 0.19\,\mathrm{mag}$
$M_V = -21.2$	$D = 760\,\mathrm{kpc}$	$D_{\mathrm{LG}} = 0.30\,\mathrm{Mpc}$
Type = Sb I–II	size $= 92' \times 197'$ (20 kpc $\times$ 44 kpc) (3)	
$M = (2–4) \times 10^{11}\,M_\odot$ (4)	$M_{\mathrm{HI}} = 5.8 \times 10^9\,M_\odot$ (5)	$M_{\mathrm{H_2}} = {\sim}3 \times 10^8\,M_\odot$ (6)
Total no. globulars $= 400 \pm 55$ (7)		
Total no. planetaries $= 8,000 \pm 1500$ (8)	Nova rate $= 29 \pm 4\,\mathrm{yr^{-1}}$ (9)	
M (nucleus) $= 7 \times 10^7\,M_\odot$ (10)		
	$12 + \log(\mathrm{O/H}) = 9.0$ (13)	

(1) de Vaucouleurs (1958) obtains $V = 3.48$ within $B = 26.8\,\mathrm{mag\,arcsec^{-2}}$.
(2) Kent (1987) finds $V = 3.28$ for $R < 97'$.
(3) Holmberg (1958).
(4) Hodge (1992, p. 109).
(5) van de Hulst, Raimond & van Woerden (1957).
(6) Dame et al. (1993).
(7) Fusi Pecci et al. (1993).
(8) Peimbert (1990).
(9) Capaccioli et al. (1989).
(10) Bacon et al. (1994).
(11) Schlegel (1998, private communication).
(12) Rubin & Ford (1970).
(13) Massey (1998b).
(14) de Vaucouleurs & Leach (1981).
(15) Rubin & D'Odorico (1969).

the formation of moderately metal-rich halo stars and clusters. The present rate of star formation in M31 is much lower than it is in the Galaxy. Furthermore, it is noted that the most luminous globular clusters in M31 (Mayall II, $\epsilon = 0.22$), in the Galaxy (ω Centauri, $\epsilon = 0.19$), and in the LMC (NGC 1835, $\epsilon = 0.21$) are also the most highly flattened. It is presently not clear why X-ray sources appear to be so much more concentrated toward the center of M31 than they are toward the center of the Galaxy. Radial velocity observations show that the rotation curve of M31 is flat beyond $R \approx 15$ kpc, indicating that the Andromeda galaxy is embedded in a dark halo with $M \propto R$. At large values of R the surface density of globular clusters appears to drop off less rapidly than does that of halo stars. The fact that star counts in the outer regions of M31 drop as $R^{1/4}$ suggests that the structure of this region was produced by an early merger of massive ancestral galaxies. It is not yet clear why stars in the nuclear region of the Andromeda galaxy, and the M31 globulars, appear to be enriched in N compared to Galactic globulars and the solar neighborhood. It also remains puzzling why none of the planetaries in the bulge of M31 are observed to be super metal rich. Finally, it would be interesting to use the *Hubble Space Telescope* in "snapshot mode" to search for the (possibly compact) missing globular clusters in the central part of the M31 bulge.

4

The Milky Way system

4.1   Introduction

We mainly owe the "discovery" of the fact that the Milky Way system is a galaxy to the work of Shapley, Lindblad, and Oort. Shapley (1918a) noted that "we may say confidently that the plane of the Milky Way is also a symmetrical plane in the great system of globular clusters." He (Shapley 1918b) also found that "The center [of this system of globular clusters], which lies in the region of the rich star clouds of Sagittarius near the boundary of Scorpio and Ophiuchus, has the coordinates R.A. = 17^h 30^m, Decl. = $-30°$." Shapley concluded that "A consideration of the foregoing results leads naturally to the conclusion that globular clusters outline the extent and arrangement of the total [G]alactic organization." Shapley's discovery may be thought of as the second Copernican revolution. Copernicus had shifted the center of the Universe from the Earth to the Sun, and Shapley took the even greater step of moving it from the solar system to the center of the Galaxy. Subsequently Lindblad (1927) and Oort (1927, 1928) were able to show that (a) the Galactic disk is in rapid differential rotation about a center that coincides with that of Shapley's globular cluster system and that (b) globular clusters and high-velocity stars rotate more slowly around the same center than does the Galactic disk. The notion that the Milky Way system is a spiral galaxy received its ultimate confirmation when Morgan, Whitford & Code (1953) were able to outline three Galactic spiral arms in the vicinity of the Sun. A year later, the existence of *global* spiral structure in the Milky Way was demonstrated by the 21-cm observations of Westerhout (1954) and Schmidt (1954). Such hydrogen line observations were able to penetrate the dense dust clouds that obscure lines of sight located at low Galactic latitudes. Finally, the wide-angle photographs by Code & Houck (1955) provided stunning visual evidence for the similarity of the Milky Way galaxy to such nearby edge-on late-type spirals as NGC 891.

Table 4.1 gives a brief summary of the evidence for the Hubble type of the Milky Way system. The presence of the large bulge seen in wide-angle photographs, and of nearly circular 21-cm spiral arms, favors the notion that the Galaxy is of type Sb. However, the high Galactic star formation rate, and the presence of giant H II complexes (such as W 49), appear to favor an Sc classification. On balance all of the presently available evidence is probably consistent with the hypothesis that the Galaxy is of Hubble stage bc. The discovery (Blitz & Spergel 1991) of a nuclear bar in the Milky Way system suggests that an S(B)bc classification type is probably appropriate. Detailed photometric evidence for the existence of this bar is provided by Stanek et al. (1996).

Table 4.1. *Hubble type of the Milky Way system*

Type of Information	Favors Sb	Favors Sc
Total amount of H I	x	
Shape of 21-cm arms	x	
Shape of optical spiral arms		x
Rate of star formation		?
Giant H II region W 49		x
Presence of large bulge	x	

One of the most striking features of wide-angle photographs of the Milky Way system is that the central bulge of the Galaxy in Sagittarius sticks out below the dust band that obscures the Galactic equatorial plane. Baade (1951) suggested that this low absorption window, which is centered on the globular cluster NGC 6522($\ell = 1°0, b = -3°9$), would be the ideal place to study stellar populations in the bulge of the Galaxy. He found 76 RR Lyrae variables in this field, which is nowadays referred to as "Baade's Window." These RR Lyrae variables exhibited a striking frequency maximum at $B \sim 17.5$. Adopting M_B (RR) = 0.0, and deriving the reddening in this direction from the integrated color of NGC 6522, Baade derived a distance $D = 8.7$ kpc to the Galactic nuclear bulge. The fact that this value is very close to the presently accepted value $D = 8.5 \pm 0.5$ kpc (see Section 4.3) is due to a fortuitous cancellation of errors. We now know that Baade's overestimate of the luminosity of the RR Lyrae variables was almost exactly counteracted by the fact that he overestimated the reddening toward the Galactic center. This overestimate was due to the fact that NGC 6522 is a cluster of intermediate metallicity, whereas Baade had assumed that it had the blue intrinsic color of metal-poor halo globulars. Finally, Morgan (1959) noted that the nuclear bulge of the Galaxy does *not*, as Baade believed, consist entirely of globular clusterlike stars of Population II. The RR Lyrae variables that Baade used to determine the distance to the nuclear bulge, in fact, only constitute a relatively minor component within the predominantly metal-rich old stellar bulge population.

Excellent recent books on the Galaxy are Blitz & Teuben's (1996) *Unsolved Problems of the Milky Way*, Gilmore, King & van der Kruit's (1989) *The Milky Way as a Galaxy*, Humphreys's (1993) *The Minnesota Lectures on the Structure and Dynamics of the Milky Way*, Majewski's (1993) *Galaxy Evolution: The Milky Way Perspective*, Morris's (1989) *The Center of the Galaxy*, and Mihalas & Binney's (1981) *Galactic Astronomy*.

4.2 The nucleus of the galaxy

The compact radio source Sgr A* is located at $\alpha = 17^h 45^m 39°9, \delta = 29° 00'$ $28''$ (J2000). Backer & Sramek (1982) find that its proper motion is consistent with the hypothesis that this source is at rest in the Galactic nucleus (i.e., the observed proper motion may be interpreted as being entirely due to reflected solar motion). Available evidence on the distance to the Galactic nucleus has been reviewed by Reid (1993), who concludes that $R_0 = 8.5 \pm 0.5$ kpc. From the *HIPPARCOS* proper motions of Cepheids, Feast & Whitelock (1997) also find $R_0 = 8.5 \pm 0.5$ kpc. A similar distance $D = 8.4 \pm 0.4$ kpc has been obtained by Paczyński & Stanek (1998) from the red giant clump in Baade's Window, in conjunction with the *HIPPARCOS* calibration of local red clump stars. However (Girardi et al. 1998, Cole 1998), it is by no means certain that such a comparison will not be affected by age and composition differences between red clump stars in the solar neighborhood and clump stars in the nuclear bulge of the Galaxy.

From *K*-band observations with the Keck 10-m telescope, Ghez et al. (1998) find that the peaks of both the stellar surface density and of the highest stellar velocity dispersion are consistent with that of the unusual radio source and black hole candidate Sgr A*. Their observations suggest that the positions of Sgr A* and of the dynamical center of the Galaxy are coincident to $\pm 0\rlap{.}''1$.

From radial velocity and proper motion observations of individual stars in close proximity to Sgr A*/Eckert & Genzel (1997) derive a mass $M = (2.45 \pm 0.14) \times 10^6 \, M_\odot$ for the region within 0.015 pc of the nucleus. This observation gives strong support to the notion that the Milky Way system resembles the Andromeda galaxy, which has an even larger black hole with a mass of $7 \times 10^7 \, M_\odot$ (Bacon et al. 1994) at its center. Particularly strong evidence for the existence of a very concentrated high mass in the Galactic center is provided by a star with a space motion $\gtrsim 1{,}500$ km s^{-1} that is located at a projected distance of $0\rlap{.}''19$ (7.4×10^{-3} pc) from the nucleus (Eckert & Genzel 1997).

Studies of the stellar population in the immediate neighborhood of the Galactic nucleus are hampered by the fact that the foreground dust absorption amounts to $A_V \sim 30$ mag. As a result it is only possible to observe the central region of the Galaxy at infrared and radio wavelengths. From $\sim 1°$ (140 pc) to $0\rlap{.}''5$ (0.02 pc) the surface brightness of the central region of the Galaxy (Genzel & Townes 1987) scales as $R^{-0.75}$, whereas it scales approximately as R^{-2} at Galactocentric distances greater than $\sim 6°$ (840 pc). It is presently not entirely certain that star formation is currently taking place within a few pc of the Galactic nucleus. Morris (1993) has suggested that black holes might acquire temporary optically thick atmospheres in collisions with the numerous red giants in the dense central region of the Galaxy. Such objects would appear as luminous blue stars with stellar winds comparable to those observed in O and WR stars. Genzel et al. (1996) note a deficiency of luminous late-type stars within $r < 5''$ of the nucleus. They attribute this effect to the large collision cross sections of luminous late-type stars. A variation on this theme has been proposed by Davies et al. (1998), who find that a large fraction of encounters between binaries and red giant stars lead to the formation of a compact binary within an extended envelope, whilst the third body is ejected. Destruction of this red giant results from the ejection of its envelope as the binary "hardens."

A 3.6-cm radio map of the central region of the Galaxy is shown in Figure 4.1. This figure shows a number of thermal sources of which G $0.7 - 0.0$, also known as Sgr B2, is probably the best known. This source is one of the most luminous H II regions in the Galaxy. Sgr B2 is unique because of the richness of its molecular line spectrum (Martin & Downes 1972), which is probably due to high metallicity of the interstellar gas at a distance of only ~ 100 pc from the nucleus. Metzger (1971) points out that Sgr B2 is unlike normal spiral-arm H II regions because of its very high internal velocity dispersion.

The central star forming disk of the Galaxy has a radius of 115 pc. Embedded within this disk is the remarkable "arches" cluster at G $0.121 + 0.017$, which contains over 100 massive young stars (Serabyn, Shupe & Figer 1998), including some with WR spectra. The total observed mass of O-type stars in this cluster is $\sim 5{,}000 \, M_\odot$. This cluster is situated at a distance of only 17 pc from the Galactic center. The total number of O stars

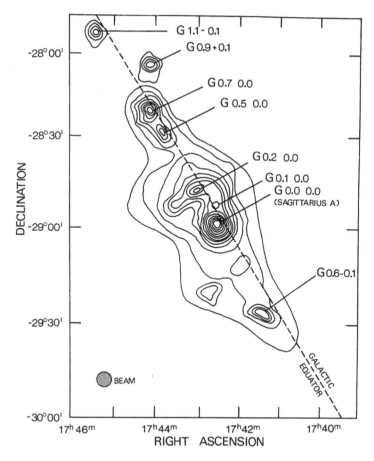

Fig. 4.1. Radio isophote maps (adapted from Downs & Maxwell 1966) of the innermost region of the Galactic Disk. Note the complex structure of the region surrounding the Galactic center. With the exception of Sgr A (and possibly of G 1.1 − 0.1) the sources shown in the figure all have thermal spectra. The source G 0.7 − 0.0 is also known as Sgr B2.

in the cluster is comparable to that in R 136, which is located at the center of the 30 Doradus complex in the LMC. However, the radius of the Arches cluster is only ∼1/3 of that of R 136. Its average density of ∼$3 \times 10^5 \, M_\odot \, pc^{-3}$ exceeds that in the core of R 136. It would be interesting to know if the mass spectrum in this cluster differs from that observed in young clusters situated in less-dense regions of the Galaxy.

The most interesting implication of the results discussed above is that the central region of the Milky Way system is still actively forming stars. It is therefore quite different from the central bulge of the Andromeda galaxy, which is devoid of any current star formation. This observation strongly suggests that the Galaxy is of later Hubble type than M31.

The Sagittarius emission complex, which is located within ∼10 pc of the Galactic center, can be divided into the shell-like nonthermal source Sgr A (East) and the thermal source Sgr A (West). The latter source coincides with the peak of the stellar 2-μm emission. An advection-dominated accretion flow onto a black hole with a mass of ∼ $2.5 \times 10^6 M_\odot$ provides an accurate description (Mahadevan 1998) of the spectral energy distribution of the nuclear source Sgr A*. Sgr A (East) is a nonthermal source that has a spectral index $\alpha = -0.76$ and shell dimensions of 8 pc × 10.5 pc (Ekers et al. 1983). It is remarkable that this supernova remnant (SNR), which is one of the three SNRs with the highest surface brightness in the Galaxy, is located only 2.5 pc from the nucleus of the Galaxy. Clearly the a priori probability that a supernova should have appeared recently so close to the Galactic nucleus, in a region that contains < 0.02% of the total Galactic baryonic mass, is exceedingly small. In this connection it is worth recalling (see Section 3.10) that S And also occurred very close to the nucleus of M31. Does this mean that the supernova rate is particularly high near spiral nuclei? Observations of extragalactic supernovae (Wang, Höflich & Wheeler 1997, van den Bergh 1997a) provide no support for the speculation that supernova rates might be unusually high near the nuclei of spiral galaxies. In fact Wang et al. arrive at the opposite conclusion that the inner 1 kpc of spirals are remarkably free of SNe Ia. However, this result might be due to the fact that photographic observations, in which the central regions of galaxy images are often "burned out," are biased against the discovery of supernovae.

The recent detection of γ rays from within $\sim 1°$ of the Galactic center by the *Compton Gamma Ray Observatory* (Melia, Yusef-Zadeh & Fatuzzo 1998) shows that the emission from this region extends to very high energies.

For a review on the Galactic center the reader is referred to Morris & Serabyn (1996).

4.3   The nuclear bulge of the galaxy

Dense absorbing clouds in the equatorial plane render the central bulge of the Milky Way galaxy almost entirely invisible at visual wavelengths. However, it shows up clearly in the *IRAS* all-sky survey (Habing et al. 1985) as a centrally peaked 15° × 20° (2.2 kpc × 2.9 kpc) concentration of 12-μm and 25-μm sources. From observations at 2.4 μm, Kent, Dame & Fazio (1991) find that the nuclear bulge of the Galaxy is an oblate spheroid with $b/a = 0.61$.[1] Most of the infrared sources are probably dust-embedded late-M giant stars. The existence of a strong metallicity gradient in the inner Galaxy (Terndrup 1988, Frogel et al. 1990, Minniti 1995) is expected to result in the formation of many very cool, dusty, metal-rich giants in the core of the Galaxy. The distribution of IR sources is therefore predicted to be more centrally concentrated than that of all evolved stars (King 1993). The metallicity gradient in the central region of the Galaxy will be enhanced by the star formation that is presently taking place (see Figure 4.1) in the inner disk within a distance of 50′ (125 pc) from the nucleus. However, the presence of a nuclear bar (Blitz & Spergel 1991) will stir up the gas and reduce the central metallicity gradient. Observations of RR Lyrae stars in the massive compact halo object (MACHO) fields in the direction of the nuclear bulge (Alcock et al. 1997a) show that a bar can only be seen in the inner fields with $\ell < 4°$ and $b > -4°$ (i.e., in those regions where the bar potential is strong enough to influence the kinematically hot RR Lyrae component). Lee (1992) has argued that these RR Lyrae stars in Baade's Window belong to the oldest subcomponent of Galactic Population II. The reason for this (Renzini & Greggio 1990) is that the collapse timescale, $\tau \sim (G\rho)^{-1/2}$, will be shorter in the dense central region of

---

[1] It is, of course, not possible from photometry to distinguish between a bar and an oblate spheroid that is viewed edge-on.

the proto-Galaxy than it is in lower density regions at larger Galactocentric distances. A potentially serious problem with this simple-minded scenario is, however, that the outer halo globular cluster NGC 2419 seems to be as old as typical metal-poor globulars in the inner halo of the Galaxy (Harris et al. 1997).

The discovery of large numbers of RR Lyrae stars in Baade's Window (Baade 1951) at first appeared to confirm Baade's hypothesis that the nuclear bulges of spiral galaxies consist of metal-poor old stars of Population II. However, Morgan (1959) later showed that the integrated light of the central bulge of M31, and of Baade's Window, was dominated by old metal-rich stars. Subsequent observations of individual red giants in the nuclear bulge of the Galaxy (Spinrad, Taylor & van den Bergh 1969, McWilliam & Rich 1994, Castro et al. 1996) confirmed the conclusion that these stars were metal rich. From an analysis of K-type giants in Baade's Window, Sadler, Rich & Terndrup (1996) find that $\langle[\text{Fe/H}]\rangle = -0.11 \pm 0.04$, with more than half of all stars having metallicities in the range $-0.4 < [\text{Fe/H}] < +0.3$. There are two reasons why the mean value for the metallicity quoted above may be lower than that of the bulk of the stars in the nuclear bulge: (1) The line of sight toward Baade's Window intersects the nuclear bulge at $Z \approx -0.6$ kpc where the metallicity is expected to be lower than it is at $Z \sim 0$ kpc and (2) the sample excludes M giants, which will, on average (Spinrad, Taylor & van den Bergh), be more metal rich than K giants. From echelle spectra obtained with the Keck telescope, Castro et al. (1996) found $[\text{Fe/H}] = +0.47 \pm 0.17$ for the star BW IV-167. This value is similar to that of the well-known nearby super metal rich star μ Leonis. From a study of the integrated light of starlight in Baade's Window, Idiart et al. (1996) find $[\text{Mg/Fe}] = +0.45$. Taken at face value this indicates that metal enrichment in the nuclear bulge was dominated by the ejecta of fast-evolving supernovae of Type II, rather than by the iron-rich material that is ejected from the more slowly evolving progenitors of supernovae of Type Ia. This in turn suggests that the bulge of the Galaxy might resemble elliptical galaxies that collapsed on a relatively short timescale. However, McWilliam & Rich (1994) find $\langle[\text{CN/Fe}]\rangle$ to be close to solar in metal-poor bulge stars. The weak CN lines in bulge stars therefore indicate that these objects must differ, in some respects, from the stars in elliptical galaxies. A similar problem was noted in Section 3.8.3, where the N abundance in M31 (and in its globulars) was found to be higher than the corresponding values in the Milky Way system. Another problem with the scenario suggested above is that McWilliam & Rich unexpectedly find $[\text{Ca/Fe}]$ and $[\text{Si/Fe}]$ in bulge stars to closely follow normal trends for Galactic *disk* stars. The data presented above indicate that the evolutionary history of the Galactic bulge was probably more complex than would be expected from a simple early collapse scenario. Possibly second-generation stars, which formed after the initial collapse of the Galaxy, provide a significant contribution to the stellar population in the Galactic nuclear bulge. Prima facie evidence for such second-generation star formation is, of course, provided by star-forming complexes like Sgr B2. A detailed discussion of the differences among disk, halo, and bulge abundances is given by Worthey (1998).

Contrary to some theoretical expectations (Morris 1993), observations of the Galactic nuclear bulge by Holtzman et al. (1998) appear to indicate that the bulge luminosity function over the range $+7 < M_V < +11$ does not differ significantly from that in the solar neighborhood. Taken at face value this result suggests that the mass spectrum of star formation does not depend strongly on environmental density.

Minniti (1996) has obtained radial velocities for metal-rich stars in a bulge field at $\ell = 8°, b = +7°$. He finds that such stars participate in Galactic rotation and have $\langle V \rangle = +66 \pm 5$ km s^{-1}, $\langle \sigma \rangle = 72 \pm 4$ km s^{-1}. In contrast, Minniti finds that metal-poor stars with [Fe/H] < -1.5 in this same field do not participate in Galactic rotation ($\langle V \rangle = -6 \pm 20$ km s^{-1}) and have a larger radial velocity dispersion ($\langle \sigma \rangle = 114 \pm 14$ km s^{-1}) than do metal-rich bulge stars. The bulge has a half-light radius $R_h \sim 200$ pc (Frogel et al. 1990), which is an order of magnitude smaller than that of the Galactic halo. Minniti (1995) concludes that the metal-rich globular clusters with $R_{GC} > 3$ kpc belong to the Galactic "Thick Disk" (however, see Section 4.7 for a different interpretation) but that those with $R_{GC} < 3$ kpc are kinematically associated with the bulge. From color–magnitude diagrams obtained with the *Hubble Space Telescope*, Ortolani et al. (1995) have concluded that the ages of the bulge clusters NGC 6528 and NGC 6553 do not differ by more than a few Gyr from those of the Thick Disk cluster 47 Tucanae. However, this inference is rendered uncertain by the fact that small differences in the helium abundance, and in the ratios of elements produced by SNe II and SNe Ia, may change the calculated ages of globular clusters by a gigayear, or more. It should also be emphasized that not all globular clusters that are presently located in the bulge are physically associated with it (i.e., they might just be falling through it). From a rather noisy color–magnitude diagram, which extends down to the main-sequence turnoff, Terndrup (1988) concluded that the bulk of the stars in Baade's Window have ages in the range 11–14 Gyr. He also found that the number of objects with ages <5 Gyr is negligible at $Z \approx -500$ pc.

Azzopardi et al. (1991) have discovered 34 carbon stars in eight fields in the Galactic nuclear bulge. The origin of these stars remains controversial. Ng (1997) has suggested that they are, in some way, related to the Sagittarius dwarf galaxy. However, Whitelock has argued that the C stars in the bulge are more metal rich than those in Sgr and therefore not related to them. Ng (1998) counters this argument by proposing that the carbon stars in the Galactic bulge were formed out of Galactic interstellar material during a collision with gas in Sagittarius, when that dSph galaxy last crossed through the Galactic plane.

NGC 6287 (Stetson & West 1994), which is a metal-poor ([Fe/H] $= -2.05$) globular cluster at a Galactocentric distance of only 1.9 kpc, might be an example of an object that has fallen in from the inner halo of the Galaxy. However, the small size ($r_h = 1.3$ pc) of this object suggests that it might, in fact, be physically associated with the nuclear bulge. If so, it could be one of the oldest objects in the Galaxy.

It is still not clear how the Galactic bulge formed. Carney, Latham & Laird (1990) and Wyse & Gilmore (1992) have suggested that it was made from low angular momentum gas that was left over after most star formation had ended in the halo. The idea that halo stars were formed from leftover halo gas, which was enriched by SNe II on a short time scale, appears to be supported by McWilliams & Rich (1994), who found that [Mg/Fe] and [Y/Fe] are (as is the case in halo stars) elevated by ≈ 0.3 dex. As had been pointed out before it is, however, not clear why [Ca/Fe] and [Si/Fe] in bulge stars should closely follow trends for normal *disk* giants. Alternatively it has been suggested (Combes et al. 1990, Pfenniger & Norman 1990, Merritt & Sellwood 1984) that galactic bulges formed from bars. However, an argument against this idea might be that bulges formed in this fashion should have erased most preexisting radial metallicity gradients. This appears to conflict with the observation (e.g., McClure 1969) that steep metallicity gradients continue to exist within some nuclear bulges.

A number of lines of evidence suggest that a short bar is embedded within the nuclear bulge of the Galaxy (Kuijken 1996). Raboud et al. (1998) have made the interesting suggestion that this bar might perturb a small fraction of stars into "hot" orbits that would allow them to wander erratically from inside the bar to outside the corotation radius. Could super metal-rich stars such as μ Leonis be objects that were ejected from the nuclear bulge into the solar neighborhood? Alcock et al. (1998b) also find evidence for a bar from observations of red giant clump stars. Only the more metal-rich RR Lyrae stars appear to belong to the barred bulge.

4.4   The Galactic disk

Van der Kruit (1986) has used the *Pioneer 10* spacecraft to measure integrated starlight in the blue and red from a position in the solar system located beyond the zodiacal dust cloud. From these observations he found the ratio of the disk scale-length h, to the vertical scale-hight h_z to be $h/hz = 17 \pm 3$. Owing to the difficulty of removing the effects of Galactic disk absorption the true uncertainty of this value may, however, be greater than the value cited above. Kent, Dame & Fazio (1991) have found $h_z = 247$ pc near the Sun. Adopting this value, rather than $h_z = 325 \pm 25$ pc used by van der Kruit, one obtains a Galactic disk scale-length of 4.2 kpc. This value is slightly smaller than the scale-length $h = 5.7$ kpc (see Section 3.6.1) that is derived from the 26′ scale-length in V light of the disk of the Andromeda galaxy. From infrared observations in a low-absorption region in the direction of the Galactic anticenter a much smaller scale-length $h = 2.5$ kpc, with an abrupt surface-brightness drop at $R_{GC} \sim 14$ kpc, has more recently been obtained by Robin, Crézé & Mohan (1992). Using 2.2-μm observations with the *COBE* satellite, Durand, Dejonghe & Acker (1996) find a scale-length $h = 2.6$ kpc; but they opine that h may be larger at visual wavelengths. If one nevertheless adopts the value $h = 4.2$ kpc, then the total luminosity of the Galactic disk becomes $L_B \approx 1.0 \times 10^{10} L_\odot$. Van der Kruit (1986) derived an integrated color $B - V = 0.95 \pm 0.15$ for the old disk component of the Galaxy.

A tabulation of published values of the Galactic disk scale-length, which range from 2.2 kpc to 5.5 kpc, is given by Robin (1997).

4.4.1   Evolution of the Galactic disk

The evolutionary relationships among the Galactic halo, the Galactic nuclear bulge, the Thick Disk, and the Thin Disk remain a subject of lively discussion and controversy. In particular it is not yet clear if the evolution from protogalaxy to disk was continuous, or if there was an extended hiatus (Berman & Suchkov 1991) between the halo and disk phases of Galactic evolution. Furthermore, it is not yet known whether the Thick Disk and Thin Disk represent distinct evolutionary phases, or if the Thick Disk (Thin Disk) was produced at the beginning (end) of a more-or-less continuous vertical collapse process. Realistic N-body calculations that take into account energy and angular momentum transfer to the Galactic halo (Barnes 1996) appear to show that the Thick Disk probably does not represent Thin Disk material that was heated by capture of infalling small satellite galaxies. Furthermore, the observation (Gratton et al. 1996) that [Mg/Fe] $\approx +0.4$ for stars in the Thick Disk, but that it decreases to [Fe/H] $= +0.2$ in the Thin Disk, appears to militate against the suggestion that the Thick Disk represents heated Thin Disk material. However, this conclusion depends somewhat on how the Thin Disk to Thick Disk transition is defined.

4.4.2   Age of the Galactic disk

The Thick Disk is observed to contain many RR Lyrae stars. Since such cluster-type variables occur in the "young" globular cluster Ruprecht 106 (Kaluzny et al. 1995), to which Richer et al. (1996) assign an age of 9.3 Gyr, it follows that the Thick Disk must have an age of at least 9 Gyr. An even larger age of 12^{+1}_{-2} Gyr has recently been obtained by Phelps (1997) for the probable open cluster Berkeley 17 at $R_{GC} \sim 11$ kpc. However, Carraro, et al. (1999) find a substantially lower age of 9 Gyr for this cluster. If one assumes that the disk formed at the end of the halo phase of Galactic evolution, then an upper limit to the age of the disk is set by the halo cluster M92, for which Bolte & Hogan (1995) derive an age of 15.8 ± 2.1 Gyr. A weaker, but entirely independent, upper limit to the age of the Galactic halo is set by the thorium abundance in the ultra metal-poor star CS 22893-052,[2] for which Sneden et al. (1996) find an age of 16 ± 6 Gyr. It should, however, be kept in mind (Phelps et al. 1994, Kaluzny et al. 1995) that there may be some overlap between the ages of the oldest Thick Disk open clusters and that of the youngest halo globular clusters. In other words the disk may have started to form before the assembly of the halo had been completed.

A quite different approach has been used by Jimenez, Flynn & Kotoneva (1998). These authors derived an upper limit of 11 ± 1 Gyr and a lower limit of 8 Gyr to the age of the disk near the Sun from the magnitudes and colors of "red clump" giants, for which accurate distances have been determined by *HIPPARCOS*. The ages of the oldest clump stars in the disk are 2–3 Gyr younger than those which these same authors compute for metal-poor globular clusters. Taken at face value, these results suggest that the formation of the Galactic disk at $R_{GC} \sim 8$ kpc must have started within 2–3 Gyr of the time when the halo started to form.

If all open clusters belong to the Thin Disk population then a lower limit to the age of the Thin Disk is set by NGC 6791, which is the oldest open cluster with a well-determined age. Garnavich et al. (1993), Kaluzny & Rucinski (1995), and Tripicco et al. (1995) find ages in the range 7–10 Gyr for this object. More recently, Chaboyer, Green & Liebert (1999) have obtained an age of 8.0 ± 0.5 Gyr for this cluster. A caveat is, however, that the motion of NGC 6791 lags circular disk motion by more than 60 km s^{-1} (Scott et al. 1995), so that its assignment to the Thin Disk population is not entirely beyond question. Peterson & Green (1998) have derived a metallicity [Fe/H] = $+0.4 \pm 0.1$ from high-dispersion spectroscopy of a blue horizontal branch proper motion member of this cluster. They find that the relative abundances of C, O, and Al to Fe are solar. However, Mg, Si, N, and Na appear to be enhanced (as they are in metal-rich galaxies). It is difficult to see how NGC 6791 could have reached such a high metallicity if it was formed in the Galactic disk ~ 10 Gyr ago. Could it perhaps have formed in the nuclear bulge and then been ejected by the Bar (or during a merger event that gave rise to the Bar)? One might imagine (Raboud et al. 1998, Lerner, Sundin & Thomasson 1999) that such a cluster could have been perturbed by the Bar (or by a merger event) into a "hot"

---

[2] Norris, Ryan & Beers (1997) note that a strong ^{13}CH feature at $\sim 4,019.0$ Å coincides with Th λ 4019.12. This might affect the use of the Th/Eu ratio as a Galactic chronometer.

orbit that allowed it to wander erratically from inside the Bar to outside the corotation region. However, it might turn out to be more difficult to accelerate a massive star cluster like NGC 6791 into a hot orbit than to accelerate an individual star, such as μ Leonis, into such an orbit. An age similar to that of NGC 6791 is obtained for matter in the solar neighborhood by Chamcham & Hendry (1996), who derive an age of at least 9 Gyr from a comparison between the calculated production ratios of the actinoid pairs $^{235}U/^{238}U$ and $^{232}Th/^{238}U$ and their presently observed abundances.

Smith (1998) has estimated the age of the local disk from the luminosity function of a sample of 152 white dwarfs that have main-sequence companions. From the luminosity of the break in their luminosity function, which results from the fact that the disk is too young to allow white dwarfs to cool to very faint luminosities, Smith derives a lower limit of $9.7^{+0.9}_{-0.8}$ Gyr for the local Galactic disk. An entirely independent lower age limit for the Thin Disk is set by the disk white dwarf ESO 439-26 (Ruiz et al. 1995) that has $M_V = +17.6 \pm 0.1$, which yields a cooling age of 6–7 Gyr. It would be of great interest to also try to obtain similar stringent age limits on the Galactic halo by observations of the luminosities of faint old halo white dwarfs. A comparison between the luminosity function of cool white dwarfs and recently calculated white dwarf cooling sequences (Leggett, Ruiz & Bergeron 1998) yields an age of 8 ± 1.5 Gyr for the disk in the solar neighborhood. However, some doubt has been thrown on such calculations by the work of Hansen (1999), which appears to show that only old white dwarfs with helium atmospheres are red, while those with hydrogen atmospheres are blue.

4.4.3 Old open clusters

The population of old open clusters in the (Thin) Disk of the Galaxy has been studied by Friel (1995). She finds no old clusters with $R_{GC} \lesssim 7$ kpc. The most probable explanation for this (van den Bergh & McClure 1980) is that such old open clusters have been destroyed by interactions with massive giant molecular clouds. The fact that old open clusters in the outermost part of the Galactic disk have *not* been disrupted appears to militate against the suggestion (Pfenniger 1998) that huge amounts of mass are present in the outer disk in the form of massive cold H I clouds. Friel concludes that old open clusters exhibit a steep radial metallicity gradient with $\langle[Fe/H]\rangle \simeq 0.0$ at $R_{GC} = 7$ kpc and $\langle[Fe/H]\rangle - 0.5$ at $R_{GC} = 12.5$ kpc. A quite different conclusion has recently been reached by Twarog, Ashman & Anthony-Twarog (1997). They find that clusters with $6.5 < R_{GC}(kpc) < 10$ all have $[Fe/H] \approx 0.0$, whereas those with $10 < R_{GC}(kpc) < 15$ have $[Fe/H] \approx -0.3$. Twarog et al. suggest that this apparent discontinuity in the metallicity of open clusters at $R_{GC} = 10$ kpc might be due to differences in Galactic evolutionary history. In particular they hypothesize that the discontinuity at $R_{GC} = 10$ kpc might represent the original edge of the Galactic disk. Figure 4.2 shows a plot of cluster metallicity [Fe/H] (B. Twarog, private communication) versus Galactocentric distance R_{GC} for the subset of clusters for which Friel (1995) has listed ages. The plot shows the following: (1) The subset of clusters with known ages exhibits the same discontinuity at 10 kpc as does the complete sample studied by Twarog et al. (1997); (2) the systematic difference in [Fe/H] between the old ($T > 2.0$ Gyr) and the young ($T < 2.0$ Gyr) clusters (in the sense old minus new) is -0.04 ± 0.05 (i.e., there is no significant difference between the metallicities of the old and young open clusters in the Twarog's sample). A strong argument against the conclusion by Twarog et al.

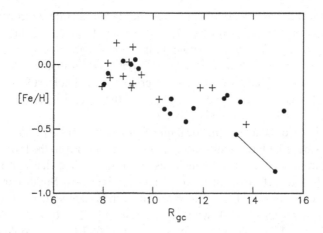

Fig. 4.2. Metallicity [Fe/H] versus Galactocentric distance R_{GC} for old ($T > 2.0$ Gyr) [plotted as •] and for young ($T < 2.0$ Gyr) [plotted as +] open clusters. Both the old and young clusters appear to exhibit a small discontinuity, at $R_{GC} = 10$ kpc. Two different solutions for the old cluster Berkeley No. 21 are joined by a thin line.

(1997) is that a plot of $12 + \log(\text{O/H})$ versus R_{GC}, for both H II regions and B stars (Henry 1998), appears to show a smooth decline of the oxygen abundance over the range $5 < R_{GC} < 15$. In their study of the age–metallicity relation Carraro, Ng & Portinari (1998) find an apparent excess of metal-rich objects with ages in the range 5–9 Gyr. Might some of these stars also represent objects that were ejected from the nucleus by the Bar (or by a merger), and subsequently settled down into the solar neighborhood? The most distant known open cluster is the 2.2–2.5 Gyr old cluster Berkeley No. 21 (Tosi et al. 1998), which is located at a Galactocentric distance $R_{GC} = 14.5$ kpc. This cluster has Fe/H $= -0.97^{+0.3}_{-0.1}$, which makes it at least as metal poor as the Small Magellanic Cloud.

Edvardsson et al. (1993) have found that the metallicity of field stars with ages <10 Gyr, and $7 \lesssim R_{GC}$ (kpc) $\lesssim 13$, exhibits no systematic dependence of element abundance on age. The most straightforward interpretation of these results is that stars in the Galactic disk formed from interstellar gas that had been enriched in heavy elements, but that was subsequently diluted by infall of more-or-less pristine gas. It should, however, be noted that the existence of a radial abundance gradient in the Galactic disk places limits on the amount of pristine gas that could have fallen into the disk. Possible evidence for *some* infall is provided by the observation (Edvardsson et al. 1993) that the scatter in [Si/Fe] is about four times smaller than that in [Fe/H]. This is exactly what would be expected if the disk gas was well mixed before the infall of clumps of almost pristine hydrogen-rich gas.

Carraro, Ng & Portinari (1998) have made the intriguing suggestion that some of the metal-rich stars with ages between 5 and 9 Gyr that are observed near the Sun might have migrated to the solar neighborhood and settled down there. Such objects might have been scattered out from the Galactic bulge when the nuclear bar formed – perhaps through some kind of merger event. It would be important to formulate suitable dynamical models to explore and check such a merger/capture scenario against observational constraints.

4.5   Young spiral arm tracers

Spiral arms are embedded within the disks of intermediate- and late-type galaxies. These arms are outlined by molecular clouds, by dust lanes, and by stellar associations. The concept of stellar associations was introduced by Ambartsumian (1949), who showed that both OB stars (O associations) and T Tauri stars (T associations) exist mainly in unstable clumps. Subsequently it was found (van den Bergh 1966b) that stars embedded in reflection nebulae are also significantly clumped (R associations). Since all types of associations are gravitationally unstable they must be young (Blaauw 1958) and disintegrate on time scales of a few $\times 10^7$ years. [Star clusters, which are stable negative energy systems, may be embedded within expanding (positive energy) associations.] Figure 4.3 shows (van den Bergh 1968c) how R associations outline the nearby Orion spiral arm. Figure 4.4 also shows that the Orion spiral arm is quite steeply inclined to the Galactic plane. The space density of R associations is greater than that of OB associations. Such R associations are therefore particularly suitable for the study of nearby spiral structure. H II regions are, as was already emphasized by Baade (1951), good spiral-arm tracers because they are produced by the radiation from luminous hot young stars. That 21-cm observations of H I outlines spiral arms so well is, at least in part, due to the fact that the OB stars in such arms photodissociate (Allen et al. 1995) the H_2 in molecular clouds. The concentration of H I along spiral arms is seen particularly well in the nearby Sb I–II spiral M81 (Rots 1975). In the case of the Milky Way system it is more difficult to disentangle the spiral structure from 21-cm observations. This is so because:

(1)   The Sun is located very close to the Galactic plane.
(2)   Individual H I clouds in the outer regions of the Galaxy may have motions that deviate from circular motion.

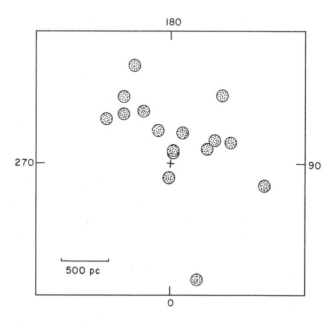

Fig. 4.3. Positions of *R* associations projected onto the Galactic plane. These associations clearly outline the Orion spiral arm. The position of the Sun is marked by a cross.

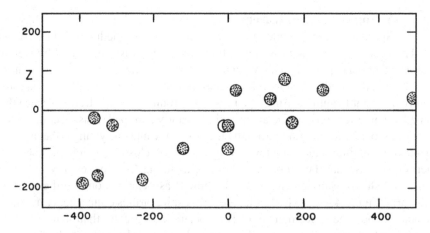

Fig. 4.4. Distance along the Orion spiral arm (measured from the point nearest to the Sun) versus height above the Galactic plane for R associations. The figure shows that the Orion arm intersects the Galactic plane at an angle of $\sim10°$.

(3) The relationship between radial velocity and distance is not single valued in the inner part of the Galaxy.

(4) Noncircular motions will be induced by the bar potential.

(5) Gas in the inner disk of the Galaxy exhibits strong systematic radial outflow.

As in the case of M31 (see Section 3.6.3), it is often difficult to combine individual Galactic spiral segments into a coherent spiral-arm pattern. However, it does appear that the large-scale spiral pattern in the Galaxy is more nearly circular than might have been expected for a spiral of type Sbc. It is also noted in passing that the region of the Galaxy with $R_{GC} > 4$ kpc exhibits little or no evidence for the existence of a bar.

4.6   The Galactic halo

Toomre (1977) wrote that "It seems almost inconceivable that there wasn't a great deal of merging of sizable bits and pieces (including quite a few lesser galaxies) early in the career of every major galaxy, no matter what it now looks like." Prima facie evidence for such mergers is provided by the Sagittarius dwarf (Ibata et al. 1994), which is presently being disrupted by, and is merging with, the Milky Way system. The debris from previous merger events may still be detectable as moving groups of halo stars (Johnston et al. 1995). Such moving groups could be used as probes of the Galactic potential in the halo (Johnston et al. 1998). From in situ measurements of the radial velocities of halo stars, Majewski, Munn & Hawley (1996) concluded that the Galactic halo is, in a dynamical sense, presently not well mixed. Additional evidence that suggests that a few close dwarf spheroidal companions to the Galaxy may have been swallowed by mergers is provided by the observation that the Galaxy presently has three dwarf spheroidal companions (Dra, Scl, UMi) with $60 < D(kpc) < 100$ but only one (Sgr) with $D < 60$ kpc. It is of interest to note that, of the nine known (Richer et al. 1996) "young" globular clusters (IC 4499, Eridanus, Arp 2, Pal 1, Pal 3, Pal 4, Pal 12, Ruprecht 106, and Terzan 7), two (Arp 2 and Ter 7) are associated with the Sagittarius system. Furthermore, Irwin (1999) suggests that Pal 2 and Pal 12 may have been tidally detached from the Sagittarius dwarf. *If young globular clusters only form in dwarf spheroidals, then these results suggest that the Sagittarius dwarf represents ~1/3 of all material that has fallen into the Galaxy.*

From the colors and metallicities of halo stars Unavene, Wyse & Gilmore (1996) conclude that less than 5% of the most metal-poor stars in the Galactic halo are young. Preston, Beers & Shectman (1994) find that the young accreted population accounts for 4–10% of the Galactic halo. The corresponding mass is $<1 \times 10^8 \, M_\odot$.

The fact that the Galactic globular clusters that are on retrograde orbits have a mean metallicity $\langle[Fe/H]\rangle = -1.59 \pm 0.07$, which is similar to that for those on direct orbits for which $\langle[Fe/H]\rangle = -1.65 \pm 0.11$ (van den Bergh 1993), suggests that most halo globulars formed in a single highly turbulent protogalaxy.

From data listed by Djorgovski (1993) it is found that the metal-poor ([Fe/H] $\leq$ -1.00) Galactic globular clusters have $R_h = 7.6$ kpc, compared to $R_h = 3.7$ kpc for the metal-rich Galactic globulars with [Fe/H] > -1.00. It is not known how these values of R_h compare to those of halo stars with various metallicities.

The radial density distributions of both RR Lyrae stars and of blue horizontal branch stars have been studied by Preston, Shectman & Beers (1991). These authors conclude that the density distribution of RR Lyrae stars suggests that the flattening of the Galactic halo decreases with increasing R_{GC}. They find that the flattening of the halo changes from an inner value $c/a \sim 0.4$ to $c/a \sim 1.0$ at $R_{GC} = 20$ kpc. Preston et al. find that the radial density distribution of halo stars may be represented by a power law of the form

$$\rho(R) \propto R^n, \tag{4.1}$$

in which $n = -3.2 \pm 0.1$ for RR Lyrae stars, and $n \approx -3.5$ for blue horizontal branch stars, which were, however, only observed over a more restricted range of R_{GC}. For blue horizontal branch stars (which are believed to belong to an old halo population) Sluis & Arnold (1998) find that $n = -3.2 \pm 0.3$. For this population they find an axial ratio $c/a = 0.52 \pm 0.11$. Preston et al. (1991) note that the horizontal branch color gradient $(B - R)/(R + V + B)$ is, at a given metallicity [Fe/H], more negative outside the solar circle than it is at smaller values of R_{GC}. In other words, halo field horizontal branch stars appear to be redder (= younger?) in the outer halo than they are closer to the Galactic center.

From their study of the Lick astrograph fields at intermediate Galactic latitudes, Suntzeff, Kinman and Kraft (1991) find that $n = -3.5$ for RR Lyrae stars. They infer that the total luminous mass of halo stars with $4 < R_{GC}$ (kpc) < 24 is $\sim 9 \times 10^8 \, M_\odot$ and that globular clusters presently account for $\sim 2\%$ of the luminous halo mass. The radial density distribution of Galactic globular clusters (Harris 1997), excluding those that are probably associated with the Sgr dwarf, is shown in Figure 4.5. From these data one finds that half of all Galactic globular clusters (Harris 1997) lie within $R_h = 5.2$ kpc. The figure indicates that the radial density distribution of Galactic globulars may not be well represented by a single power law. Over the range $2 < R$(kpc) < 8 the data are adequately represented by a relation of the form $\rho(R) \propto R^{-2.2}$. The drop-off below this relation for clusters within 2 kpc of the Galactic nucleus, no doubt, largely results from our failure to discover some clusters in the heavily absorbed central region of the Galaxy. However, destruction of clusters by tidal friction (Tremaine, Ostriker & Spitzer 1975) might also have contributed to the paucity of Galactic globular clusters at $R_{GC} < 2$ kpc.

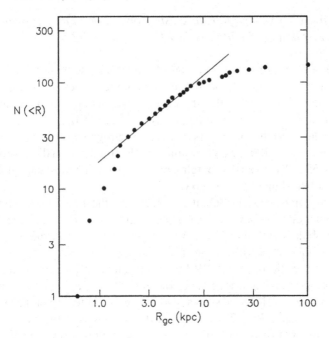

Fig. 4.5. Radial density distribution of Galactic globular clusters. The figure shows that the data cannot be represented by a single power-law density distribution. The line represents a distribution with $n = -2.2$. The deficiency of clusters with $R_{GC} < 2$ kpc may be due to incompleteness and to removal of clusters by dynamical friction.

Figure 4.5 also shows that the exponent n in Eq. 4.1 becomes progressively more negative toward greater values of R_{GC}.

Wilhelm (1996) finds that observations of halo blue horizontal branch stars in the solar neighborhood yield a direct rotational velocity $\langle V_{rot} \rangle = +40 \pm 17$ km s^{-1}, whereas similar objects with [Fe/H] < -1.6 and $|Z| > 4$ kpc have an average retrograde velocity $\langle V_{rot} \rangle = -93 \pm 36$ km s^{-1}. In other words it appears that the inner halo has a small net direct motion, whereas the outer halo seems to exhibit a significant net retrograde motion. The conclusion that stars in the outer Galactic halo have a net retrograde motion is confirmed by the study of Lowell proper motion stars by Carney et al. (1996). For a complete discussion of this question the reader is referred to Majewski (1992). From proper motions near the North Galactic Pole Majewski finds a rotation, relative to the local standard of rest, $V_{LSR} = -275 \pm 16$ km s^{-1}. Assuming $V_{rot} = -220$ km s^{-1} for the local standard of rest, this gives a retrograde motion of ~ -55 km s^{-1} for the high halo. Majewski also notes that the stars in the outer halo do not appear to exhibit a vertical metallicity gradient. Such a gradient would, of course, have been expected if the outer Galactic halo had been pressure supported during its collapse. From a study of the kinematics of blue horizontal branch field stars Sommer-Larsen et al. (1997) find that the radial component of the stellar velocity ellipsoid decreases from $\sigma_r \simeq 140 \pm 10$ km s^{-1} near the Sun to an asymptotic value $\sigma_r = 89 \pm 19$ km s^{-1} at large R_{GC}. This decrease in σ_r is matched by a corresponding increase in the tangential component σ_t of the velocity dispersion. Sommer-Larsen et al. argue that this result is consistent with that expected if the outer halo was built up by capture of proto-Galactic fragments. Chen

(1998) has used *HIPPARCOS* parallaxes and proper motions to show that the stars in the Galactic halo are not well mixed dynamically. In addition Grillmair (1997) has found evidence for tidal tails associated with sixteen Galactic globular clusters and with four globulars in M31. The rate of mass loss from Galactic satellites has been calculated by Johnston, Sigurdsson & Hernquist (1998). Johnston et al. (1998) have shown how such tidal streams, formed from satellite galaxies and globular clusters, can be used to measure the Galactic halo potential. The shape of this Galactic halo potential might, in principle (Olling & Merrifield 1998), also be determined from the manner in which the Galactic gas layer flares at large values of R_{GC}.

Von Hippel (1998) has shown that the white dwarfs produced by massive stars that evolved long ago can account for 8–9% of the inner halo mass, if stars were formed with a Salpeter luminosity function with a mass spectrum of slope $\alpha = -2.35$. For a shallower slope $\alpha = -2.0$ they could account for 15–17% of the mass of the inner halo. Furthermore, Alcock et al. (1998c) find that MACHOS with individual masses in the range $10^{-7} < M/M_\odot < 10^{-3}$ cannot account for more than 25% of the dark halo mass. Zaritsky (1998) has calculated the mass of the Galactic halo from the radial velocities of distant globular clusters and dwarf companions of the Galaxy. He finds a total Galactic mass (out to a distance of ~200 Mpc) of $\gtrsim 1.2 \times 10^{12} M$.

4.7 Galactic globular clusters

Traditionally (Kinman 1959, Zinn 1985) the Galactic globular cluster system is regarded as consisting of a metal-rich disk and metal-poor halo. However, Figure 4.6 suggests that the metal-rich ([Fe/H] ≥ -1.00) cluster population is more properly regarded as being concentrated toward the Galactic nucleus. Contrary to expectation the figure shows little evidence for a metal-rich disk component. A second striking feature of Figure 4.6 is that the bulge component also shows up clearly among the metal-poor

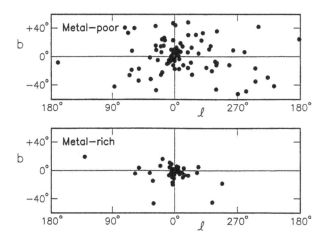

Fig. 4.6. Distribution of metal-poor ([Fe/H] < -1.00) globular clusters on the sky is shown in upper panel, and that of metal-rich ([Fe/H] > -1.00) clusters in the lower panel. Metal-rich clusters exhibit a strong concentration toward the Galactic center. Contrary to expectation, the metal-rich clusters do not appear to outline the Galactic disk. Clusters that are probably associated with the Sagittarius dwarf have been omitted from the plot.

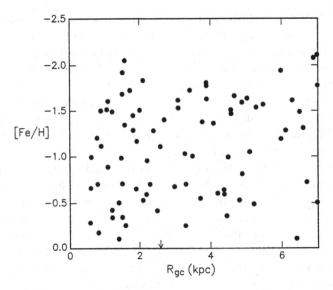

Fig. 4.7. Relation between globular cluster metallicity and Galactocentric distance. The figure shows a radial metallicity gradient in addition to a large dispersion in [Fe/H] at each value of R_{GC}.

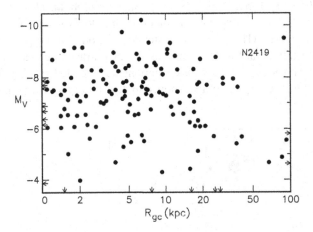

Fig. 4.8. Luminosities of globular clusters as a function of Galactocentric distance. Note how faint most of the clusters with $R_{GC} > 40$ kpc are. Faint clusters at small Galactocentric distances were probably destroyed over a Hubble time by tidal forces and by Disk and bulge shocks. Clusters thought to be associated with the Sagittarius dwarf have been omitted from the plot.

clusters. Figure 4.7, which is based on the compilation of Harris (1997), shows that (1) the metallicity dispersion at any value of R_{GC} is large and that (2) the mean metallicity of Galactic globular clusters decreases with increasing Galactocentric distance. Figure 4.8 gives a plot of the luminosities of globular clusters as a function of Galactocentric distance. The figure shows that for $R_{GC} > 10$ kpc, $\langle M_V \rangle$ becomes fainter with increasing Galactocentric distance. The lack of faint globulars at small values of R_{GC} might be due to cluster destruction by bulge and disk shocks. However, the lack of luminous clusters (NGC 2419 excluded) at large radii must be an intrinsic effect because such massive clusters, if they ever existed, would have survived to the present day.

As has already been pointed out in Section 4.6 half of all presently known Galactic globular clusters have $R_{GC} < 5.2$ kpc. Furthermore, 92 out of 141 (65%) of Galactic globular clusters with known values of R_{GC} have $R_{GC} < 8.2$ kpc (i.e., they lie interior to the Sun). In other words the majority of globular clusters are, contrary to a common misconception, not members of the outer halo of the Galaxy. Barbuy, Bica & Ortolani (1998) have estimated that ~15 globular clusters located behind the Galactic bulge remain to be discovered. By extrapolating the radial distribution of clusters with $R_{GC} > 4$ kpc, to smaller Galactocentric distances, Capuzzo-Dolcetta & Vignola (1997) find that 56 Galactic globulars remain undiscovered or were swallowed by the nucleus via dynamical friction.

Vesperini (1998) has made a detailed study of the disintegration of Galactic globular clusters due to two-body relaxation, disk shocking, and dynamical friction. They conclude that the Galaxy originally contained ~300 globular clusters, having a total mass of ~$9 \times 10^7 M_\odot$. It is found that both an initially log-normal mass distribution *and* an initial power-law mass spectrum will evolve into the presently observed log-normal mass distribution of clusters. Thus it is (almost) impossible to recover the faint end of the original luminosity distribution of globular clusters from present-day observations.

Van den Bergh (1993) has pointed out that the metal-poor Galactic globular clusters with $-2.0\langle$[Fe/H]$\rangle - 1.0$ can be divided into two population components on the basis of their position in a plot of metallicity versus horizontal branch population gradient $(B-R)/(B+V+R)$, in which B, V, and R are the number of blue, variable, and red stars, respectively. Lee, Demarque & Zinn (1994) independently arrived at a similar conclusion. The members of van den Bergh's α population fall below the curve (isochrone?) shown in Figures 4.9a, b, while the members of his β population scatter along it. Figure 4.9a

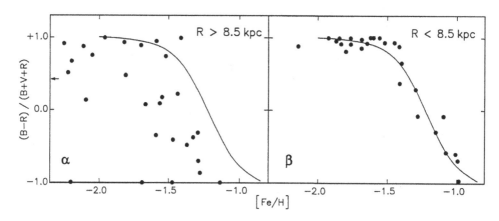

Fig. 4.9. Comparison between plots of $(B - V)/(B + V + R)$ versus [Fe/H] for globular clusters with $R_{GC} < 8.5$ kpc and with $R_{GC} < 8.5$ kpc. It is speculated that the α population, which dominates at $R_{GC} > 8.5$ kpc, was accreted by the Galaxy after it had collapsed. Clusters of the β population probably formed while the main collapse of the protoGalaxy was taking place.

shows that the majority of the metal-poor globular clusters with $R_{GC} > 8.5$ kpc fall below the plotted curve (isochrone?) and therefore belong to van den Bergh's α population. However, Figure 4.9b shows that almost all globulars with $R_{GC} < 8.5$ kpc lie along this curve and therefore belong to van den Bergh's α population. Lee (1992) and van den Bergh have speculated that the members of the β population are relatively young objects that have been captured by the Galaxy, whereas the members of the β population may have formed during the large-scale collapse of the Galaxy envisioned by Eggen, Lynden-Bell & Sandage (1962).

Recent papers by Stetson et al. (1989), Buonanno et al. (1993), Ferraro et al. (1995), Richer et al. (1996), Rosenberg et al. (1997), and by Stetson et al. (1999) appear to show that the globular clusters Arp 2, Eridanus, Palomar 1, Palomar 3, Palomar 4, Palomar 12, Ruprecht 106, Terzan 7, and IC 4499 have ages that may be up to 3 Gyr younger than those of the majority of globulars having similar metallicities. In Figure 4.10 these young clusters are shown as filled circles. Inspection of this figure shows that such young clusters are distributed differently in the M_V versus R_{GC} plane than are other globulars. All young clusters are seen to be located in the outer halo at $R_{GC} > 15.0$ kpc. Moreover, all such young clusters are seen to have below-average luminosities. The fact that the very old outer halo globular NGC 2419 (Harris et al. 1997) is very luminous ($M_V = -9.53$), while the young outer halo clusters are faint, suggests that the *peak of the globular cluster*

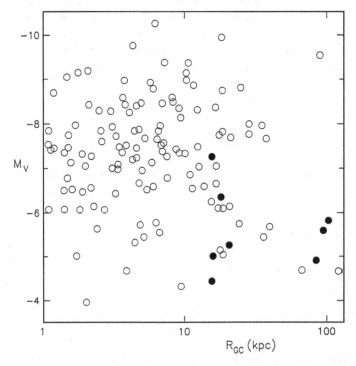

Fig. 4.10. Plot of M_V versus R_{GC} for Galactic globular clusters. The figure shows that "young" globular clusters (which are shown as filled circles) *all* have $R_{GC} > 15$ kpc and that they tend to have below-average luminosities. The young globular cluster Pal 1 at $M_V = -1.88$ and $R_{GC} = 15.9$ kpc is not plotted in the figure.

luminosity function shifted rapidly to lower luminosities. If this conclusion is correct then the last generation of globular clusters to form in the outer halo had luminosities that are almost as low as those of typical old open clusters. It should, however, be emphasized that the tentative conclusion that the luminosities with which globular clusters formed decreased rapidly with time only applies to the one quarter of all Galactic globulars that are located at $R_{GC} > 15$ kpc. There is no evidence for any relation between age and luminosity for the bulk of the Galactic globular clusters that are situated at $R_{GC} < 15$ kpc.

4.8 Galactic supernovae

Information on the frequency and distribution of Galactic supernova may be obtained from a few historical supernovae that have occurred during the past two millennia, and from the optical and radio remnants of supernovae that have occurred during the past ~100,000 years. Because of the insidious effects of dust absorption only radio observations can be used to probe the large-scale distribution of Galactic SNe. Important constraints on the total past supernova rate, and on the ratio of SNe Ia to SNe II, can be derived from the present oxygen abundance and O/Fe ratio in young stars and gaseous nebulae.

Figure 4.11 (van den Bergh 1988a) shows a plot of the distribution of the most luminous (high surface brightness) supernova remnants on the sky. Also shown for comparison are the distribution of CO emission, *IRAS* 100-μm infrared emission, and *Cos B* γ-rays. The figure shows that Galactic supernova remnants are strongly concentrated in the region of the disk that is located within 60° of the Galactic center. The reasons for this are that (1) young SNe II will preferentially form from the densest gas clouds in the Galactic disk and that (2) radio supernova remnants in dense environments will expand more slowly, and hence live longer, than those formed in low-density regions. The apparent absence of supernova remnants associated with the Galactic nuclear bulge is, no doubt, almost entirely due to the fact that few (if any) young stars are forming at high $|Z|$ in the bulge and because the low gas density in the bulge would allow the remnants of SNe Ia that do form there to expand out of existence on a short timescale. Inspection of Figure 4.11 shows that, within the nuclear disk, the distribution of SNRs does not depend strongly on Galactic longitude. This indicates that the outer region of the nuclear disk is more densely populated than its inner part (i.e., the nuclear disk is really a nuclear annulus).

The data in Table 4.1 suggest that the Galaxy is probably spiral of a type intermediate between Hubble stages Sb and Sb. The supernova rate in such a galaxy may be derived by comparison with similar extragalactic objects. Van den Bergh & Tammann (1991) report supernova rates of $0.49\,h^2$, $0.27\,h^2$, and $1.35\,h^2$ SNu[3] for supernova of Types Ia, Ib, and II, respectively, in galaxies of types Sab and Sb. The corresponding rates in spirals of types Sbc–Sd were found to be $0.49\,h^2$, $0.77\,h^2$, and $3.93\,h^2$ SNu, respectively, for Types Ia, Ib, and II. The following supernova rates are therefore expected in the Milky Way system: Type Ia: $0.5\,h^2$ SNu, Type Ib: $0.5\,h^2$ SNu, and Type II: $2.6\,h^2$ SNu, for a total of $3.6\,h^2$ SNu. According to van der Kruit (1989) the total luminosity of the Galaxy is $L_B = 1.9 \times 10^{10} L_\odot (B)$. Combining this luminosity with the rates cited above, and assuming $h^2 = 0.5$ and $R_\odot = 8.5$ kpc, yields expected Galactic supernova rates of 0.5 SN Ia, 0.5 SN Ib, and 2.5 SN II per century. How do these values compare with

---

[3] One SNu is a rate of one supernova per century per $10^{10} L_\odot (B)$, and $h = H_o(\mathrm{km\ s^{-1} Mpc^{-1}})/100$.

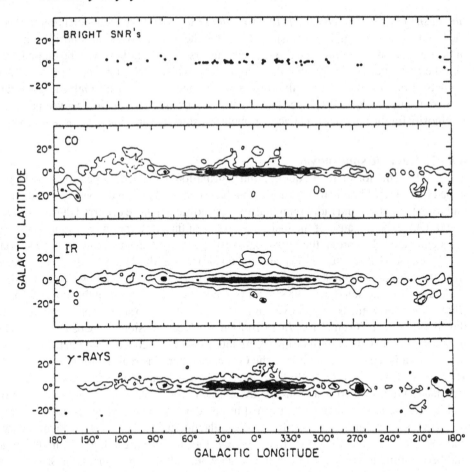

Fig. 4.11. Comparison among the distributions of bright (high surface-brightness) radio supernova remnants, CO emission, *IRAS* 100-μm emission, and *Cos B* γ-rays in the energy range 70 MeV to 5 GeV. All four distributions reveal the presence of the same thin nuclear disk of molecular hydrogen gas.

the evidence provided by historical supernova observations and nucleosynthesis? From a comparison of solar system abundances with models for nucleosyntheses, Nagataki, Hashimoto & Sato (1998) estimate that $7 \lesssim N(\text{SNe II})/N(\text{SNe Ia}) \lesssim 18$. This is slightly, but perhaps not significantly, larger than the ratio $N(\text{SNe II})/N(\text{SNe Ia}) = 5$ that was derived from historical Galactic supernovae. From the observations discussed by Clark & Stephenson (1977), together with more recent information (van den Bergh 1990), one obtains the data listed in Table 4.2. This table is restricted to the historical supernovae of the second millennium, because information for those SNe that occurred during the first millennium is less reliable and almost certainly more incomplete.[4] If it is assumed that the data in Table 4.2 are complete for supernovae that occurred within 3 kpc of the

---

[4] The recent discovery of gamma rays produced by ^{44}Ti (Iyudin et al. 1998) from Cas A, together with a probable nearby SNR in Vela, suggests that we may soon have a much more complete sample of Galactic supernova remnants.

Table 4.2. *Supernovae of the second millennium*[a]

Year	Name	D(kpc)	Type
1006	Lupus	1.4	Ia
1054	Crab	2.0	II
1181?	3C58	2.6	II?
1318?	Vela[b]	0.2	?
1572	Tycho	2.5	Ib?
1604	Kepler	4.2	Ib/II?
1658 ± 3	Cas A	2.8	Ib

[a] van den Bergh (1990).
[b] Iyudin et al. (1998).

Sun during the second millenium, then the local supernova rates are: 1 SN Ia, 2 SN Ib, and 2 SNe II per millennium. According to the Galactic model of Ratnatunga & van den Bergh (1989) the region within 3 kpc of the Sun produces 1/35 of all disk stars, and it is therefore expected to contain 1/35 of all Galactic core-collapse supernovae. From the data given above one would therefore expect the total Galactic rate for both SNe Ib and for SNe II to be $2 \times 35/10 = 7$ per century. This value is higher, but perhaps not significantly higher, than the rates of 0.5 SNe Ib and 2.4 SNe II per century that were predicted from the supernova rates observed in extragalactic spirals. The observed discrepancy could be reduced by a factor of 1.77 by assuming that the Galactic data are, in fact, complete to a distance of 4 kpc, rather than only 3 kpc. The discrepancy could also be reduced by a similar factor if one assumes that the scale-length for star formation in the Galactic disk is 3 kpc, rather than the value of 4 kpc adopted by Ratnatunga & van den Bergh. Finally, the predicted supernova rate could be reduced by a factor of 2 if one were to adopt $h^2 = 0.25$ rather than $h^2 = 0.5$. It is noted in passing that van den Bergh (1991c) has pointed out that the local mass spectrum of star formation may not contain enough massive stars to sustain the high frequency of core-collapse supernovae predicted by the model of Ratnatunga & van den Bergh (1989).

The large amount of star forming activity near the center of the Galaxy is also attested to by the supernova remnant (SNR) Sgr A (East) at 2.5 pc from the Galactic nucleus and by the SNR candidate G 359.87 + 0.18 (Yusef-Zadeh, Cotton & Reynolds 1998) at a distance of 32 pc from the nucleus.

4.9   Novae

The distribution of novae on the sky (Payne-Gaposchkin 1957) shows a strong concentration toward the Galactic nuclear bulge, in addition to a contribution from a disk component. In addition at least one nova (T Scorpii 1860) is known to have occurred in a globular cluster (M80). This shows that the Galactic nova population has disk, bulge, and halo components. In this respect the distribution of novae resembles that of planetary nebulae (see Section 4.10). Liller & Mayer (1987) have used a homogeneous sample of novae discovered during the period 1978–1986 to derive a Galactic nova rate of 73 ± 24 yr^{-1}. This value is more than an order of magnitude higher than the raw discovery rate of 3.7 yr^{-1}. The difference between the observed and calculated true nova rates is due to large, and uncertain, corrections for incompleteness resulting from Galactic absorption. It is particularly surprising that this calculated Galactic nova rate is higher than the rate 29 ± 4 yr^{-1} that Capaccioli et al. (1989) have derived for M31, which is more luminous than the Galaxy. As was previously pointed out in Section 3.11, this difference might be due to the fact that the rate of star formation in the disk of the Andromeda galaxy is presently much lower than it is in the disk of the Milky Way. A nova rate lower than that of Liller & Mayer has, however, been derived by Della Valle & Livio (1994), who find a total Galactic rate of only 20 yr^{-1}, and by Shafter (1997), who derived a rate of $\sim35 \pm 11$ yr^{-1}. A *disk* nova rate of 5 ± 2.5 yr^{-1} is obtained by Della Valle & Duerbeck (1993). Perhaps the most sophisticated modeling of the occurrence rate of Galactic novae is by Hatano et al. (1997b). These authors use a Monte Carlo technique, in conjunction with a simple model for the distribution of dust in the Galaxy, to derive a Galactic nova rate of 41 ± 29 yr^{-1}. Yungelson, Livio & Tutukov (1997) arrive at the conclusion that the current rate of nova production in spiral and irregular galaxies is mainly determined by the present disk star formation rate, while that in elliptical galaxies is primarily determined by galactic mass, for which K-band (2.0–2.4 μm) luminosity can be used as a proxy. Yungelson et al. associate the slow novae in the Galactic bulge with an old stellar population, which would be expected to typically have relatively low white dwarf masses. These authors argue that the hypothesis that nova speed, defined as the time taken to decline by two magnitudes from maximum, correlates with the age of the stellar population with which they are associated receives some support from the observation that more fast and very fast novae occur in the Milky Way than in the Andromeda galaxy. However, it is not yet clear to what extent this conclusion might be affected by selection effects, which favor the discovery of fast luminous Galactic novae.

Della Valle & Livio (1998) find that novae belonging to the He/N class are fast, bright, and tend to be concentrated close to the Galactic plane. They typically have a scale-height $\lesssim100$ pc. In contrast, Fe II novae are slow and dim and occur up to, or beyond, ~1 kpc (e.g., RW UMi 1956) from the Galactic plane. Although novae process only $\lesssim0.3\%$ of all interstellar matter (Gehrz et al. 1998), they are important sources of ^{7}Li, ^{15}N, ^{17}O, and of the radioactive nucleids ^{22}Na and ^{26}Al. By resolving ^{26}Al from ^{27}Al in cosmic radiation, Simpson & Connell (1998) find a Galactic confinement time (partly in the Galactic magnetic halo) for cosmic radiation of 19 ± 3 Myr and an average density of ~0.26 cm^{-3}.

4.10   Planetary nebulae

The distribution of planetary nebulae on the sky (Minkowski 1965) shows a strong concentration toward the Galactic bulge with a surface density maximum near Baade's low-absorption window at $\ell = 1°0$, $b = -3°9$. A distinct disk component is also seen to be present. The discovery of the planetary nebula K 648 in the metal-poor globular cluster M15 (Pease 1928) shows that the Galactic population of planetary nebulae also contains a weak halo component. Peimbert (1990) estimates that the Galaxy contains $7,200 \pm 1,800$ planetary nebulae and that their present formation rate is ~0.3 yr^{-1}. Carbon, nitrogen, and oxygen are formed by p- and α-capture reactions in the interiors of stars. Subsequently they are partly released into the interstellar medium by planetary nebulae and stellar winds, and by ejection from novae and supernovae. Some constraints on the relative importance of these processes might be derived from observations of the ^{14}N/^{15}N ratio in various environments (Chin et al. 1999).

Table 4.3. *Comparison between M31 and the Galaxy*

	Andromeda Galaxy	Milky Way System
Type	Sb I–II	S(B)bc I–II
L_V	$2.6 \times 10^{10} L_\odot$	$2: \times 10^{10} L_\odot^a$
D	760 ± 35 kpc	8.5 ± 0.5 kpc
M_{HI}	$5.8 \times 10^9 M_\odot$	$4 \times 10^9 M_\odot$
M(nucleus)	$7 \times 10^7 M_\odot$	$2.5 \times 10^6 M_\odot$
Bulge light	$25\%^a$	$12\%^a$
R_e (bulge)	2.4 kpc	2.5 kpc
Bulge dispersion	$\sigma = 155$ km s^{-1}	$\sigma = 130$ km s^{-1}
Rotational velocity	~ 260 km s^{-1a}	~ 220 km s^{-1a}
No. globulars	400 ± 55	160 ± 20

[a]van der Kruit (1989).

4.11   Comparison with M31

M31 and the Galaxy are the two most luminous objects in the Local Group. A comparison between some of the main characteristics of these two galaxies is shown in Table 4.3. This table shows the following:

(1) The nuclear bulge of M31 is significantly more luminous than the Galactic bulge.
(2) The effective radii of the central bulges of M31 and the Galaxy are similar.
(3) The velocity dispersion in the Andromeda bulge is larger than that of the Milky Way.
(4) M_{HI}, the scale-length, and luminosity of the disks of the Galaxy and of M31 are comparable.
(5) The rotational velocity of M31 is greater than that of the Galaxy.
(6) The rate of star formation, and hence L_{IR}, is significantly greater for the Milky Way system than it is for the Andromeda galaxy.

4.12   Summary and desiderata

Available evidence suggests that the Milky Way system is a spiral of type S(B)bc. In addition to an actively star-forming disk it contains an inner disk of radius ~ 115 pc in which star formation is presently taking place. This disk is centered on a nuclear black hole with a mass of 2–3 million solar masses. In this respect the Galaxy differs from M31 in which the nuclear bulge presently appears to be devoid of any star formation. It is not yet clear if the presence of the supernova remnant Sgr A (East), located at a projected distance of only 2.5 kpc from the Galactic nucleus, and of S And near the nucleus of M31, result from exceptionally high supernova rates near galactic nuclei.

The fact that individual stars in the Galactic bulge have high [Mg/Fe] values is consistent with a scenario in which enrichment of the bulge took place on a short timescale by ejecta from supernovae of Type II. It is, however, not clear why these same stars appear to exhibit [Ca/Fe] and [Si/Fe] ratios that are similar to those of disk stars.

Most presently available evidence is consistent with a scenario in which the main body of the Galaxy formed via a rapid ELS-like collapse (Eggen, Lynden-Bell & Sandage 1962), with stars and clusters of the outer halo having been accreted over a longer period.

Table 4.4. *Data on the Milky Way system*

$\alpha(2000) = 17^h 45^m 39^s.9$ (1)		$\delta(2000) = -29°00'28''$ (1)
$\ell = 0°00$		$b = 0°00$
$V_r = +16$ (13)		
$(m - M)_0 = 14.51 \pm 0.14$ (2)		$A_V \sim 30$ mag
$M_V = -20.9?$	$D = 8.0 \pm 0.5$ kpc (2)	$D_{LG} = 0.46$ Mpc
type = S(B)bc I–II:		
$M = (1.8–3.7) \times 10^{11} M_\odot$ (3)		$M_{HI} = 4. \times 10^9$ (4)
$M_{HII} = 8.4 \times 10^7 M_\odot$ (5)		$M_{H_2} \sim 3 \times 10^8 M_\odot$ (6)
Total no. globulars		R_e (globulars) = 5.2 kpc (8)
= 160 ± 20 (7)		
Total no. planetaries		Nova rate = 20 yr^{-1} (10)
= 7,200 ± 1800 (9)		
Total star formation rate		
= 0.35 $M_\odot$ yr^{-1} (11)		
M (nucleus) = (2.45 ± 0.14)		
×$10^6 M_\odot$ (12)		
[Fe/H] = +0.06^a (14)		$12 + \log(O/H) = 8.7$
		(solar neighborhood) (15)

^{a}Young stars in solar neighborhood.
  (1) Backer & Sramek (1982).
  (2) Reid (1993).
  (3) Hodge (1992, p. 109).
  (4) van der Kruit (1989).
  (5) Gottesman, Davies & Reddish (1966).
  (6) Dame et al. 1993.
  (7) Harris (1991).
  (8) See Section 4.4.
  (9) Peimbert (1990).
(10) Delle Valle & Livio (1994).
(11) Walterbos & Braun (1994).
(12) Eckert & Genzel (1997).
(13) Grebel (1997).
(14) Meusinger, Reimann & Stecklum (1991).
(15) Massey (1998b).

On such a picture the Galaxy formed inside out, with the central bulge being the oldest. It is not yet clear how to reconcile this scenario with the observation that the ancient globular cluster NGC 2419 (which is situated at $R_{GC} \approx 90$ kpc) is as old as the oldest metal-poor globular clusters near the center of the Galaxy. Another point that remains unresolved is whether there was, or was not, a gap between the disk and halo phases of Galactic formation. On balance, presently available evidence suggests that the disk near the Sun was formed ~11 Gyr ago and that the halo started to form 2–3 Gyr earlier. The nature and evolution of the dark matter corona, in which the Milky Way system is embedded, remains a topic of deep interest and active research. The observation that many old open clusters have survived in the outer Galactic disk appears to militate against the hypothesis that significant amounts of dark matter are concentrated in massive equatorial cloud like structures at large Galactic radii. It would be interesting to know if super metal-rich stars such as μ Leonis, and old open clusters such as NGC 6791, might have been formed in the Galactic bulge and were subsequently expelled.

A summary of data on the Galaxy is given in Table 4.4.

5

The Triangulum galaxy (M33)

5.1   Introduction

The galaxy M33 (= NGC 598) is the third-brightest member of the Local Group. It is a late-type spiral of type Sc II–III. The large angular size of the Triangulum galaxy, and its intermediate inclination $i = 56°$ (Zaritsky, Elston & Hill 1989), make it particularly suitable for studies of spiral structure and stellar content (see Figure 5.1). Only M31, the Magellanic Clouds, and the tidally disrupted Sagittarius dwarf have larger angular diameters than the Triangulum galaxy. The spiral nature of this galaxy was first hinted at by visual observations made by the Earl of Rosse (1850).

The modern era of exploration of M33 began with the independent discovery of variable stars in this object by Duncan (1922) and by Wolf (1923). Neither of these papers show any indication that these authors anticipated the revolutionary impact that the discovery of variable stars in nearby "nebulae" would soon have on Man's view of the Universe. In the words of Hubble (1926) "The nature of nebulae and their place in the scheme of the universe have been favorite subjects of controversy since the very dawn of telescopic observations." Hubble writes that his investigation "followed naturally upon the partial resolution of Messier 33 into swarms of actual stars." He concluded that "The data lead to a conception of the object as an isolated system of stars and nebulae, lying far outside the limits of the [G]alactic system." In his paper Hubble was able to show that 35 of the variables in M33 were classical Cepheids, thus demonstrating, beyond reasonable doubt, that galaxies were "island universes," and ending the great debate (Heatherington 1972, Hoskin 1976) on the nature of spiral nebulae. Additionally, Hubble (1926) was able to show that the Cepheid distance to M33 was entirely consistent with those he obtained from the novae and star clusters.

Early work on M33 has been surveyed by Gordon (1969), and a more up-to-date review of the literature on this galaxy is given in van den Bergh (1991a). A good bibliography, complete to 1973, is provided by Brosche, Einasto & Rummel (1974). The nomenclature for objects in M33 has been reviewed by Lortet (1986). A summary of observational parameters for the Triangulum galaxy is provided in Table 5.1.

5.2   Reddening and distance

According to Sandage (1963b) UBV photometry by Johnson & Sandage shows a reddening of $E(B - V) = 0.09$ in the direction of M33. This value agrees well with $E(B - V) = 0.08 \pm 0.03$ that Schmidt-Kaler obtained from a comparison between the UBV colors of open clusters in the Galaxy and in M33. From BVRI photometry of

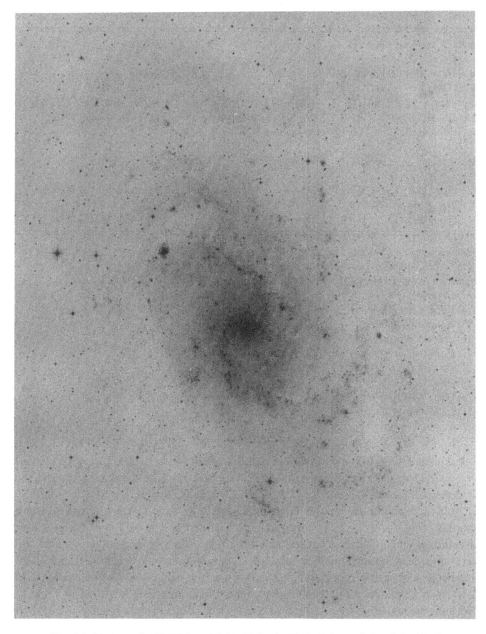

Fig. 5.1. Photograph of M33 in red light obtained with the Palomar 1.2-m Schmidt telescope. The bright H II region in the NE part of this galaxy is NGC 604.

Cepheids in the Triangulum galaxy Freedman et al. (1991) find $E(B - V) = 0.10 \pm 0.09$. Using intermediate-band photometry of Galactic foreground field stars in the direction of M33 Johnson & Joner (1987) obtain $E(B - V) = 0.08$. Assuming the standard ratio of hydrogen column density to absorption Burstein & Heiles (1984) find a Galactic foreground absorption $A_B = 0.18$ mag, corresponding to $E(B - V) = 0.04$. In the

Table 5.1. *Data on M33*

$\alpha(2000) = 01^h 33^m 50.9$ (1)		$\delta(2000) = +30° 39' 37''$ (1)
$\ell = 133.61$		$b = -31.33$
$V_r = -181.0 \pm 1.3$ km s^{-1} (2)		$i = 56°$ (2)
$V = 5.85$ (3)	$B - V = 0.65$ (3)	$U - B = 0.00$ (3)
$(m - M)_0 = 24.5 \pm 0.2$ (4)	$E(B - V) = 0.07$ (4)	A_V(foreground) $= 0.22$ mag (4)
$M_V = -18.87$ (4)	$D = 830 \pm 20$ kpc (4)	$D_{LG} = 0.38$ Mpc
Type = Sc II–III	Optical size = $53' \times 83'$(13 kpc $\times$20 kpc) (5)	
Major axis = $23° \pm 1°$ (6)	H I size = $74' \times 135'$(18 kpc $\times$33 kpc) (7)	
$M^a = (0.8–1.4) \times 10^{10} M_\odot$ (2)		$M_{H I} = 2.6 \times 10^9 M_\odot$ (8)
Total no. globulars = 25 (9)		
Total no. planetaries ≥ 58 (10)	Nova rate ≈ 5yr^{-1} (11)	$12 + \log$ (O/H) = 8.4 (13) (12)

^{a}Excludes possible dark halo.
(1) de Vaucouleurs & Leach (1981).
(2) Deul & van der Hulst (1987).
(3) Jacobsson (1970).
(4) van den Bergh (1991a).
(5) Holmberg (1958).
(6) de Vaucouleurs (1959a).
(7) Corbelli et al. (1989).
(8) Corbelli & Schneider (1997).
(9) Christian (1993).
(10) Ford (1983).
(11) Della Valle et al. (1994).
(12) Massey (1998b).

subsequent discussion it will be assumed that $E(B - V) = 0.07$, corresponding to $A_B = 0.22$ mag in the direction of M33. With these values the total magnitude and integrated colors of M33 are $V_0 = 6.28$, $(B - V)_0 = 0.58$, and $(U - B)_0 = -0.05$.

In his classical paper on the Triangulum galaxy Hubble (1926) derived a distance modulus $(m - M)_0 = 22.1$, corresponding to a distance of 263 kpc. We now know that this distance modulus was too small by more than two magnitudes because of a serious error in the zero-point of the Cepheid period–luminosity relation (Baade 1954), and because of scale errors (Sandage 1983) in the faint magnitude sequences used at Mt. Wilson. A compilation of recent distance determinations to M33 is given in Table 5.2. Note that the first (discordant) entry in the table is based on old photographic observations, to which uncertain scale corrections had been applied. The last entry, which is based on H_2O maser observations, represents a new technique that holds great promise. Detached eclipsing variables are expected to yield an essentially geometric distance determination of M33 in the not too distant future. The remaining nine entries yield a formal mean value $\langle (m - M)_0 \rangle = 24.60 \pm 0.05$, corresponding to a distance of 830 kpc. Note that this distance is slightly larger than, but comparable to, the distance of 760 kpc adopted in Section 3.2 for M31. This small difference in distance, together with their separation on only 14.9 on the sky, supports the notion that M31 and M33 form part of a relatively compact subclustering within the Local Group.

Table 5.2. *Recent distance determinations to M33*

Method	Reference	$(m - M)_0$
Cepheids m_{pg}	Christian & Schommer(1987)	$24.05^{a,b} \pm 0.18$
Cepheids I	Mould (1987)	$24.82^{a,b} \pm 0.15$
Cepheids H	Madore et al. (1985)	$24.50^a \pm 0.2$
Cepheids BVRI	Freedman et al.(1991)	$24.64^b \pm 0.13$
Red giants	Wilson et al. (1990)	24.60 ± 0.30
Tip giant branch	Salaris & Cassisi (1998)	24.97 ± 0.18
Red giants	Mould & Kristian (1986)	24.80 ± 0.30
LP variables	Kinman et al. (1987)	$24.55^a \pm 0.10$
LP variables	Mould et al. (1990a)	24.52 ± 0.17
RR Lyrae	Pritchet (1988)	24.60 ± 0.23
Novae	Della Valle (1988)	24.38^a
H_2O masers	Greenhill et al. (1993)	$23–25$

[a]Following Freedman (1988a) a value $A_V = 0.40$ mag was adopted for the sum of the Galactic foreground absorption and the internal absorption. The corresponding absorption values at other wavelengths are $A_B = 0.52$ mag, $A_I = 0.19$ mag, and $A_K = 0.04$ mag.
[b]A distance modulus $(m - M)_0 = 18.5 \pm 0.1$ was assumed for the LMC.

5.3   The nucleus of M33

The Triangulum galaxy consists of four principal components: (1) an exponential disk, (2) a halo that contains RR Lyrae variables and clusters, which, on the basis of their colors and kinematics, are globulars, (3) a semi-stellar nucleus, and (4) *possibly* a small nuclear bulge. An image of the nucleus of M33, obtained in subarcsecond seeing with the *Canada–France–Hawaii* telescope, has been published by Kormendy & McClure (1993). A color image of the M33 nucleus obtained with the *Hubble Space Telescope* is shown in Lauer et al. (1998). De Vaucouleurs & Leach (1981) find the following position (B1950) for the nucleus: $\alpha = 01^h 31^m 01\overset{s}{.}67$, $\delta = +30° 24' 15''.0$. Photometry of the nuclear region of the Triangulum galaxy is listed in Table 5.3. Adopting V(nucleus) = 13.95, and $(m - M)_V = 24.72$, yields $M_V = -10.77$. This value is more luminous than that of any Galactic globular cluster. Kormendy & McClure (1993) find that the

Table 5.3. *Photometry of the nucleous of M33*

Diaphragm	V	$B - V$	$U - B$	Reference
6''.4	. . .	0.65	. . .	Sandage (1963a)
10.4	14.54^a	0.68	$+0.05$	Walker (1964)
13.9	14.16^a	0.62	-0.06	Walker (1964)
24	13.68^a	0.56	$+0.01$	Walker (1964)
nucleus	13.85^b			Nieto & Aurière (1982)
nucleus	13.95^b			Kormendy & McClure (1993)

[a]Referred to the average brightness of the nebula one diaphragm diameter east and west of the nucleus.
[b]Total magnitude of nucleus integrated under brightness profile. $B - V = 0.65$ assumed.

half-light radius of the nucleus is $r_h = 0.''34\,(1.4\,\text{pc})$ and that the velocity dispersion in the nucleus is only $21 \pm 3\,\text{km s}^{-1}$. From these data they obtain a seeing-corrected central mass-to-light ratio $M/L_V \lesssim 0.4$ (in solar units). Because this mass-to-light ratio is so low, the nucleus of M33 cannot contain a very massive black hole. Kormendy & McClure conclude that a strict upper limit to the mass of such a central black hole is $5 \times 10^4\,M_\odot$. An even lower limit of $2 \times 10^4\,M_\odot$ is derived by Lauer et al. (1998). These authors speculate that the absence of a massive black hole in M33 is due to the fact that the central part of this galaxy does not (like M31, M32, and the Milky Way) have a central spheroid in which such a black hole could grow. Surface photometry of the central region of M33 with the *Hubble Space Telescope* by Lauer et al. shows an extremely compact nucleus. The brightness profile of this nucleus falls rapidly with a power-law slope of $\gamma = -1.5$ for $0.''08 < r < 0.''25$, and then steepens to $\gamma = -1.9$ for $0.''25 < r < 0.''8$. The profile of the nuclear component begins to fall below that of the M33 disk at $r \sim 2''$. Lauer et al. find a core radius of $r_c = 0.''034 \pm 0.''006\,(0.14\,\text{pc})$. Possibly, collisions resulting from the very high density of stars would induce some new star formation in the nucleus.

Wilson et al. (1988) have observed three molecular clouds near the nucleus of M33. For the cloud closest to the nucleus to be stable against tidal disruption by the nucleus the mass in the inner 100 pc of M33 must be $<3 \times 10^7\,M_\odot$. This value is comfortably above the mass of the nucleus of the Triangulum galaxy that has been derived above. The nucleus of M33 contains an X-ray source with a 0.1–6 keV luminosity of $\sim 1.5 \times 10^{39}\,\text{erg s}^{-1}$ (Trinchieri, Fabbiano & Peres 1988; Peres et al. 1989), which makes it the most powerful steady X-ray source in the Local Group. The central X-ray source is ten times brighter than any of the other sources in M33. The X-ray spectrum of this nuclear source is significantly softer than that of typical active galactic nuclei (Takano et al. 1994). Observations by Dubus et al. (1998) show that the source M33 X-8, which is coincident with the optical nucleus of M33, is variable and exhibits a period of ~ 108 days. This shows that (most) of the emission from M33 X-8 must arise from a single object, probably a binary with a black hole primary of $\sim 10\,M_\odot$. In addition to the central source there is also a diffuse X-ray emission component associated with the disk of this galaxy. The hard (>2 keV) component of this diffuse emission probably represents the integrated light produced by accreting systems and young supernova remnants. The soft (<1 keV) component of the disk X-ray emission is likely due to the integrated emission of the stellar coronae, old supernova remnants, and, perhaps, hot diffuse gas in the disk of M33.

Spectroscopically (van den Bergh 1976) the nucleus of M33 is clearly composite, with a late A-type being indicated by K/(H + Hε), F2–F4, by $\lambda 4226/\text{H}\gamma$, and F3–F4 by CH/Hγ. The observed spectrum of the semi-stellar nucleus of the Triangulum galaxy might be accounted for in two entirely different ways: (1) It could be produced by a young metal-rich population, or (2) it might be accounted for by an old globular cluster–like population that is *very* metal poor. Van den Bergh (1976) showed that the $\lambda 4325$ line of Fe I was stronger in M33 than in it is in the spectra of very metal-poor globular clusters. This result militates against the hypothesis that the nucleus consists of metal-poor stars. The interpretation that the nucleus consists of young metal-rich stars is strongly supported by the infrared spectra of Schmidt, Bica & Alloin (1990). O'Connell (1983) has interpreted the observed spectrum of the semi-stellar nucleus in terms of nuclear star formation at a mean rate of $\sim 3 \times 10^{-4} \, M_\odot \, \text{yr}^{-1}$, which has continued over the past 1 Gyr. A somewhat different scenario is advocated by Ciani, D'Odorico & Benvenuti (1984) and Gordon et al. (1999), who used the *International Ultraviolet Explorer* (*IUE*) satellite to obtain spectra over the range $\lambda\lambda 1200 - 3000$, together with published multicolor photometry, to derive a population model in which the nucleus of M33 underwent a burst of star formation with $M_V \sim -15.5$ (at age 10 Myr). This nucleus now appears to be embedded in a clumpy shell with $\tau_V \sim 2 \pm 1$. The dust in this shell has a strong 2175 Å bump (i.e., it differs from that in most starburst galaxies).

According to Rubin & Ford (1986) the nuclear spectrum of M33 is unique because it shows [N II] $\lambda 6548$ and $\lambda 6583$ in emission while Hα is seen in absorption. This observation suggests that the nitrogen-to-hydrogen ratio is unusually high for the gas that is situated within the nucleus of the Triangulum galaxy. Rubin & Ford point out that nuclear Hα absorption is likely to be overwhelmed by disk Hα emission in the spectra of more distant galaxies. This suggestion should be checked by high-resolution observations with the *Hubble Space Telescope*. Garnett et al. (1997) find similar central oxygen abundances and radial oxygen gradients for the Local Group galaxy M33 (Sc II–IV) and for the M81 group galaxy NGC 2403 (Sc III).

5.4 Central bulge and halo of M33

The existence of a tiny nuclear bulge within the central part of the disk of M33 remains controversial. Such a bulge was first reported by Patterson (1940), but it is not clearly seen in the UBV photometry of de Vaucouleurs (1959a). However, Boulesteix et al. (1979) do find such a bulge with a luminosity $\sim 1\%$ of that of the M33 exponential disk. They claim that this small nuclear bulge exhibits an $r^{1/4}$ law profile with an effective radius $r_e = 2\overset{.}{.}75$. Possible evidence for such a nuclear bulge is also seen in the VRI photometry of Guidoni, Messi & Natali (1981). However, more recently Kent (1987) has cast doubt on the reality of such a bulge. He writes "Inside 3′ the profile rises above the exponential, suggesting the presence of a bulge component. However, close examination of the galaxy images shows that this region is still dominated by the spiral structure of the disk." From *I*-band CCD frames and *IRAS* 12-μm data Bothun (1991) finds a conservative upper limit of $M_V = -15$ for the integrated luminosity of the bulge of the Triangulum galaxy. Bothun concludes that "the luminosity profile of M33 does exhibit a clear excess of light relative to the exponential disk interior to $R = 0.5$ kpc." Using H-band observations Minniti, Olszewski & Rieke (1993) have found a bulge interior to $R = 0.5$ kpc. However, they note that the brightest bulge stars are substantially brighter (and hence presumably younger) than those that are seen in the Galactic bulge. This conclusion is confirmed by McLean & Liu (1996). Mighell & Rich (1995) find that the "Population II" stars, which are more centrally concentrated than the stars of Population I, are probably part of an intermediate-age population. However, McLean & Liu (1996) have used IJK photometry of the central region of M33 to conclude that "our study does not therefore reveal a bulge component." The presence or absence of such a bulge component is important because Schommer et al. (1991) and Christian (1993) have discovered 25 halo globular clusters in M33. These halo clusters are found to form a nonrotating system with a velocity dispersion of $\sim 70 \, \text{km s}^{-1}$.

If one were to assume that the halo and bulge of M33 constitute a single entity with $M_V > -15$, then this bulge/halo of the Triangulum galaxy would have a specific cluster frequency of $S^1 > 25$.[1]

Observations in the V and I bands by Mighell & Rich (1995), who used the planetary camera of the *Hubble Space Telescope*, have recently clarified the nature of the stellar population near the nucleus of the Triangulum galaxy. Their results show the presence of three distinct population components: (1) A young population of blue main sequence stars, red supergiants, and candidate Cepheids of intermediate color, (2) a population of infrared-bright giants that were probably formed during an intermediate-age burst of star formation, and (3) old stars of Population II, which exhibit a rather large abundance spread and comprise objects with metallicities ranging from metal-poor like M15 to relatively metal-rich stars resembling those in 47 Tucanae. In this respect the nuclear Population II differs from the halo Population II (Mould & Kristian 1986), which only contains metal-poor stars. Mighell & Rich find that these nuclear Population II stars are more centrally concentrated than are the Population I stars.

Additional evidence for the existence of a halo surrounding M33 is provided by the discovery of seven RR Lyrae stars in the Triangulum galaxy by Pritchet & van den Bergh (Pritchet 1988). For these objects $\langle B \rangle = 25.79 \pm 0.15$, which yields a distance for M33 that is consistent with that derived from Cepheids. A halo field, at a projected distance of 5 kpc from the nucleus of M33, has been studied by Mould & Kristian (1986). This field is so far from the center of M33 that internal absorption effects can probably be neglected. After correcting for foreground absorption the I versus $V - I$ diagram of Mould & Kristian shows a rather small dispersion. For their adopted distance modulus of $(m - M)_o = 24.8 \pm 0.3$ the color–magnitude diagram of the M33 halo is reasonably well fitted by the giant branches of metal-poor Galactic globular clusters. According to Lequeux (1999) $V - I$ photometry of individual halo stars shows that they mostly have metallicities in the range $-1.6 < [\text{Fe/H}] < -0.7$.

5.5   Disk of M33

Detailed UBV photometry of M33 has been published by de Vaucouleurs (1959a). He found that most of the luminosity of the Triangulum galaxy resides in an exponential disk. This disk has a B scale-length of $7\rlap{.}'9$ (1.9 kpc). Within the accuracy of the data the U and V scale-lengths do not differ from that in B. A slightly larger disk scale-length $9\rlap{.}'6$ (2.3 kpc) has more recently been measured by Kent (1987). According to Lequeux (1999) the disk of M33, as outlined by red giants, has a sharp outer edge. De Vaucouleurs notes that this disk has an inclination of $55° \pm 1°$, while the position angle of the major axis is $23° \pm 1°$. These photometric values agree well with the inclination $i = 56° \pm 1°$ and major axis position angle $23° \pm 1°$ that Zaritsky, Elston & Hill (1989) derived from a study of the positions and radial velocities of 55 H II regions. The M33 rotation curve measured from H II regions agrees reasonably well with the one that Rogstad, Wright & Lockhart (1976) obtained from H I observations, except for a small difference at large radii where the H I velocities are marginally larger than those derived from H II regions.

---

[1] The specific globular cluster frequency S (Harris & van den Bergh 1981) is defined as $S = N \times 10^{0.4(M_V+15)}$, in which N is the number of globular clusters in a galaxy with absolute magnitude M_V. In other words S is the number of globular clusters per $M_V = -15$ of parent galaxy light.

Neutral hydrogen observations (Corbelli, Schneider & Salpeter 1989) show that the gaseous envelope of M33 has dimensions of $74' \times 135'$ (18 kpc $\times$ 33 kpc), which is significantly larger than the optical dimensions of $53' \times 83'$ (13 kpc $\times$ 20 kpc) (Holmberg 1958). Corbelli & Schneider (1997) have fit a tilted disk model to 21-cm line observations of M33 obtained with the Arecibo radio telescope. They find that the outer hydrogen disk of the Triangulum galaxy is tilted by $\sim 30°$ with respect to its inner disk. If we adopt a distance of 830 kpc their observations yield a total hydrogen mass of $2.6 \times 10^9 \, M_\odot$. From the rotation curve of M33, which extends out to $\sim 70'$ (17 kpc), and a distance of 830 kpc, one finds a total mass of $5.5 \times 10^{10} \, M_\odot$. The sharp outer cutoff (Kenney 1990) of the hydrogen envelope of the Triangulum galaxy is believed to be due to ionization by the intergalactic radiation field.

Observations by Freedman (1985) show that the upper ends of the stellar luminosity functions are similar in M31, M33, and in the LMC. Humphreys & Sandage (1980) find that the brightest individual blue and red supergiants in M33 have $M_V = -9.4$ and $M_V = -8.2$, respectively.

Madore, van den Bergh & Rogstad (1974) have compared counts of young stars in the U-band with the 21-cm H I observations by Rogstad et al. (1976). They find different linear relations between the number of stars (and hence the rate of star formation) and the H I surface density in the inner and outer regions of M33. Assuming a relation of the form

$$\sigma_s \propto (N_{H_I})^n, \tag{5.1}$$

in which σ_s is the stellar surface density and N_{H_I} is the number of hydrogen atoms cm^{-2} along the line of sight, Madore et al. find $n = 0.6$ in the inner region and $n = 2.4$ in the outer region of M33. Similarly they find that

$$\sigma_{H_{II}} \propto (N_{H_I})^n, \tag{5.2}$$

in which $n = 0.9$ in the inner region and $n = 2.6$ in the outer region. These relations, which are plotted in Figure 5.2 and Figure 5.3, respectively, show that the rate of star and H II region formation depends much more critically on the H I surface density in the outer area of the Triangulum galaxy, than it does in the inner regions of this system. At a given surface density of gas the star formation rate is highest in the inner region of M33. The most plausible explanation for this result appears to be that the H I gas layer is much thinner in the inner part of M33 than it is in the outer part of this galaxy. A similar phenomenon is observed in the Milky Way (Kerr 1969), in which the H I layer also thickens toward larger R_{GC}.

Rogstad et al. (1976) show that the outermost part of the disk of the Triangulum galaxy has been strongly warped. These authors note that they had previously found a similar strong warping of the outer disk of M83 (= NGC 5236). In the latter case a tidal interaction with its companion galaxy NGC 5253, which took place ~ 1–2 Gyr ago (Rogstad, Lockhart & Wright 1994), may have been responsible for the observed warping. However, M33 has no nearby companion that could have warped its outer disk recently.

Both observations of stars (McCarthy et al. 1995, Monteverdi et al. 1997) and of H II regions show that there is a radial abundance gradient in the disk of M33. For H II regions Vílchez et al. (1988) find a steep O/H gradient, a shallower S/H gradient, and no N/O gradient. Monteverdi et al. find an oxygen abundance gradient of -0.16 dex kpc^{-1}, which agrees well with the value -0.13 dex kpc^{-1} derived by Vílchez

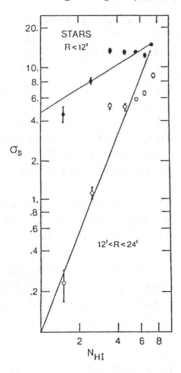

Fig. 5.2. Relation between surface density of neutral hydrogen (expressed in units of $2.73 \times 10^{20} \mathrm{cm}^{-2}$) and the surface density of young stars. Data taken from Madore, van den Bergh & Rogstad (1974). At a given surface density of hydrogen the rate of star formation is seen to be greatest in the innermost region of M33, where the gas layer is thin.

et al. However, Zaritsky et al. (1989) find a somewhat shallower slope of -0.09 ± 0.02 dex kpc^{-1} for O/H. A similar gradient of -0.09 ± 0.05 dex kpc^{-1} was derived from supernova remnants by Smith et al. (1993). The existence of a radial gradient in the ratio of blue-to-red supergiants in the Triangulum galaxy (Walker 1964) remains controversial. Humphreys & Sandage (1980) and Ivanov (1998) appear to confirm its reality, whereas Freedman (1985) and Wilson (1990) do not.

The rate at which massive stars lose matter in radiation-driven stellar winds is expected to increase with metallicity (Smith 1988). A massive star is expected to become a Wolf–Rayet (WR) object when most of its hydrogen atmosphere has been removed by such winds. A metal-rich star will therefore become a WR star sooner than a more metal-poor star. Such rapid mass loss will also speed up the removal of the outer nitrogen-rich layers of Wolf–Rayet stars. Since M33 has a radial abundance gradient one would therefore expect the WC-to-WN ratio, in which WC refers to carbon-type WR stars, and WN are nitrogen-type Wolf–Rayet stars, to decrease with increasing galactocentric distance. This expectation is confirmed by the observations of Massey & Conti (1983). According to Schild, Smith & Willis (1990) the linewidths of early WC stars in M33 also appear to correlate loosely with ambient metallicity and with galactocentric distance.

Eight fields in M33 have been searched for WR stars by Massey & Johnson (1998). These authors find that the Triangulum galaxy fits the relation

$$N(WC)/N(WN) = 0.96[12 + \log(O/H)] - 7.82, \tag{5.3}$$

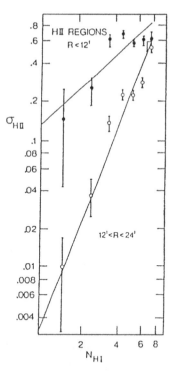

Fig. 5.3. Relation between surface density of neutral hydrogen gas and the surface number density of H II regions in the disk of M33. The figure shows that more H II regions form at a given hydrogen surface density in the inner region of M33, where the hydrogen layer is thin and the hydrogen density high, than in the outer region of this galaxy. Data from Madore, van den Bergh and Rogstad (1974).

which is also obeyed by most other Local Group galaxies. The fact that the scatter in this relation is only 0.08 in WC/WN provides strong support for a scenario in which the evolution of massive stars is strongly influenced by the effects of mass loss that is a function of metallicity.

5.6   Star clusters

A catalog of 250 nonstellar objects in M33 has been published by Christian & Schommer (1982). These authors also provided BVR photometry for some 60 clusters, in addition to references to previous work on the star clusters in the Triangulum galaxy. A catalog of 185 nonstellar objects in the OB associations of M33 (which are discussed in more detail in Section 5.7) has been given by Ivanov (1992). These objects are proba- bly a mixture of concentrated clusters, groupings of OB stars, and compact H II regions. Mochejska et al. (1998) have recently listed 35 additional cluster candidates in M33. Pho- tometry of individual star clusters in the Triangulum galaxy has been reported by Cohen, Persson & Searle (1984) and by Christian & Schommer (1988). The color–magnitude diagrams for the integrated colors and magnitudes of clusters in the Triangulum galaxy (Christian & Schommer 1988) and in the Large Magellanic Cloud (van den Bergh 1981b) are compared in Figure 5.4. Intercomparison of these figures shows a number of striking differences. In the first place the Large Cloud is seen to contain brighter blue clusters than

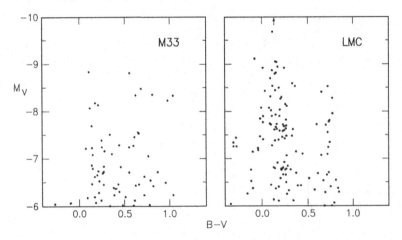

Fig. 5.4. Color–magnitude diagrams for clusters in M33 and in the LMC. The Large Cloud is seen to contain more bright clusters. Furthermore, the color gap between blue and red clusters is more pronounced in the LMC than it is in M33.

does M33. These luminous populous star clusters in the LMC have sometimes, perhaps misleadingly, been referred to as "blue globular clusters." Baade (1963, p. 110) writes "Such a cluster is NGC 1866 in the Large Magellanic Cloud, a very unusual cluster, in which the brightest stars are of the order of -3^M, and so rich that everybody who looks at a photograph swears it is a globular cluster." The presence of these objects hints at a possible difference in the mass spectrum of star formation in spiral and irregular galaxies. In M33 only S 135 ($M_V = -8.8$, $B - V = 0.11$) and C 39 ($M_V = -8.8$, $B - V = 0.56$) seem to rival the LMC populous clusters. Secondly the LMC is seen to exhibit a more pronounced color gap between blue and red clusters than does M33. This probably indicates that the Triangulum galaxy contains a larger population of intermediate-age star clusters than does the LMC (i.e., the Triangulum galaxy has not suffered a burst of star formation similar to that which has occurred recently in the Large Cloud). However, larger reddening of some clusters may also have contributed to filling in the color gap in the Triangulum galaxy. From observations with the *Hubble Space Telescope* Chandar (1998) has formed an unbiased sample of 60 clusters in the disk of M33. These observations suggest that star clusters formed continuously from $\sim 4 \times 10^6$ to $\sim 1 \times 10^{10}$ yr ago.

The red clusters with $B - V > 0.6$ in M33 appear to be true globular clusters. Cohen et al. (1984) were able to show that the integrated spectra of four of these objects indicated that they were more metal-rich than the Galactic globular 47 Tuc, which has [Fe/H] $= -0.7$. However, Christian & Schommer (1983) conclude from their spectra that none of the old clusters in M33, for which they obtained spectra, were as metal poor as the Galactic globular M15 ($=$ NGC 7078), which has [Fe/H] $= -2.2$. Brodie & Huchra (1991) found a mean metallicity $\langle$[Fe/H]$\rangle = -1.55 \pm 0.37$ for the 22 clusters that they observed in M33. This mean metallicity is lower than the value $\langle$[Fe/H]$\rangle = -1.21 \pm 0.02$ that Brodie & Huchra found for M31. The difference is in the sense that the most luminous parent galaxies tend to have the most metal-rich globular cluster systems (van den Bergh 1975). Finally, Schommer et al. (1991) have shown (see Figure 5.5) that the reddest M33

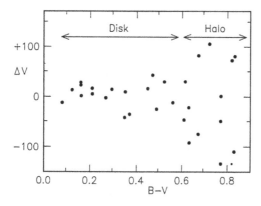

Fig. 5.5. Velocity difference $\Delta V = V$(cluster) $- V$(disk) in km s^{-1} versus cluster color from data by Schommer et al. (1991). The plot shows that red clusters with $B - V > 0.6$ kinematically belong to a halo population.

clusters with $B - V > 0.6$ have halo kinematics with a velocity dispersion $\sim$70 km s^{-1}, whereas the bluer clusters with $B - V < 0.6$ have disk kinematics. Schommer et al. (1991) show that the young blue clusters in M33 rotate with the H I and H II gas. Intermediate-age clusters have motions similar to those of the young clusters, with a marginally lower rotational velocity and higher velocity dispersion. According to Harris (1991), the M33 globular cluster system has $\langle M_V \rangle = -7.0 \pm 0.2$, with $\sigma = 1.2$ mag, which is similar to that of other globular cluster systems.

Sarajedini et al. (1998) have used the *Hubble Space Telescope* to obtain color–magnitude diagrams for 10 objects, which, on the basis of their red integrated colors $[(B - V) > 0.6]$, are true globular clusters. This identification is supported by the fact that these clusters appear to have metallicities as low as [Fe/H]$= -1.6$. From the morphology of their color–magnitude diagrams Sarajedini et al. find that the average age of these globular clusters appears to be a few gigayears younger than that of Galactic halo globular clusters, which in turn have ages similar to those of the globulars in the LMC (Olsen et al. 1998, Johnson et al. 1999). The unexpected result of this work is that *the M33 globulars, which exhibit halo kinematics, are relatively young. On the other hand the LMC globulars are old, but exhibit disk kinematics* (Schommer et al. 1992). These results do not fit comfortably within any currently fashionable scenarios for galaxy evolution.

5.7 Associations and spiral structure

The Triangulum galaxy is a late-type spiral of DDO type Sc II–III. This classification indicates that this object has rather open, and not very well-developed, spiral structure. NGC 2403 [see panel 36 of the *Hubble Atlas of Galaxies* (Sandage 1961)] in the nearby M81 group is a close morphological match to the Triangulum galaxy. Inspection of Figure 5.1 shows that M33 has two main spiral arms, both of which have a rather patchy and discontinuous structure. A number of stellar interarm features are also present. The spiral arms are most clearly outlined by H II regions (Boulesteix et al. 1974). They are less clearly visible as concentrations of loose star clouds and OB associations. Because of the high surface density of stellar images such associations are much more difficult to identify in M33 than they are in M31. Many of the associations in M33 appear

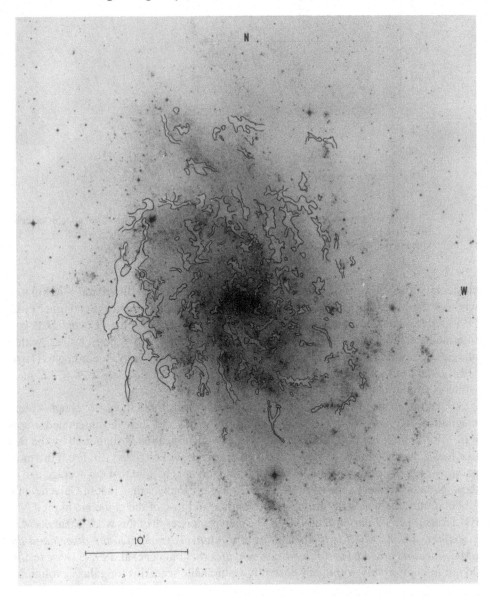

Fig. 5.6. Distribution of dust clouds in M33 according to Humphreys & Sandage (1980). Note that the spiral structure in M33 does not appear to be very clearly outlined by dust lanes. (Reproduced with permission.)

to form larger clumpings with dimensions ~0.5 kpc, which are probably similar to the "constellations" in the LMC (McKibben Nail & Shapley 1953, van den Bergh 1981b). The spiral arms in the Triangulum galaxy do not show up well in the distribution of dust patches (see Figure 5.6) or in the distribution of H I over the face of this spiral (Wright, Warner & Baldwin 1972; Newton 1980). The positions of 1,156 candidate OB stars, cataloged by Ivanov, Freedman & Madore 1993), also do not outline the principal spiral arms very clearly. However, they show up a bit more distinctly in the distribution of OB

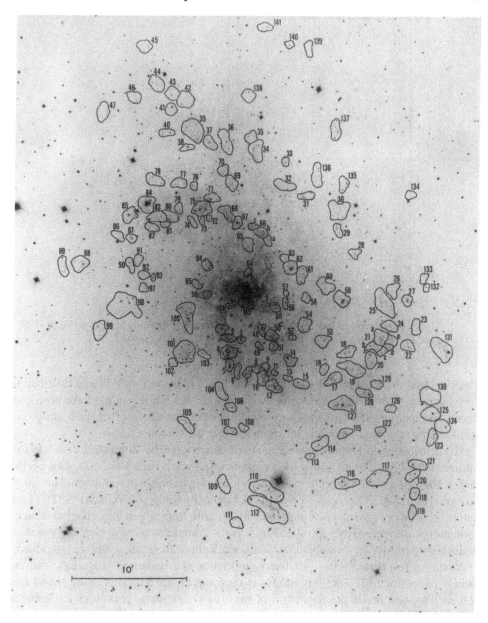

Fig. 5.7. Distribution of 143 OB associations in M33 according to Humphreys & Sandage (1980). The main spiral arms and inter-arm features in the outer regions of the Triangulum galaxy are clearly outlined by these OB associations. (Reproduced with permission.)

associations (see Figure 5.7) that have been isolated by Humphreys & Sandage (1980). From the distribution of OB associations in M33, van den Bergh (1991a) finds that the scale-length for associations (see Figure 5.8) with $R < 8$ kpc is $9\rlap.'9$ (2.4 kpc), which is similar to the $9\rlap.'6$ scale-length found by Kent from surface photometry but longer than the $7\rlap.'8$ scale-length obtained by de Vaucouleurs (1959a). Deul & den Hartog (1990) have

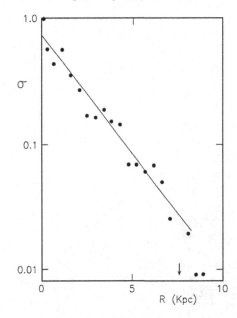

Fig. 5.8. Normalized and rectified surface density of associations in M33 from Ivanov (1987). For $R < 8$ kpc the observed density distribution has a scale-length of 9.9.

observed that the locations of small ($D < 200$ pc) holes in the M33 H I distribution correlate well with the positions of OB associations. Such holes might have been produced by supernova-driven winds, or they could be due to the fact that the hydrogen in star-forming regions is mainly in molecular, rather than atomic, form.

The distribution of H II regions, which has been mapped by Boulesteix et al. (1974), outlines the southern spiral arm of M33 most clearly, even though NGC 604, which is the largest H II region in the Triangulum galaxy, is located in its northern spiral arm. The southern spiral arm of M33 has been studied in great detail by Reagen & Wilson (1993). These authors find an age gradient *perpendicular* to this arm, from which they conclude that its formation was triggered by a spiral density wave. A complete survey of Cepheid variables of different periods (ages) might throw additional light on the evolution of this spiral arm.

A comparison between the radial surface brightness distribution of starlight and the radial distribution of H II regions and H I is shown in Figure 5.9. This figure shows that H II regions, and hence the formation of massive young stars, is presently somewhat less centrally concentrated than is the integrated light of M33. This implies that star formation was once more concentrated to the nuclear region than it is at present. Figure 5.9 also shows that H I, and hence the location of future star formation, is less centrally concentrated than are the H II regions that define the present locations of massive star formation in M33.

Wilson & Scoville (1989) have studied CO emission in the central part of the Triangulum nebula. From these observations the central 3.9 kpc^2 of M33 is found to contain $3.4 \times 10^7 M_\odot$ of molecular hydrogen in addition to $1.3 \times 10^7 M_\odot$ of atomic hydrogen (Wilson et al. 1991). Assuming that 2/3 of all gas (by mass) is in the form of hydrogen, it then follows that the total amount of gas in the central 3.9 kpc^2 of M33 is

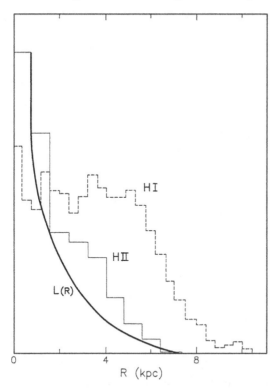

Fig. 5.9. Comparison between the radial distribution of the integrated light of M33 and that of H II regions and of H I gas. The figure, which was adapted from Boulesteix et al. (1974), shows that star formation is gradually moving outwards in this galaxy.

$7.5 \times 10^7 \, M_\odot$. Using a mean reddening $E_{B-V} = 0.3$ one then finds that the central 3.9 kpc^2 of the Triangulum nebula emits 4.8×10^{39} erg s^{-1} in Hα. Adopting a modified Miller–Scalo luminosity function (Kennicutt 1983), one obtains a total rate of formation for stars with $M_\odot > 0.1 \, M_\odot$ of $0.043 M_\odot$ yr^{-1}. (The corresponding rate for stars with $M_\odot > 10 \, M_\odot$ is $0.0068 \, M_\odot$ yr^{-1}.) The timescale for exhaustion of all gas in the central 3.9 kpc^2 of M33 is therefore 1.7×10^9 yr for stars with $M_\odot > 0.1 \, M_\odot$. This timescale would be approximately doubled if recycling of gas by evolving stars is taken into account. The evolutionary timescale would, of course, be longer for a "top-heavy" mass spectrum of star formation (Scalo 1998). If only stars with $M_\odot > 10 \, M_\odot$ are formed, the timescale for gas depletion becomes 11×10^9 yr. These results indicate that the *gas depletion timescale in the central region of M33 is significantly shorter than the Hubble time*. A similar problem in the Galactic disk near the Sun was first noticed by van den Bergh (1957) and is discussed in detail by Sandage (1986b).

5.8   H II regions in the Triangulum Galaxy

Finding charts and positions of 432 compact Hα emitting regions have been published by Calzetti et al. (1995). A catalog of larger H II regions in M33 has been given by Courtès et al. (1987). Hα fluxes and sizes for most of these emission regions are presented by Wyder, Hodge & Skelton (1997). These authors find that the bright end of the differential luminosity function of Hα emission regions may be represented by a power law with exponent -2.40 ± 0.15. The Hα luminosity function turns over at a luminosity of $\sim 10^{36.4}$ erg s^{-1} ($D = 830$ kpc assumed), probably due to the effects of incompleteness. Wyder et al. also find that the diameter distribution of H II regions in the Triangulum galaxy is approximately represented by the exponential distribution given in Eq. 3.4 with $D_0 \approx (32 \pm 1)$ pc.

The largest H II region in M33 is NGC 604. With a diameter $D = 265$ pc it exceeds the size of the 30 Doradus nebula in the LMC, for which $D = 228$ pc (van den Bergh 1968a). NGC 604 has an H I mass of $1.6 \times 10^6 \, M_\odot$, an H II mass of $\sim 10^6 \, M_\odot$, and an Hα luminosity of 4.5×10^{39} erg s^{-1} (Sabalisck et al. 1995). The gas in NGC 604 exhibits a large velocity dispersion, no doubt due to strong stellar winds, "champagne flows," and supernova shells (Benvenuti et al. 1979). Embedded within this nebulosity (Drissen, Moffat & Shara 1993) are numerous WR and early O-type stars. These stars appear to form a number of associations. In this respect NGC 604 differs from 30 Doradus in the Large Magellanic Cloud, which is dominated by a single supercluster. For NGC 604 Drissen et al. find that the slope of the upper end of the mass spectrum between $15 \, M_\odot$ and $60 \, M_\odot$ is not as steep as that found for star forming regions in the Galaxy and in the LMC [but see cautionary comments by Scalo (1998)]. Wilson & Matthews (1995) find that some CO clouds have managed to survive the intense recent burst of star formation in NGC 604. Most of the emission regions in M33 were produced by giant associations of OB stars. However, two huge Hα bubbles, each with a diameter of ~ 130 pc, turn out to be centered on a single O9 star (Oey & Massey 1994). As has previously been pointed out in Section 3.6.5, it is unexpectedly found that the largest H II regions in the Sc galaxy M33 are somewhat smaller than are the largest emission regions in the Sb galaxy M31. Arp & Brueckel (1973) have suggested that this effect is due to the fact that clumps of smaller H II regions in M31 may have been classified as single large emission objects.

5.9   Supernovae and supernova remnants

No supernovae have been seen in the Triangulum galaxy during historical times. According to van den Bergh & Tammann (1991) the expected supernova rate in Sbc–Sd galaxies is $5.2 \, h^2$ SNu. With a luminosity of $\sim 3 \times 10^9 \, L_\odot(B)$, and $h^2 \approx 0.5$, this yields an expected supernova rate of ~ 0.8 per century. So the lack of any recent SNe in M33 is not unexpected. From an optical survey of M33 SNRs Long et al. (1990) were only able to place very broad limits on the frequency of supernovae in M33. They concluded that the number of known optical SNRs in M33 suggests a supernova rate of between 0.3 and 3 per century. From the frequency distribution of SNR diameters van den Bergh (1991a) estimates a frequency of one SN per 325 yr for $E_0/n = 10^{51}$ ergs cm^3, where E_0 is the explosion energy and n is the ambient density of hydrogen atoms.

Mathewson & Clark (1972) were able to show that the supernova remnants in the LMC had higher [S II]/Hα ratios than those in H II regions. This is so because of strong [S II] emission from shocked dense gas behind the shock fronts in SNRs. Using this technique D'Odorico, Dopita & Benvenuti (1980) were able to identify 19 supernova remnants in M33. For a complete listing of optical work on the SNRs in M33 the reader is referred to Smith et al. (1993), which presents spectra of 42 supernova remnants (and SNR candidates) in this galaxy. All of these objects are found to have [S II]/H$\alpha \geq 0.4$, which is the canonical dividing line between SNRs and H II regions. Strong [O I], which is a diagnostic for strongly shocked gas, is also observed in the majority of the M33 SNRs. Smith et al. find little evidence for a systematic variation of line intensity ratios with SNR diameter, and hence with evolutionary stage or age. Furthermore, Lumsden & Puxley (1995) find no indication of a peak in [Fe II] emission from SNRs having a narrow range of ages. Finally, the data obtained by Smith et al. give no evidence for any relation between SNR diameter and galactocentric distance. Smith et al. (1993) find a radial gradient in [N/Hα] for SNRs, in the sense that N is significantly greater at $R = 0.5$ kpc than it is at $R = 4.5$ kpc. Significant radial gradients are also found for [N/H] and for [O/H]. However, for reasons that are not yet well understood, the [N/H] values in SNRs are, at any radius, about 0.3 dex larger than those which Kwitter & Aller (1981) find in H II regions. Similarly the [O/H] values found in SNRs are, at any radius, systematically smaller by ~0.15 dex than are those found in H II regions by Kwitter & Aller. Smith et al. observe a radial oxygen abundance gradient of -0.09 ± 0.05 dex kpc^{-1} in the Disk of M33.

All 52 optical SNRs in M33 that were known in 1992 have been inspected on high-resolution radio surveys at 4.84 GHz and 1.42 GHz by Duric et al. (1993), who were able to identify 26 of them at radio wavelengths. They obtained the following results: (1) The cumulative radio luminosity function of the radio SNRs may be represented by a power law with index -0.8. (2) The radio power of the brightest SNR in M33 is about seven times smaller than that of the Galactic supernova remnant Cassiopeia A. (3) The known SNRs in M33 collectively account for <5% of the total radio power of M33 at a wavelength of 20 cm. Nevertheless, if one allows for the residence time of electrons in the disk, the energetic charged particles produced in SNRs are a plausible source for all of the nonthermal radiation emitted by the Triangulum galaxy. Duric et al. derive a residence time of 8×10^6 yr for the cosmic ray electrons in M33, compared to an estimate of 2×10^7 yr in the Galaxy. Benvenuti et al. (1979) have discovered a supernova shell in the giant H II region NGC 604. A more detailed study of NGC 604 has been published by Yang et al. (1996). These authors find that this H II region has a complex morphology, with evidence for numerous filaments and shell-like structures. Most of these shells are observed to be expanding, with some of them having expansion velocities >100 km s^{-1}. The suggestion that some of these shells are supernova remnants is supported by the presence of X-ray emission in NGC 604. Gordon et al. (1993) have also found a SNR in the H II region NGC 592. High-resolution images of two SNRs in M33 have been published by Blair & Davidsen (1993).

The most recent survey of supernova remnants in M33 is by Gordon et al. (1998), who used interference filter images and spectra to catalog 98 SNR candidates. For this sample these authors find that the cumulative distribution of SNR diameters is well represented by a Sedov–Taylor expansion model, but not by a free expansion model. From the Sedov–Taylor model the mean interval between supernovae in M33 is found to be one every 360 years. This value is in good agreement with van den Bergh (1991a), who estimated a frequency of one supernova every 325 years. Gordon et al. find a good correlation between [N II]/Hα and galactocentric distance. This correlation, no doubt, can be attributed to the strong radial abundance gradient in the Triangulum galaxy. The large dispersion about the mean [N II]/Hα relation is probably due to differences in shock velocities. A plot of the distribution of the 98 SNR candidates over the face of M33 shows a weak correlation with the positions of spiral arm segments. There also appears to be an even weaker correlation between the positions of SNRs and features in the X-ray contour map of Long (1996). Dubus et al. (1998) find that the X-ray source M33 X-7 has an 3.45 day eclipse period and a 0.31 second pulse period. The observations suggest that M33 X-7 is a neutron star with a 15–40 $M_\odot$ binary companion.

Van der Hulst (1997) has used high-resolution 21-cm observations to obtain position–velocity maps for the disk of M33. On such maps he has found a small high-velocity feature produced by a hydrogen cloud with $\sim 1 \times 10^5 \, M_\odot$ that deviates from the general disk motion by ~ 50 km s^{-1}. The total kinetic energy of this feature is $\sim 5 \times 10^{51}$ ergs, which is equivalent to the energy of a small number of supernovae. It therefore seems plausible that such features may have been accelerated out of the disk of M33 by supernovae.

5.10   Variable stars

Novae, Cepheids, and Hubble–Sandage objects were among the first variable stars found in the Triangulum galaxy. As has already been noted in Section 5.1, the first variables (other than novae) in M33 were discovered independently by Duncan (1922) and by Wolf (1923). Hubble (1926) was able to show that 35 of the variables in M33 were classical Cepheids. Owing to problems with photometric scales at faint magnitudes (Sandage & Carlson 1983, Christian & Schommer 1987) and difficulties with internal absorption, Hubble's pioneering studies of the Cepheids in the Triangulum galaxy remained unsurpassed for half a century. References to some more recent observations of Cepheids in M33 are listed in Table 5.2. This table shows that the modern Cepheid observations yield a distance to the Triangulum galaxy that is consistent with that derived from long-period variables, RR Lyrae stars, and the maximum magnitude versus rate-of-decline relation of novae. Unfortunately the presently available data on Cepheids are not yet numerous (or homogeneous) enough to study the distribution of Cepheids of different period (age) relative to the M33 spiral arms.

Mould et al. (1990a) found that the long-period variables in M33 scatter more about their adopted period–luminosity relation than do those in the LMC. The reason for this may be that M33, because of its higher luminosity and metallicity, exhibits a larger range of absorption values than does the Large Magellanic Cloud. Alternatively this luminosity dispersion might be due to a larger range in metallicity of the variables in M33, or to stronger patchy dust absorption in the Triangulum galaxy. A large metallicity range is expected because M33 exhibits a clear-cut metallicity gradient (see Section 5.9), whereas the presence of the Bar in the LMC should keep the gas so stirred up that a metallicity gradient cannot be established.

Van den Bergh, Herbst & Kowal (1975) found that the brightest red supergiant variable star in M33 has $V(\text{max}) \simeq 17.0$, corresponding to $M_V(\text{max}) = -7.7$. Such bright red variables appear to be preferentially found in the outer regions of the Triangulum galaxy. The light curves of the blue Hubble–Sandage variables in M33 have been discussed in some detail by Sharov (1990). Della Valle et al. (1994) have used the "control time" method to derive a nova rate of 4.6 ± 0.9 yr^{-1} for M33. This rate is compared with those which these authors derive for other galaxies in Table 5.4. This table shows that the nova rate per $10^{10} L_\odot (H)$ appears to be higher in M33 and in the LMC, which have actively star forming disks, than it is in galaxies that are dominated by an older stellar population. Taken at face value the results of Della Valle et al. suggest that the majority of novae in M33, and in the LMC, are probably associated with an intermediate-age

Table 5.4. *Comparison of nova rates[a]*

Galaxy	Nova rate per year	Rate per $10^{10} L_\odot$ (H)
LMC	2.5 ± 0.5	14.7 ± 3
M33	4.6 ± 0.9	13.4 ± 2.8
NGC 5128	28 ± 7	5.2 ± 1.3
M31	29 ± 4	4.0 ± 0.6
M81	24 ± 8	3.3 ± 1.1

[a] Data from Della Valle et al. (1994).

stellar population. Half of the 10 novae listed by Della Valle et al. (1994) occur within $r_e = 14'$ (3.4 kpc).

5.11 Summary and desiderata

The Triangulum galaxy consists of a nucleus, a disk, and a halo. The existence of a nuclear bulge and of an outer dark corona have not yet been established with certainty. If the nucleus of M33 contains a black hole it has $M < 2 \times 10^4 \, M_\odot$. The relation between the column density of hydrogen atoms and the rate of star formation suggests that the thickness of the M33 gas layer increases with radius. Observations of both H II regions and of supernova remnants show that there is a pronounced abundance gradient in the disk. However, the source of a systematic difference between abundances derived from H II regions and those from SNRs remains obscure. In the inner disk the N/H ratio is high. It would be interesting to use the high-resolution spectrograph of the *Hubble Space Telescope* to see if other spiral galaxies also exhibit such high N/H ratios close to their nuclei. The timescale for the exhaustion of gas by star formation is found to be only 2–3 Gyr in the inner disk of M33. Radial velocity observations of old red clusters prove the existence of a halo population of globular clusters. Color–magnitude diagrams for individual stars in these clusters suggest that they are relatively young, even though they have halo kinematics. This contrasts with the situation in the LMC in which the globular clusters are very old but appear to have disk kinematics. The reality of a small bulge component in the inner region of M33 remains controversial. The radial velocity observations of young blue clusters in M33 show that these objects exhibit disk kinematics. It would be very interesting to use CCD surveys to obtain a complete census of the Cepheid variables in the Triangulum galaxy. Such a survey should allow one to study the relation between the periods (ages) of Cepheids and their location relative to spiral arms.

Radio observations show that the outer disk of M33 is strongly warped. Tidal effects produced by M31 are too small to account for this warp. The neutral hydrogen layer in M33 exhibits a sharp outer cutoff, so that it is not possible to use 21-cm observations to measure the rotation curve of the Triangulum galaxy at large radii. As a result it is not known if M33 is embedded within a massive dark corona. Hernandez & Gilmore (1997) find that 93.5 ± 2.5% of the total mass of typical late-type spirals resides in such dark envelopes.

6

The Large Magellanic Cloud

6.1   Introduction

To the naked eye the Magellanic Clouds appear as detached portions of the Milky Way. They have probably been known to the inhabitants of the southern hemisphere for thousands of years. In the north the Large Cloud was, like M31, already known to al-Sufi in the 10th Century. The Large Magellanic Cloud (see Table 6.1) is the largest and the brightest (external) galaxy in the sky. Because of this proximity ($D \approx 50$ kpc), its stellar content can be studied in more detail than that of any other external galaxy. Observations that require a 5-m class telescope in M31 can be carried out with a 0.5-m telescope in the LMC. The Large Cloud belongs to the barred subtype of Hubble's irregular class. Its DDO classification is Ir III–IV [i.e., it has a morphology intermediate between that of giant (III) and subgiant (IV) galaxies]. Based on the presence of a faint streamer of nebulosity, which extends from $\alpha = 5^{\rm h}$, $\delta = -73°$ to $\alpha = 3^{\rm h}5$, $\delta = -55°$ (de Vaucouleurs 1954a,b, 1955), the Large Cloud has often been described as a late-type *spiral*. However (de Vaucouleurs & Freeman 1972), this spiral arm–like feature actually appears to be a faint streamer of Galactic foreground nebulosity. This feature in the constellation Mensa has also been observed in the infrared. From multicolor photometry and high-dispersion photometry of 38 stars in the vicinity of this filament Penprase et al. (1998) conclude that it produces a mean reddening $E(B - V) \sim 0.17$ and is located at a distance of 230 ± 30 pc. The absence of a semi-stellar nucleus in the LMC also militates against the hypothesis that the Large Cloud is a late-type *spiral* (van den Bergh 1997b).

Ever since the pioneering work on Cepheids by Henrietta Leavitt (1907)[1] the Large Cloud has been a convenient "stepping stone" on the long and treacherous path leading to the extragalactic distance scale. More recently the MACHO project has also made it possible to compile a well-defined sample of variable stars in the LMC (Alcock et al. 1997c, Welch et al. 1997). This work, on a complete sample of 1,477 LMC variables, is beginning to make a major contribution to our understanding of the population content and evolution of the Large Cloud. Finally, the great discovery by Butcher (1977) "that the bulk of star formation began 3–5×10^9 years ago in the LMC, rather than 10×10^9 years ago as in the Galaxy," is starting to affect our ideas on how galaxies evolve, and

---

[1] Volume 60 of the *Harvard Annals* is dated 1908. However, the Harvard Annual Report for the period ending 1907 September 30 indicates that Volume 60, No. 4 had already been distributed – so historically 1907 is the correct date (Gingrich 1997, private communication).

Table 6.1. *Data on the Large Magellanic Cloud*

$\alpha(2000) = 5^h\ 19\overset{m}{.}6$ (1)		$\delta(2000) = -69°\ 27\overset{'}{.}1$ (1)
$\ell = 280\overset{\circ}{.}19$		$b = -33\overset{\circ}{.}29$
$V_r = +275\ \mathrm{km\,s^{-1}}$ (2)	$i = 27{-}45°$ (3)	p.a major axis $= 170°$ (1)
$V = 0.4$ (2)	$B - V = 0.52 \pm 0.03$ (6)	$R_h = 3\overset{\circ}{.}03 \pm 0\overset{\circ}{.}05$ (6)
$(m - M)_0 = 18.50 \pm 0.05$ (4)	$E(B - V) = 0.13$ (5)	$A_V = 0.40$ mag (5)
$M_V = -18.5$	$D = 50$ kpc	$D_{\mathrm{LG}} = 0.48$
Type: Ir III–IV	Disk scale-length	
	$= 101' \pm 3'$ (1.5 kpc) (6)	
$M = (0.6{-}2.0) \times 10^{10}\ M_\odot$ (2)	$M_{\mathrm{H\,I}} = 7.0 \times 10^8\ M_\odot$ (2)	
$M_{\mathrm{H}_2} = (1.0 \pm 0.3) \times 10^8\ M_\odot$ (7)		
Total no. globulars $= 13$ (8)		
Total no. planetaries $= 265$ (9)	Nova rate $\simeq 0.7 \pm 0.2\ \mathrm{yr^{-1}}$ (10)	
[Fe/H] $= -0.30 \pm 0.04$ (11)	$12 + \log(\mathrm{O/H}) = 8.37 \pm 0.09$ (12)	

(1) Centroid of yellow light isophotes (de Vaucouleurs & Freeman 1972).
(2) Westerlund (1997, p. 28).
(3) Westerlund (1997, p. 30).
(4) See Table 6.2.
(5) See Section 6.2.1.
(6) Bothun & Thompson (1988).
(7) Israel (1998).
(8) Suntzeff (1992).
(9) Morgan (1994).
(10) See Section 6.6.3.
(11) Luck et al. (1998).
(12) Kurt & Dufour (1998).

on how they might (in some cases) transform themselves from one Hubble type into another.

An excellent and very complete discussion of the LMC, with a detailed bibliography that is complete to mid-1995, is provided in Westerlund's (1997) book *The Magellanic Clouds*. A beautiful atlas that identifies clusters, variable stars, and emission regions in the LMC has been published by Hodge and Wright (1967). Individual bright LMC members and Galactic foreground stars are identified on charts by Fehrenbach, Duflot & Petit (1970). Figure 6.1 shows a clear separation between the Galactic foreground stars centered at $\langle V_r \rangle = +26$ km s^{-1} and members of the Large Cloud with $\langle V_r \rangle = +311$ kms^{-1}. Only for a small number of objects with radial velocities in the range $+120$ km s^{-1} to $+220$ km s^{-1} is the membership assignment ambiguous. A spectacular atlas of narrow-band Hα + [N II] photographs of the LMC and SMC has been published by Davies, Elliot & Meaburn (1976).

Recent books on the Magellanic Clouds are *IAU Symposium No. 108* (van den Bergh & de Boer 1984), *IAU Symposium No. 148* (Haynes & Milne 1991), and *IAU Symposium No. 190* (Chu et al. 1999). A complete listing of reviews and proceedings on the Magellanic Clouds is given in Appendix 2 of Westerlund (1997). A digital photometric survey of the LMC, which will eventually provide UBVI photometry of up to 25 million stars, is presently being undertaken by Zaritsky, Harris & Thompson (1997)

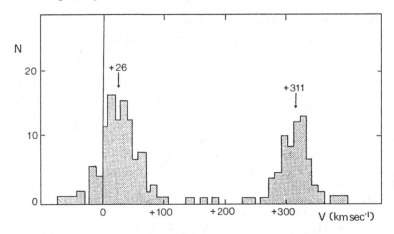

Fig. 6.1. Distribution of radial velocities of bright stars in the direction of the Large Magellanic Cloud according to Fehrenbach et al. (1970). The data show a clear-cut separation between LMC members with $\langle V \rangle = +311$ km s^{-1} and Galactic foreground objects with $\langle V \rangle = +26$ km s^{-1}.

6.2   Distance and reddening

The Large Magellanic Cloud is arguably the most important rung on the extragalactic distance ladder. Measurements of its distance and reddening are therefore critical for the determination of the extragalactic distance scale. A possible complication has been introduced by Zaritsky & Lin (1997), who found a vertical extension of the red clump stars in the color–magnitude diagram of the LMC, which they interpreted as being due to a concentration of stars closer to us than the main body of the Large Cloud. However, more recently Beaulieu & Sackett (1998) and Gallart (1998) have argued that this small concentration of stars just above the red clump might, in fact, be due to a short-lived evolutionary phase of old stars in the LMC. The interpretation by Beaulieu & Sackett is supported by the color–magnitude diagram of the Fornax dwarf spheroidal by Stetson, Hesser & Smecker-Hane (1998), which shows a feature above the red giant clump similar to that seen in the Large Cloud. Alcock et al. (1997c) have used the MACHO database to search for RR Lyrae stars that might be associated with a dwarf spheroidal galaxy in front of the LMC. Their search for foreground RR Lyrae variables yielded 20 stars whose distance distribution followed the expected Galactic halo density profile; that is, there data provide no support for the existence of an intervening galaxy that might contribute to the microlensing of LMC stars. Additional arguments against the existence of a galaxy in front of the LMC have been summarized by Gould (1998).

6.2.1   Reddening of the large cloud

From a study of 1,409 individual stars, Oestreicher, Gochermann & Schmidt-Kaler (1995) find that the foreground reddening in the direction of the LMC is quite variable, with values ranging from 0.00 mag to 0.15 mag. From five-color photometry of foreground stars of types A and F in the direction of the Large Cloud, McNamara & Feltz (1980) obtained $\langle E(B - V) \rangle = 0.034 \pm 0.004$. A detailed discussion of both the internal reddening in, and foreground reddening toward, the LMC has been given by Bessell

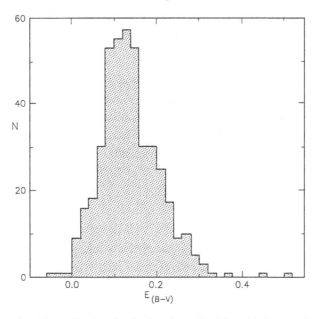

Fig. 6.2. Reddening distribution determined for 414 OB stars in the LMC by Massey et al. (1995).

(1991). More recently Massey et al. (1995) have used UBV photometry of 414 early-type stars in the Large Cloud (see Figure 6.2) to obtain $\langle E(B-V)\rangle = 0.13$, corresponding to $A_V = 0.40$ mag. However, from a bigger sample of 2,069 OB stars Harris, Zaritsky & Thompson (1997) find a larger value of $\langle E(B-V)\rangle = 0.20$. The reason for the difference between the mean reddening values found by Massey et al. and by Harris et al. might be that Massey et al. used a constant V cutoff for their sample, which would exclude the most highly reddened stars. Inspection of Figure 6.2 shows that the observed reddening distribution of OB stars is asymmetrical with a long wing extending to $E(B-V) \simeq 0.4$. Since OB stars are young they might be more closely associated with gas and dust clouds than older stars. It is therefore possible that the mean reddening for LMC Cepheids might be slightly lower than the values that Massey et al. and Harris et al. obtained for OB stars. This problem is presently being attacked by Madore (1997, private communication), who is obtaining UBV photometry of early-type stars near 40 long-period Cepheids in the LMC. Oestreicher & Schmidt-Kaler (1996) have combined spectroscopy and photometry of 1,507 luminous stars of types O, B, and A to study the distribution of reddening values in the Large Cloud. For their sample they find a mean foreground reddening $\langle E(B-V)\rangle = 0.16$. They fit their reddening distribution with a model that consists of a component with $\langle E(B-V)\rangle = 0.04$ due to small clouds and a second component with $\langle E(B-V)\rangle = 0.4$ that is due to reddening in large dust clouds. Oestreicher & Schmidt-Kaler note that $E(B-V)$ values in the range 0.4 to 0.8 are only observed in stars brighter than $V_0 = 12.3$. This result suggests that their observational sample of OBA stars was strongly biased against heavily reddened objects.

The extinction of emission regions in the Large Cloud has been studied by Ye (1998), who finds that most H II regions in the LMC have extinction properties similar to those of Galactic ones. Galaxies (Gurwell & Hodge 1990) and objects with active galactic nuclei (Crampton et al. 1997) that are located behind the LMC could be used to probe the interstellar medium in the Large Cloud.

6.2.2 Distance to the Large Cloud

A very detailed review of distance determinations of the LMC published prior to mid-1995 is given in Westerlund (1997, pp. 6–20). From observations of Cepheids, RR Lyrae stars, Mira variables, OB stars, red stars, SN 1987A, and novae Westerlund finds an unweighted mean distance modulus of $\langle (m - M)_o \rangle = 18.48 \pm 0.04$, corresponding to a distance of 49.7 ± 0.9 kpc. An excellent review of distance determinations to the Magellanic Clouds is given by Walker (1998), who concludes that $\langle (m - M)_o \rangle = 18.55 \pm 0.10$. Recent distance determinations are discussed in more detail below.

Cepheids

Feast & Catchpole (1997) have used *HIPPARCOS* observations of Galactic Cepheids to set the zero-point of the Cepheid period–luminosity relation. Application of this relation to the Cepheids in the LMC yields a Large Cloud distance modulus $(m - M)_o = 18.70 \pm 0.10$. A possible problem with this determination is that only a small number of Cepheids have parallaxes that significantly exceed their errors. In fact most of the weight of the solution comes from the (somewhat peculiar) Cepheid Polaris (α UMi). An additional difficulty is that it is not easy to disentangle the effects of metallicity and reddening of Cepheids (Madore & Freedman 1998, Sekiguchi & Fukugita 1997). From observations of Cepheids at different radial distances in M101, Kennicutt et al. (1998) conclude that the change in the distance moduli of Cepheids varies with metallicity as $\delta(m - M)_o/\delta[O/H] = -0.24 \pm 0.16$ mag dex^{-1}; that is, the change of Cepheid luminosity with metallicity is only marginally significant. Oudmaijer, Groenewegen & Schrijver (1998) have rediscussed the Galactic calibration of Cepheids observed with the *HIPPARCOS* satellite by taking the Lutz–Kelker bias into account. The resulting distance modulus of the LMC is $(m - M)_o = 18.56 \pm 0.08$, which is 0.14 mag smaller than the value obtained by Feast & Catchpole (1997). Using a maximum-likelihood technique, Luri et al. (1998) find a Cepheid period–luminosity relation from *HIPPARCOS* data that is 0.37 ± 0.20 mag fainter than that found by Feast & Catchpole. The corresponding LMC distance modulus is $(m - M)_o = 18.33 \pm 0.20$. However, a reanalysis of the *HIPPARCOS* data by Baumgardt et al. (1999) yields $(m - M)_o = 18.77 \pm 0.17$.

Using the infrared Barnes–Evans surface brightness technique Gieren, Fouqué & Gómez (1998) find an LMC distance modulus of $(m - M)_o = 18.46 \pm 0.06$. Wood, Arnold & Sebo (1997) have used nonlinear pulsation models to simulate the light curve of the "bump" Cepheid HV 905. Such models constrain the mass, luminosity, and effective temperature of this object. Comparison of the predicted values of the luminosity and color of HV 905 with existing V and I photometry yields a LMC distance modulus of $(m - M)_o = 18.51 \pm 0.05$. Some doubt is thrown on the reliability of the Cepheid distance scale by the observation (Tanvir 1999) that the luminosity calibration provided by the Cepheids in Galactic clusters and associations suggests an age dependence of the main-sequence fitting distance. However, the numbers involved are small, and the assumed cluster/association membership of some Cepheids might be questioned.

RR Lyrae variables

Reid (1997) has fitted the main sequence of the Galactic globular cluster NGC 6397 to the *HIPPARCOS* parallaxes of nearby subdwarfs to determine the luminosity of its horizontal branch. Fitting this horizontal branch to that of the LMC globulars NGC 1466 and NGC 2257 yields an LMC distance modulus of $(m - M)_0 = 18.71 \pm 0.06$. It is noted in passing that this fit yields $M_V(\text{RR}) \sim +0.25$. Such a bright value is in serious conflict with $M_V(\text{RR}) \sim +0.7$, which Fernley et al. (1998a,b) derive from the statistical parallaxes of RR Lyrae stars by using their *HIPPARCOS* proper motions.[2] Moreover, Reid (1997) points out that the high luminosity of the LMC RR Lyrae variables derived from the *HIPPARCOS* distance of NGC 6397 yields an unacceptably large distance to the Galactic nucleus. These difficulties could, of course, be avoided by making the ad hoc assumption (Gratton 1998) that field RR Lyrae variables of a given metallicity are systematically fainter than their counterparts in globular clusters. However, Catelan (1998) has recently shown that the period–temperature distributions for field and cluster RR Lyrae variables are essentially indistinguishable. This suggests that there are no significant luminosity differences between the RR Lyrae variables in clusters and in the field. It is concluded that the value of $M_V(\text{RR})$ remains uncertain at the ~ 0.2 mag level. However, if there is indeed no significant systematic luminosity difference between field and cluster RR Lyrae stars, then it is difficult to understand why *HIPPARCOS* proper motions of RR Lyrae variables yield $M_V(\text{RR}) = +0.74 \pm 0.12$ (Popowski & Gould 1998), whereas cluster main sequence fitting to local subdwarfs of similar metallicity yields $M_V(\text{RR}) = +0.44 \pm 0.08$ (Reid 1997) or $M_V(\text{RR}) = +0.49 \pm 0.04$ (Gratton 1998). Excellent discussions of this problem are given by Hendry (1997), Layden (1999), and Gould & Popowski (1998). One might, following Dostoevsky (1880, p. 4), conclude that "Being at a loss to resolve these questions, I am resolved to leave them without any resolution." Alternatively, following Soderblom et al. (1998), one might assume that the $M_V(\text{RR})$ of globular clusters has been affected by some, as yet not understood, systematic error of the *HIPPARCOS* parallaxes. Some support for this view is provided by the bizarre distance determination of the Pleiades cluster obtained by *HIPPARCOS* (Narayanan & Gould 1999).

For the RR Lyrae variables in seven LMC globular clusters Walker (1992) obtains $\langle V \rangle_0 = 18.95 \pm 0.04$ and $\langle [\text{Fe/H}] \rangle = -1.93$. With a LMC distance modulus of $(m - M)_0 = 18.5 \pm 0.1$ this yields $M_V(\text{RR}) = +0.45 \pm 0.11$ at $[\text{Fe/H}] = -1.93$. Adopting a slope of $+0.18 \pm 0.03$ for the RR Lyrae M_V versus $[\text{Fe/H}]$ relation (Fernley et al. 1998b) implies that the LMC RR Lyrae variables would have had $M_V(\text{RR}) = +0.52 \pm 0.11$ at $[\text{Fe/H}] = -1.53$, which is the mean metallicity of Galactic field RR Lyrae variables. Using *HIPPARCOS* proper motions, Fernley et al. (1998a) find that these Galactic field RR Lyrae have $\langle M_V \rangle = +0.77 \pm 0.15$. These two results differ by 0.25 ± 0.19 mag, in the sense that the statistical parallaxes give an LMC distance modulus that appears to be too small by about a quarter of a magnitude. Could this difference be due to a

---

[2] Carney, Lee & Habgood (1998) point out that HD 17072, which has the best-determined *HIPPARCOS* parallax of any metal-poor ($[\text{Fe/H}] = -1.17$, $[\alpha/\text{Fe}] = +0.33$) horizontal branch star, has $M_V = +0.97 \pm 0.15$, which supports the fainter luminosities for such stars found from statistical parallaxes. From six horizontal branch stars that have $\pi/\epsilon(\pi) \gtrsim 4$, de Boer (1999) finds $M_V = +0.71$ at $(B - V)_0 = 0.20$ and $[\text{Fe/H}] = -1.5$.

systematic difference between cluster and RR Lyrae variables? For RR Lyrae field stars near three LMC clusters Walker finds $\langle V_0 \rangle = 19.05 \pm 0.04$ and $\langle[\text{Fe/H}]\rangle = -1.73$. With the $M_V(\text{RR})$ versus [Fe/H] slope of Fernley et al. (1998b) this implies that the LMC field RR Lyrae stars would have had $\langle V_0 \rangle = 19.01 \pm 0.04$ at $[\text{Fe/H}] = -1.93$, which does not differ significantly from the value $\langle V_0 \rangle = 18.95 \pm 0.04$ that is observed for the LMC cluster RR Lyrae variables. It is therefore concluded that the Large Cloud does not provide evidence for a systematic difference between field and cluster RR Lyrae stars.

Using observations of multimode RR Lyrae stars obtained during the MACHO project, in conjunction with pulsation theory, Alcock et al. (1997b) find an LMC distance modulus of $(m - M)_0 = 18.48 \pm 0.19$.

McNamara (1997b) has used five-color observations of SX Phoenicis stars, together with a zero-point derived from *HIPPARCOS* parallaxes, to derive a semempirical period–luminosity relation for SX Phe and large-amplitude δ Scuti stars. This relation was then employed to determine the distances to globular clusters that contain such variables. From this indirect calibration of M_V (RR) McNamara obtains a LMC distance modulus of $(m - M)_0 = 18.57 \pm 0.03$.

Red clump stars

Udalski et al. (1998) have used the assumption that the luminosity of red clump stars is independent of their age and metallicity, in conjunction with *HIPPARCOS* parallaxes of red clump stars in the solar neighborhood, to derive a discrepant Large Cloud distance modulus of $(m - M)_0 = 18.08 \pm 0.03$. A similar result has been published by Stanek, Zaritsky & Harris (1998). Even neglecting the possible systematic errors of *HIPPARCOS* parallaxes that were alluded to above, this determination is suspect because of evidence that M_V (clump) is a function of both age and metallicity (Alves & Sarajedini 1998, Cole 1998, Girardi et al. 1998). After allowing for such population effects Cole finds that $(m - M)_0 = 18.36 \pm 0.17$ for the Large Cloud. From a comparison of the apparent magnitude of red clump stars in the LMC, with the absolute magnitudes of stars in the red giant clumps in the metal-poor clusters NGC 2420 and NGC 2506 that are located in the outer disk of the Galaxy, Twarog, Anthony-Twarog & Bricker (1999) find that $(m - M)_0 = 18.42 \pm 0.16$. An additional caveat that should be kept in mind, when using red clumps for distance estimates, is that red clump stars in Baade's Window appear to be 0.2 mag redder in $(V - I)_0$ than those near the Sun (Paczyński 1998).

Mira variables

Van Leeuwen et al. (1997) have employed the *HIPPARCOS* satellite to obtain parallaxes for 16 Galactic Mira stars. These were then used to calibrate the period–luminosity relation for Mira stars in the K-band and in M_{bol}. Application of these relationships to the Mira stars in the LMC yields $(m - M)_0 = 18.60 \pm 0.05$ from M_K and $(m - M)_0 = 18.47 \pm 0.05$ in M_{bol}. By fitting the period–luminosity relation of 115 carbon-rich long-period variables observed with *HIPPARCOS* to the period–luminosity relation for similar objects in the LMC, Bergeat, Knapik & Rutily (1998) obtain a distance modulus of $(m - M)_0 = 18.50 \pm 0.17$.

M-type supergiants

From a comparison of line indices for a small number of Galactic calibrators with a few supergiants in the LMC, Schmidt-Kaler & Oestreicher (1998) obtain $(m - M)_0 = 18.34 \pm 0.09$. However, it is not yet known how the systematic metallicity difference between the supergiants in the Large Cloud and in the Galaxy might have affected this result.

Young clusters

Fischer et al. (1998) have fit the main sequence of the 10^8 year old cluster NGC 2157 to model isochrones to obtain an LMC distance modulus of $(m - M)_0 = 18.4$. From a comparison of 100 O-type stars in 30 Doradus with similar stars in the Galaxy, Walborn & Blades (1997) find $(m - M)_0 \approx 18.55$.

Supernova 1987A

Panagia (1998) has combined *IUE* light curves of SN 1987A with *Hubble Space Telescope* imaging to derive an absolute size of the SNR ring of $(6.23 \pm 0.08) \times 10^{17}$ cm and an angular size of $R = 0{.}''808 \pm 0{.}''017$. The corresponding distance $D(1987A) = 51.4 \pm 1.2$ kpc. Allowing for displacement of SN 1987A from the center of the Large Cloud yields an LMC distance of 52.0 ± 1.3 kpc, corresponding to a true distance modulus of $(m - M)_0 = 18.58 \pm 0.05$. This result is, however, inconsistent with $(m - M)_0 < 18.37 \pm 0.04$, which was obtained by Gould & Uza (1998) from the "light echo" times to the near and far side of the ring around SN 1987A. This discrepancy serves to remind us that this technique, while elegant, may be subject to systematic errors resulting from asymmetric geometry.

Eclipsing binaries

Detached eclipsing binaries are, in principle, capable of providing distances that depend only on geometrical factors. Guinan et al. (1999) have employed this technique to derive a distance modulus for the of LMC of $(m - M)_0 = 18.30 \pm 0.07$, using the detached eclipsing variable HV 2274. It would clearly be of great importance to extend such detailed observations to a large number of detached eclipsing variables in the Clouds of Magellan.

A summary of recent distance determinations to the Large Cloud is given in Table 6.2. An unweighted mean of these 18 values gives $\langle(m - M)_0\rangle = 18.49 \pm 0.01$. In the subsequent discussion a distance modulus of 18.50 ± 0.05, corresponding to a distance of 50.1 ± 1.2 kpc, will be adopted for the LMC.

6.3   Global properties

Westerlund (1997, p. 21) writes that "The history of the Magellanic Clouds is still veiled in obscurity." However, some hints regarding their evolution are provided by the present structure and morphology of the LMC and SMC. The Large Cloud is a disk that is seen almost pole-on. According to de Vaucouleurs & Freeman (1972), $i = 27° \pm 2°$ and $r_e = 3°0$ (2.6 kpc). From wide-field imaging of the LMC Bothun & Thompson (1988) find that the diameter of the $B = 25$ mag arcsec^{-1} isophote is $9°1 \pm 0°3$ (7.9 kpc). They

Table 6.2. *Recent distance determinations to the LMC*

Method	$(m - M)_o$	Author(s)	Comments
Cepheids	18.46 ± 0.06	Gieren et al. (1998)	Barnes–Evans technique
Cepheids	18.51 ± 0.05	Wood et al. (1997)	Bump Cepheid HV 905
Cepheids	18.70 ± 0.10	Feast & Catchpole (1997)	*P–L* calibrated by *HIPPARCOS*
Cepheids	18.56 ± 0.08	Oudmaijer et al. (1998)	Lutz–Kelker corrected P-L
RR Lyrae	18.71 ± 0.06	Reid (1997)	M_V (RR) from *HIPPARCOS* subdwarfs
RR Lyrae	18.48 ± 0.19	Alcock et al. (1997b)	Multimode RR + pulsation theory
RG tip	18.64 ± 0.14	Salaris & Cassisi (1998)	Tip of red giant branch
Red clump	18.08 ± 0.12	Udalski et al. (1998)	Using *HIPPARCOS* calibration
Red clump	18.36 ± 0.17	Cole (1998)	Recalibrated M_V (clump)
SX Phe	18.56 ± 0.10	McNamara (1997b)	M_V (RR) via *P–L* of SX Phe variables
Miras	18.60 ± 0.05	van Leeuwen et al. (1997)	*P–L* in M_K
Miras	18.47 ± 0.05	van Leeuwen et al. (1997)	*P–L* in M_{bol}
Miras	18.50 ± 0.17	Bergeat et al. (1998)	Carbon LPVs
Open cluster	$18.40 \pm 0.05 - 0.1$	Fischer et al. (1998)	Young cluster main sequence fitting
Planetaries	18.44 ± 0.18	Jacoby (1997)	PN luminosity function
Eclipsing variable	18.30 ± 0.07	Guinan et al. (1999)	Geometrical
O stars	18.55	Walborn & Blades (1997)	Galactic calibration
SN 1987A	18.58 ± 0.05	Panagia (1998)	Light echo
SN 1987A	$<18.37 \pm 0.04$	Gould & Uza (1998)	Light echo

derive a disk scale-length of $101' \pm 3'$ (1.5 kpc). Asymmetrically embedded within this disk is a prominent bar with a length of $\sim 3°$. Multi-color photographs of the Large Cloud (Walker, Blanco & Kunkel 1969) show that the Bar is mainly outlined by intermediate-age stars, whereas young clusters and associations are mostly located in the outer disk. Far-ultraviolet observations with the *Hubble Space Telescope* (Brosche et al. 1999) show that the UV surface density of stars in the LMC is a few times higher than it is in the Galactic disk near the Sun. The distribution of neutral hydrogen gas in the LMC (McGee & Milton 1964) is shown in Figure 6.3. Note the very clumpy distribution of gas and the fact that the largest concentration of neutral hydrogen (in which the 30 Doradus complex is embedded) is located just beyond the end of the Bar. Table 6.3 shows that the center of mass of the neutral hydrogen gas in the Large Cloud is displaced significantly from the center of the LMC defined by optical isophotes and the center of symmetry of the Large Cloud rotation curve. In other words it appears that the gas in the Large Cloud is "sloshing around" in the potential well of the LMC.

Table 6.3. *Population centroids in the Large Cloud*[a]

Population	α	(2000)	δ
Yellow light isophotes	05[h]20[m]	−69°27′	
Optical center of Bar	05 24	−69 45	
H I rotation center	05 21	−69 14	
H I center of mass	05 34	−68 28	
All star clusters	05 32	−69 17	
Globular clusters[b]	05 17	−69 32:	
Population I rotation center	05 20	−68 45	
Planetary nebulae	05 24	−68 58	
Novae[c]	05 21	−69 10	

[a] Adapted from Westerlund (1997, p. 33).
[b] See Section 6.4.3.
[c] Includes LMC 1997.

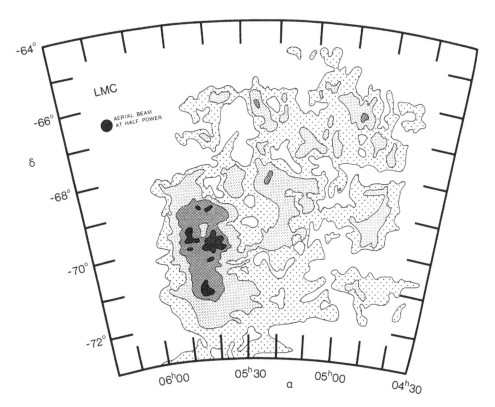

Fig. 6.3. Distribution of neutral hydrogen gas in the LMC adapted from McGee & Milton (1966). The figure shows that the hydrogen in the Large Cloud has a very irregular and clumpy distribution. The largest hydrogen cloud (in which the 30 Doradus complex is embedded) is located just beyond the end of the LMC Bar.

6.4   Star clusters

6.4.1   Introduction

The first mention of star clusters in the Large Cloud is by Herschel (1847). Numerous LMC clusters are listed in the NGC catalog (Dyer 1895). Lyngå & Westerlund (1963) find that the LMC cluster system has dimensions of $11° \times 15°$(9.5 kpc $\times$ 12.9 kpc) and is centered at $\alpha = 5^h \, 31^m \, 49^s$, $\delta = -69° \, 17'$. These authors find that the Large Cloud cluster system has an axis ratio of about 1.0 to 1.4 and that the position angle of its major axis is $7° \pm 10°$. Plots of the distribution of the Large Cloud clusters on the sky have been published by Kontizas et al. (1990), by Irwin (1991), and by Bica et al. (1996, 1999). An up-to-date listing of cluster surveys in the LMC is given in Westerlund (1997, p. 46). Most of the bright clusters in the Large Cloud may be conveniently located on the charts published by Hodge & Wright (1967). Van den Bergh (1991d) has given a review on star clusters in the Clouds of Magellan in his Russell lecture.

6.4.2   Populous Young blue clusters

Just before the Second World War, Baade (1963, p. 110) asked Thackeray in South Africa to obtain photographs of the populous cluster NGC 1866 in the Large Cloud. Surprisingly, Thackeray found that the brightest stars in this cluster were blue, whereas the brightest stars in Galactic globular clusters are red. Subsequently, Gascoigne & Kron (1952) made the unexpected discovery that the luminous star clusters in the LMC fall into two distinct color groups. They correctly assumed that the luminous red clusters, such as NGC 1846, were broadly similar to the globular clusters of Baade's Population II. The existence of populous blue clusters such as NGC 1866, which have no Galactic counterpart, was, however, a puzzling surprise. Shapley (1956) proposed that such populous blue clusters should be dubbed "circular clusters." It now appears (van den Bergh & Lafontaine 1984) that the existence of populous clusters, such as NGC 1866, is due to the fact that the mass spectrum with which young open clusters are formed in the LMC is less deficient in high-mass objects than is the mass spectrum with which open clusters form in the Galaxy and M31. The mass spectrum with which clusters were formed in the main body of the proto-Galaxy $>10^{10}$ years ago may have been weighted even more strongly toward high-mass objects. However, lower mass "young" globular clusters appear to have formed in the *outer* halo of the Galaxy $\sim 10^{10}$ years ago (van den Bergh 1998a).

Fischer et al. (1998) have found clear evidence for mass segregation in the luminous young LMC cluster NGC 2157. The cluster mass spectrum is found to be steeper (have more faint stars) in its outer region than it does in its core. The age of NGC 2157 is $\sim 1 \times 10^8$ yr, which is only about 1/10 of its relaxation time t_{rh}. It follows that most of the mass segregation in this cluster cannot be the result of relaxation. This suggests that some mergers between cloudlets may have occurred in the cluster core before protostellar collapse. Elson et al. (1998a) have found that the binary star population increases toward the center of the Large Cloud cluster NGC 1818. These authors observe the fraction of binaries to be $\sim 20 \pm 5\%$ in the outer parts of this cluster, compared to $\sim 35 \pm 5\%$ in its core region. Elson et al. find that this observed radial change in the fraction of binary stars is consistent with dynamical segregation and need not be primordial.

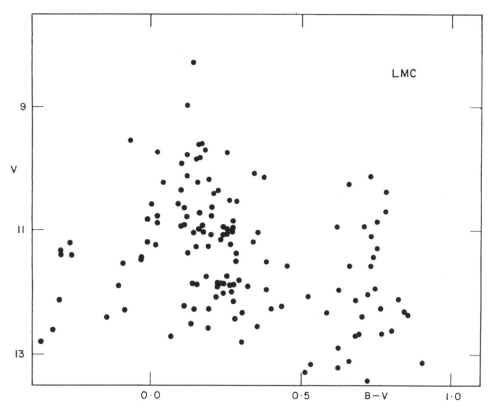

Fig. 6.4. Color–magnitude diagram for clusters in the LMC. Note the color gap between the brightest red and blue clusters.

6.4.3   *LMC globular clusters*

The color gap between the most luminous blue and red clusters in the LMC is clearly seen in the integrated color–magnitude diagram for Large Cloud clusters (van den Bergh 1981b) that is shown in Figure 6.4. It is now clear that the sharp dichotomy between blue and red LMC clusters is, at least in part, due to the wide age gap between true globular clusters, with ages $\gtrsim 12$ Gyr and [Fe/H] < -1.5, and open clusters, with ages $\lesssim 4$ Gyr and [Fe/H] $\gtrsim -1.0$ (Da Costa 1991, Geisler et al. 1997). Sarajedini (1998b) has found three clusters (NGC 2121, NGC 2155, and SL 663) that have ages of ~ 4 Gyr and [Fe/H] ~ -1. These may represent objects that formed as the recent great burst of cluster formation started to ramp up. Only the cluster ESO 121-SC03, which is located in the outer part of the LMC (see Figure 6.5) and has an age of ~ 9 Gyr, is known to fall in the age gap between ~ 4 Gyr and ~ 12 Gyr.

A listing of the 13 true globular clusters associated with the LMC (Suntzeff 1992) is given in Table 6.4. A plot of the positions of these clusters on the sky is shown in Figure 6.5. This figure shows that the globular clusters in the LMC are centered just south of the LMC Bar near the centroid of the outer H I isophotes. The centroid of the LMC globular cluster system is located at $\alpha = 5^{\mathrm{h}}17^{\mathrm{m}} \pm 12^{\mathrm{m}}$, $\delta = 69.5 \pm 1.5$ (equinox 2000).

Table 6.4. *Globular clusters associated with the Large Cloud*[a]

Cluster	α (2000) δ		R^b	$V_r(kms^{-1})$	V	ϵ^c	[Fe/H]	n(RR)
NGC 1466	3h 44m4	−71°36′	8:4	200	11.6	0.09	−1.85	38
NGC 1754	4 54.4	−70 27	2.6	236	11.4	0.06	−1.54	...
NGC 1786	4 58.8	−67 45	2.5	264	10.9	0.02	−1.87	9
NGC 1835	5 05.8	−69 24	1.4	188	9.5	0.21	−1.79	31
NGC 1841	4 44.9	−84 00	14.9	214	11.0	...	−2.11	22
NGC 1898	5 16.7	−69 36	0.6	210	11.1	...	−1.37	...
NGC 1916	5 17.5	−69 22	0.2	278	9.9	...	−2.08	...
NGC 2005	5 30.0	−69 44	0.9	270	11.2	0.14	−1.92	...
NGC 2019	5 31.7	−70 10	1.3	269	10.7	0.07	−1.81	0
NGC 2210	6 11.4	−69 07	4.4	343	10.4	0.07	−1.97	12
NGC 2257	6 29.9	−64 10	8.4	302	11.5	...	−1.8	31
Hodge 11	6 14.4	−69 51	4.7	246	12.1	0.11	−2.06	0
Reticulum	4 36.2	−58 50	11.4	243	12.8	...	−1.71	29

[a]Data mainly from Schommer et al. (1992) and Suntzeff (1992).
[b]Measured from H I rotation center at $\alpha = 5^h20^m40^s$, $\delta = -69°$ 14.2 (J2000).
[c]Cluster ellipticity.

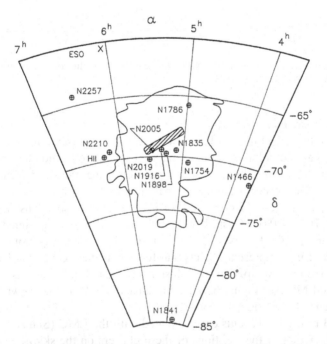

Fig. 6.5. Distribution of globular clusters in the Large Cloud. These clusters are seen to be centered just south of the LMC Bar. The Reticulum cluster at $\alpha = 04^h 36^m 14^s$, $\delta = -58°$ 50′ 00″ (J 2000) lies just outside the boundary of the figure. The position of the ∼10-Gyr-old cluster ESO121 SC03 is marked by a cross.

The radius containing half of all globular clusters in projection is $r_e = 2°5$, corresponding to 2.2 kpc. The data in Table 6.4 show no clear-cut evidence for a radial abundance gradient among the globular clusters in the LMC. However, it is noted that NGC 1841 is both the most metal-poor globular and the one that is most distant from the center of the LMC. An additional hint at the possible existence of a radial abundance gradient is provided by the observation that NGC 1898, which is the most metal-rich globular cluster in the Large Cloud, is located only 0°6 from its center. Because the number of globular clusters in the LMC is so small it will probably never be possible to establish the presence (or absence) of a radial abundance gradient among them with certainty. With $M_V = -18.5$ and n(globulars) $= 13$ the specific globular cluster frequency in the Large Cloud is $S = 0.5$. This value (Harris 1991) is quite similar to that observed in other late-type galaxies.

The globular cluster radial velocities V_i and distances from the H I rotation center R_i that are listed in Table 6.4 may be used to calculate the mass of the LMC via the "projected mass estimator" of Bahcall & Tremaine (1981). These authors suggest that, in the absence of specific information on the orbital eccentricities of N individual objects, the best mass estimator is

$$M = (24/\pi N G) \sum_{i=1}^{N} V_i^2 R_i \qquad (6.1)$$

More recently Heisler, Tremaine & Bahcall (1985) have recommended that the factor 24 in Eq. 6.1 above be replaced by 32, for a situation in which the orbits of clusters are isotropic. Substituting the data from Table 6.1 into the revised version of Eq. 6.1 yields a mass of $\sim 17 \times 10^9 M_\odot$ for the LMC. If one excludes the four most distant clusters, then one finds a mass of $\sim 11 \times 10^9 M_\odot$ for the region of the Large Cloud that is located within a distance of 5 kpc from the H I rotation center. This may be compared to a mass of $(6.2 \pm 0.9) \times 10^9 M_\odot$ that Kunkel et al. (1997) find for the same region from the radial velocities of 759 carbon stars. Cowley & Hartwick (1991) have studied CH stars in the LMC. In the Milky Way system such stars are excellent tracers of the Galactic halo. They find that these objects cover a region with a diameter of $\sim 13°$ (11 kpc). Cowley & Hartwick find that the velocity distribution of these CH stars has a rather sharp cutoff at high velocities and an extended low-velocity tail. They speculate that this asymmetry is due to the fact that the CH stars in the Large Cloud belong to two distinct population components, one of which is associated with halo globular clusters and the other with the LMC Disk. However, this conclusion appears to be at odds with that of Schommer et al. (1992), who find that the Large Cloud globulars themselves appear to form a Disk system. These authors conclude that the oldest clusters rotate with an amplitude comparable to that of young disk objects and that their (poorly determined) velocity dispersion is small. Taken at face value this result suggests that the globular clusters in the Large Cloud form a disk, rather than a halo. However, this conclusion is intrinsically uncertain because the number of globular clusters in the LMC is so small. It would therefore be important to check this conclusion by obtaining radial velocity observations of individual LMC Population II objects such as RR Lyrae stars.

Bica et al. (1996) show that the spatial distribution of the oldest LMC clusters has a major axis that is ~ 3 times larger than that of the youngest clusters. In other words, the region in which Large Cloud clusters form has shrunk with time. Bica et al. also find interesting evidence for a possible change in the inclination of the LMC cluster system with time. For the system of old clusters these authors find an axial ratio of 1.0 to 1.40 ± 0.14, corresponding to an inclination of $i = 45° \pm 4°$. However, these authors find that young clusters with ages <30 Myr appear to have a distribution with an axial ratio of 1.0 to 1.0, which suggests that they form a flattened system that is viewed almost pole-on. Bica et al. also find that the centroid of these clusters drifted by ∼1° on the sky over time. Dottori et al. (1996) have studied the migration of centers of star formation by comparing the distributions of clusters having ages of 0–10 Myr, 10–30 Myr and 30–70 Myr. A more detailed discussion of the history of star and cluster formation in the Large Cloud will be given in Section 6.6.

Color–magnitude diagrams for the six LMC globular clusters NGC 1754, NGC 1835, NGC 1898, NGC 1916, NGC 2005, and NGC 2019 have been obtained with the *Hubble Space Telescope* by Olsen et al. (1998) and are shown in Figure 6.6. The clusters NGC 1466, NGC 2257, and Hodge 11 have been observed by Johnson et al. (1998). The morphologies of their color–magnitude diagrams match those of Galactic globular clusters of similar metallicity. All six clusters studied by Olsen et al. have well-developed and quite blue horizontal branches, four of which have stars on both sides of the RR Lyrae "gap." On average, the LMC globular clusters are found to have the same ages as Galactic globulars of the same metallicity to within 1.0 ± 0.2 Gyr. It has previously been pointed out that Schommer et al. (1992) found the LMC globular clusters to form a disk system. If this conclusion is correct, and if the LMC and Galactic halo globulars have the same age, then *the Large Cloud would already have collapsed to a disk at the time when the Milky Way was still forming halo clusters.* As has already been noted in Section 5.6 the situation in the Large Cloud, where the globulars are very old and appear to lie in a disk, differs from that in M33, in which the globulars populate a halo – even though they do not appear to be very old. It remains a profound mystery why the LMC should have old disk globulars, while M33 appears to be surrounded by relatively young halo globular clusters. In this connection it is also of interest to note that the outer halo cluster NGC 2419, located at a Galactocentric distance of 90 kpc, is as old as the clusters in the inner halo of the Galaxy (Harris et al. 1997). These results appear to indicate that Baade's Population II started to form more or less simultaneously in different regions of space. There is, however, one puzzling systematic difference between globular (and open!) clusters in the Galaxy and the Magellanic Clouds. This is that the clusters of all ages in the Clouds are, on average, much more highly flattened than their Galactic counterparts (Geisler & Hodge 1980, van den Bergh & Morbey 1984a). This effect is illustrated in Figure 6.7, which shows the intermediate-age LMC cluster NGC 1978, which has $\epsilon = 0.30$. One might speculate that the difference between the mean flattening of Cloud clusters and of clusters in the Galaxy is related to the apparent difference in the frequency of cluster pairs in these systems. In the Clouds physical cluster pairs (e.g., Dieball & Grebel 1998) appear to be quite common, whereas such pairs seem to be rare in the Galaxy.

6.5   Young clusters and associations

6.5.1   *Associations and superassociations*

Star formation can take place on a variety of scales. The largest star forming complexes in the Large Cloud have been dubbed "constellations" by McKibben

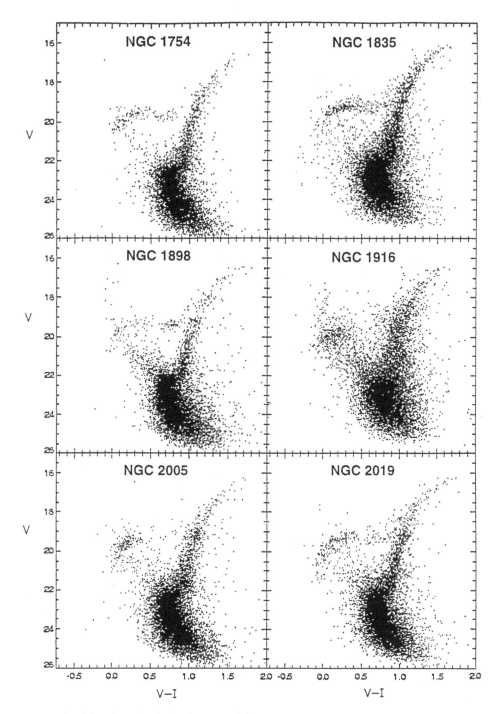

Fig. 6.6. Color–magnitude diagrams of six LMC globular clusters obtained with the *Hubble Space Telescope*. Adapted from Olsen et al. (1998). Note that these clusters all have significant numbers of stars to the blue of the RR Lyrae "gap."

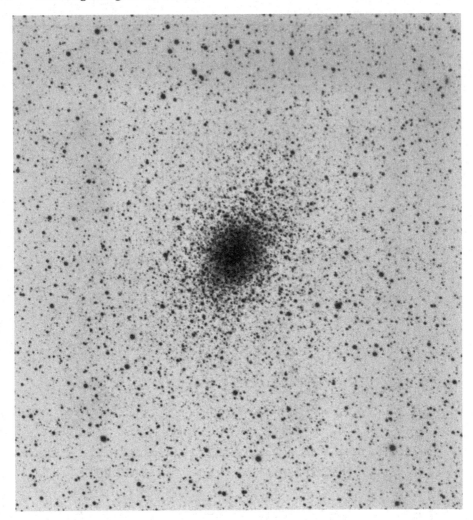

Fig. 6.7. Photograph in yellow light, obtained with the 4-m Blanco telescope, of the highly flattened intermediate-age Large Cloud cluster NGC 1978. (Reproduced with the kind permission of E.W. Olszewski.)

Nail & Shapley (1953). Star forming zones in the LMC that are smaller than constellations, but larger than typical OB associations, have been called "star clouds" by Lucke & Hodge (1970). Efremov (1995) and Magnier et al. (1997) refer to groupings of OB stars with sizes <100 pc as associations and call those with dimensions ~500 pc "complexes." Figure 6.8 shows the distribution of constellations in the LMC, which van den Bergh (1981b) has outlined by blinking multicolor Schmidt plates of the Large Cloud. Constellations I–V are those previously identified by McKibben Nail & Shapley, and constellations VI–IX are from van den Bergh (1981b). The well-known 30 Doradus complex has not been given a Roman numeral designation. Some confusion may result from the fact that Martin et al. (1976) have assigned Roman numerals I–VIII to designate supergiant emission shells in the Large Magellanic Cloud.

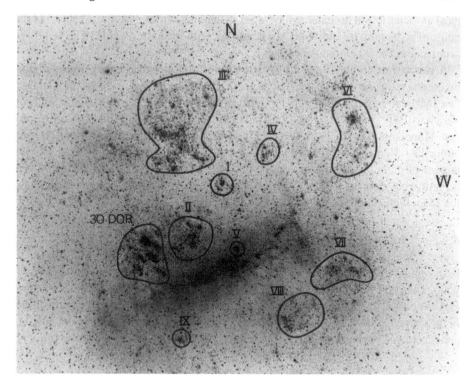

Fig. 6.8. Major LMC star forming complexes known as "constellations" outlined on the basis of inspection of multicolor Schmidt plates. The majority of these constellations are seen to be located north of the LMC Bar.

At present 30 Dor and Constellation III appear to be the most active regions of star formation in the LMC. Efremov & Elmegreen (1998) have discussed the evolutionary history of Constellation III, LMC 4, which contains a 600-pc-long arc of young stars and clusters. The form of this arc suggests that the presently observed gas was uniformly swept up by a central source of pressure. They point out that there are six ~30-Myr-old A-type supergiants, and a Cepheid of similar age, located near the center of curvature of this arc. These objects might be related to the source of the pressure that formed this feature. Alternatively (Efremov, Elmegreen & Hodge 1998; Loeb & Perna 1998), large arcs of star clusters, H I supershells, and giant dust rings might have been formed by γ-ray bursts. Such a shell can only form if the disk thickness of the parent galaxy exceeds the perturbation diameter of the high-pressure region. Otherwise the high-pressure gas will escape into the galactic halo.

Some associations contain multiple compact cores. Observationally it is often difficult to distinguish small expanding *positive energy* cores of associations from stable *negative energy* young star clusters that are frequently embedded within associations. Efremov & Elmegreen (1998) point out that small regions form stars quickly, whereas larger regions (which often contain smaller cores) form stars over a more extended period. Lucke & Hodge (1970) were able to identify 122 OB associations in the Large Cloud. From color–magnitude diagrams of association members Lucke (1974) concluded that most associations in his catalog have ages in the range from 3 Myr to 13 Myr. He found a range of reddening for individual associations, with the largest value $\langle E(B-V)\rangle = 0.25$ occurring in the 30 Doradus complex.

A number of authors have drawn attention to the fact that binary clusters appear to be common in the Magellanic Clouds, even though they are rare in the Galaxy. The fact that some of these Magellanic cluster pairs exhibit tidal tails (Leon, Bergond & Vallenari 1999) shows that they constitute physically linked systems. Leon et al. propose that Magellanic cluster pairs were formed by tidal captures within the cores of giant molecular clouds. (A similar mechanism was proposed by van den Bergh 1996b for the merger of globular clusters within the potential wells of dSph galaxies.) However, it is presently not clear why binary clusters appear to be so much rarer in the Galaxy than in the Magellanic Clouds.

6.5.2   Evolution of cluster distribution

Van den Bergh (1981b) made very rough age estimates of LMC clusters on the basis of their integrated $U - B$ and $B - V$ colors. The distribution of very young (<4 Myr), young (4–200 Myr), intermediate-age (0.2–1 Gyr) and old (>1 Gyr) clusters is shown in Figures 6.9–6.12. The very young clusters are seen to be widely dispersed over the face of the Large Cloud. Young clusters have both a widely dispersed component and one that appears to be concentrated in the Bar. Intermediate-age clusters seem to be concentrated

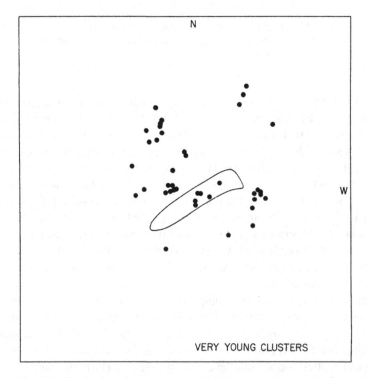

Fig. 6.9. Distribution of very young (age <4 Myr) clusters over the face of the Large Cloud. These objects are widely distributed over the LMC and exhibit no concentration to the Bar.

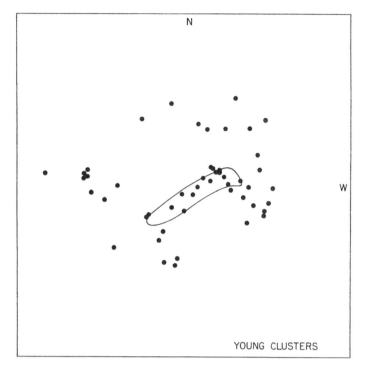

Fig. 6.10. Distribution of young (4–200 Myr) clusters over the face of the LMC. A significant fraction of these clusters are seen to be concentrated in, or near, the Bar of the Large Cloud.

near, but not in, the Bar. Finally, the old clusters are even more widely dispersed over the face of the LMC than are the very young clusters. It is particularly noteworthy that none of the very old clusters appears to be associated with the LMC Bar. This suggests that the Bar is an intermediate-age structure. This conclusion is supported by Figure 6.13, which shows the distribution of Cepheids of different periods (Gaposchkin 1943) over the face of the LMC. The corresponding Cepheid ages were taken from Tammann (1969). The figure shows little concentration of Cepheids in the Bar for ages <20 Myr. In contrast, Cepheids with periods of 40–70 Myr appear to exhibit a strong concentration to the LMC Bar. Observational selection effects are expected to have the opposite effect because faint old Cepheids would be difficult to discover in the crowded Bar region, whereas it would have been easy to discover luminous young Cepheids in the Bar. This conclusion, which is derived from the distribution of Cepheids, is strengthened and confirmed by Dottori et al. (1996), who find that very young clusters with ages <30 Myr exhibit no concentration to the LMC Bar. However, they note that about half of the slightly older clusters, with ages between 30 and 70 Myr, appear to be associated with the Bar. Taken at face value these results by Dottori et al. (1996) suggest that cluster formation in the Bar started between 30 Myr and 70 Myr ago.[3] Grebel & Brandner (1998) have also used Cepheids to study the

---

[3] In this connection it is of interest to note that Abraham et al. (1999) find that the fraction of all galaxies that are barred drops precipitously beyond redshifts of $z \sim 0.5$. This suggests that the Large Cloud might provide an example of ubiquitous late-onset bar formation.

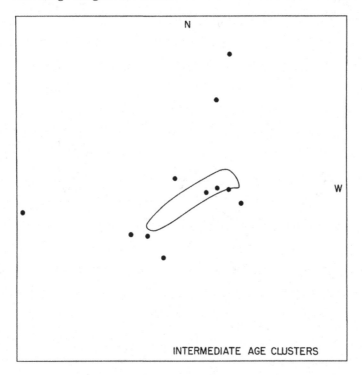

Fig. 6.11. Distribution of intermediate-age (0.2–1 Gyr) clusters over the face of the LMC. A significant fraction of these clusters appear to be located near the Bar of the Large Cloud.

recent star formation history of the LMC. They conclude that star formation in the LMC occurred stochastically across the face of the Large Cloud and that the clumpy distribution of LMC gas prevented self-propagating star formation on a large scale. These authors also concluded that loci of active star formation migrated within the Bar. The most active regions seem to have occurred in the northeastern part of the Bar 150–230 Myr ago. Star formation finally appears to have engulfed the entire Bar region 50–150 Myr ago.

De Boer et al. (1998a) have recently proposed that star formation may be triggered as gas is compressed in a bow shock formed at the leading edge of the LMC. Such a scenario would favor star formation at the SE edge of the Large Cloud, where the cumulative effects of the LMC's space motion and rotation are largest. However, inspection of Figures 6.9 and 6.10 shows that this is actually the region where the smallest number of young clusters is observed. Grebel & Brandner (1998) also find that the distribution of populations with differing ages over the face of the LMC does not support the hypothesis of de Boer et al. These results suggest that the gas density in the outer regions of the Galactic halo is too small to produce compression in the bow shock resulting from the relative motion of $\sim$465 km s^{-1} between the putative gas in the halo and that in the LMC. However, some small-scale shock-induced star formation, near the edge of the Large Cloud, probably cannot yet be entirely excluded.

6.5.3   30 Doradus and R 136

The 30 Doradus region, which is associated with the largest H I complex in the LMC (see Figure 6.3), is presently the single most active area of star formation in the

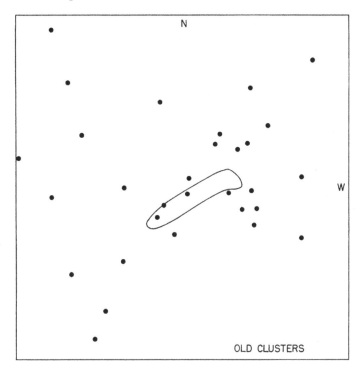

Fig. 6.12. Distribution of old (age >1 Gyr) clusters over the face of the LMC. Old clusters are distributed even more widely over the Large Cloud than are very young clusters (i.e., the region of star and cluster formation in the LMC appears to have shrunk with time). No old clusters seem to be associated with the Bar, suggesting that this structure is less than 1 Gyr old.

Large Cloud. From observations with the *Hubble Space Telescope* Massey & Hunter (1998) find that the young supercluster R 136, which is located near the center of the 30 Dor complex, contains more massive O3 stars than are known in all of the rest of the LMC. Their observations suggest that star formation in this region started 4–5 Myr ago and continued until very high mass stars formed. These massive stars shut down star formation <1–2 Myr ago. Unexpectedly Massey & Hunter find that the mass spectrum of star formation in the cluster R 136 has a slope $\Gamma^4 = -1.3$ to -1.4 over the range 3 $M_\odot$ to 120 $M_\odot$, which is similar to the slope $\Gamma = -1.35$ that Salpeter (1955) obtained for the solar neighborhood (in which the majority of massive stars were formed in small groups and clusters). The stellar density in R 136[5] is ~200 times greater than that in typical OB associations.[6] Taken at face value these results suggest that the mass spectrum of star formation is independent of the density and mass of star forming regions. For a more pessimistic assessment of the constancy of Γ the reader is referred to Scalo (1998). Malumuth & Heap (1994) find that the central region of the 30 Dor complex has a core

[4] The slope of the mass spectrum f is defined as $\Gamma = d \log F(\log M)/d \log M$.
[5] R numbers refer to the Radcliffe catalog of the brightest stars in the Magellanic Clouds (Feast, Thackeray & Wesselink 1960).
[6] It would be of interest to use a future very large telescope in space to derive the luminosity function of individual stars in extreme starburst galaxies, such as those in Arp 220 (Downes & Solomon 1998), in which the rate of star formation exceeds that in 30 Doradus by a factor of ~1,000.

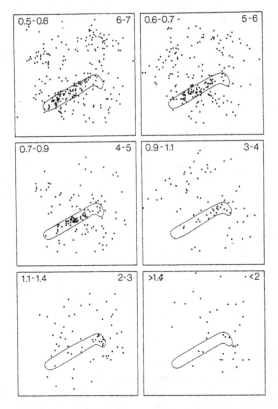

Fig. 6.13. Distribution of Cepheids of various ages over the face of the LMC. The ranges of log P(days) plotted in each panel are given in the upper left-hand corner of each figure. The corresponding age ranges, in units of 1×10^7 yr, are given in the upper right-hand corner of each panel. Active star formation seems to have been taking place in the Bar 40 to 70 Myr ago.

radius $R_c = 110''$ (27 pc) and a total mass $M \approx 1.7 \times 10^4 \, M_\odot$. It is noted in passing that this mass is significantly smaller than that of typical globular clusters. (The mass difference between 30 Dor and typical globulars becomes even larger when mass losses due to stellar evolution are taken into account.) The compact massive young cluster NGC 3603 (Sher 1965,Crowther & Dessart 1998) is the closest known Galactic analog to the supercluster R 136 (Moffat, Drissen & Shara 1994). In this connection it is puzzling that Eisenhauer et al. (1998) find that a mass distribution with a slope $\Gamma = -0.73$ gives a good representation of the mass spectrum of NGC 3603 over the range 1–30 $M_\odot$. However, these authors caution that this value should be regarded as an upper limit, since systematic effects might have resulted in an overestimate of Γ.

The kinematics of the gas associated with the 30 Doradus region is complex (Chu & Kennicutt 1994). Its outer parts have a smooth velocity field with a turbulent velocity of 30–40 km s^{-1}. In the central 9' (130 pc) core the velocity field is dominated by a large number of expanding structures with dimensions ranging from 1 pc to 100 pc, which exhibit expansion velocities that range from 100 km s^{-1} to 300 km s^{-1}. These large fast-expanding shells are coincident with extended X-ray sources. They were probably produced by the combined effects of supernovae and stellar winds from OB associations. Ultraviolet observations with the *Hubble Space Telescope* by Prinja & Crowther (1998) show that the O3 supergiant R 136a-608 holds the record for the highest terminal wind velocity ($\sim$3,640 km s^{-1}) of any star so far observed in the Local Group.

Scowen et al. (1998) have imaged the central 20 pc of the 30 Doradus region in three different emission lines. They conclude that most of the emission in 30 Dor originates in a thin zone located between the hot interior of the nebula and the surrounding dense cool molecular material. The emitting region is a photoionized, photoevaporative flow. Scowen et al. find that the ram pressure in this flow, which is derived from the thermal pressure at the surface of the column, balances the pressure in the interior of the nebula derived from previous X-ray observations.

Rubio et al. (1998) have discovered numerous infrared sources in (or near) the bright nebular filaments to the west and northeast of R 136. The positions of these sources appear to be intimately connected with the nebular microstructure, as well as with early O-type stars embedded in nebular knots. These observations suggest that a new generation of star formation is being triggered by the massive central cluster in the interstellar material on its periphery. Rubio et al. speculate that 30 Dor will evolve into a giant H II shell ionized by a younger population, which surrounds an evacuated central cavity containing an older association. Additional support for the notion that R 136 has triggered extensive "second-generation" star formation within the 30 Dor nebula is provided by Walborn et al. (1999). The stellar winds from the hot stars in R 136 release energy at a rate of $\sim$3 $\times$ 10^{39} erg s^{-1}. Over the lifetime of R 136 the integrated energy release via stellar winds will be $\sim$10^{53} ergs (Wang 1999). To this should be added the energy contributed by dozens of exploding supernovae.

The distribution of dust clouds, which are visible in absorption on a deep blue exposure of the LMC (van den Bergh 1974a), is shown in Figure 6.14. This figure shows that the most conspicuous dark clouds are located near the giant HI complex with which 30 Doradus is associated. It is particularly noteworthy that no major dust lanes appear to be embedded in the Bar of the Large Cloud.

6.6 Variable stars

Almost every type of variable star that is known in the Milky Way system has also been found in the Large Magellanic Cloud. Among the most luminous stars in the LMC, long-term variability is so ubiquitous that listings of the ten brightest stars in the Large Cloud differ substantially from decade to decade. The modern era of variable star research in the Magellanic Clouds began with Henrietta Leavitt's (1907) discovery that "It is worthy of note that in Table VI the brightest [Cepheid] variables have the longest periods." Another highlight was provided by the discovery (Thackeray & Wesselink 1953) of RR Lyrae variables in the SMC globular cluster NGC 121 and in the LMC cluster NGC 1466. The variables in both of these clusters were found to be fainter than expected. This provided support for Baade's (1954) conclusion that the distance scale in the Universe was twice as great as had previously been assumed. Finally, the discovery of the supernova 1987A in the Large Cloud, and the burst of neutrinos associated with it, provided us with a ringside seat to one of the most spectacular events in recent astronomical history.

6.6.1 Cepheids

A discussion of the statistical properties of over a thousand Cepheids that were known in each of the Magellanic Clouds in the early 1970s is given by Payne-Gaposchkin (1971b). She draws particular attention to the difference in period–frequency relations

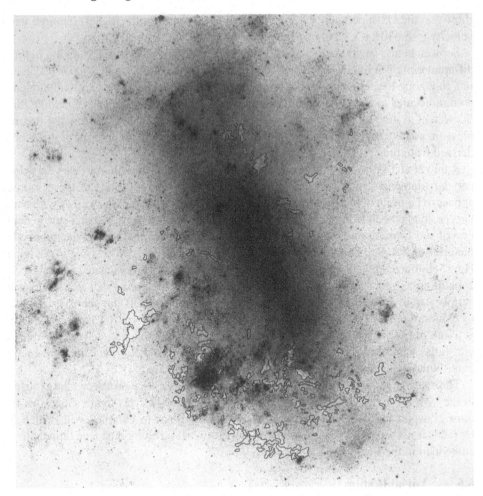

Fig. 6.14. Twenty-minute blue (103aO + GG 13) exposure of the Large Magellanic Cloud obtained with the Curtis Schmidt telescope on Cerro Tololo. Most dust clouds are seen to be associated with the giant H I complex that also contains the 30 Doradus super association.

between the LMC and the SMC. Figure 6.15 shows a progression with the peak frequency shifting from short periods in the SMC, to intermediate periods in the LMC, and finally to relatively long periods in the Galaxy. It seems highly probable that this trend is due to the fact that the mean metallicities of young stars are [Fe/H] $\approx$ −0.6, −0.3, and 0.0 in the SMC, LMC, and Milky Way, respectively. The excess of short-period Cepheids in the SMC, over that found in the LMC, is particularly striking (van den Bergh 1968a, p. 31) when the frequencies of Cepheids per unit area are compared. Furthermore, the pulsation amplitudes of SMC Cepheids with log P(days) $\approx$ 0.4 is greater than it is in the LMC and much greater than it is in the Galaxy. Among Galactic Cepheids short periods predominate outside the solar circle, while long periods are found to be most common at smaller Galactocentric distances (van den Bergh 1958). The frequency distribution of Cepheids in the Galactic anticenter direction resembles that seen in the LMC. These

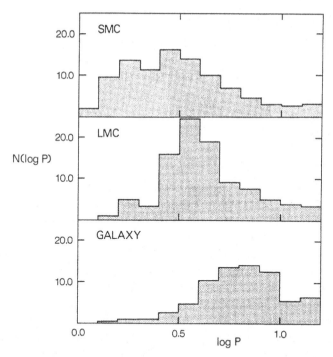

Fig. 6.15. Comparison of the frequency distributions of known Cepheids in the LMC, the SMC, and the Galaxy. The observed trend is almost certainly due to the fact that the Small Cloud has a lower metallicity than does the Large Cloud, while the mean metallicity of the Galaxy is higher than that of the LMC.

observations led van den Bergh (1958) to postulate the existence of a radial abundance gradient in the Galaxy. It is now known that this dependence of Cepheid periods on [Fe/H] is due to the fact that metallicity affects the extension of the blue loops of evolutionary tracks in the color–magnitude diagram. For high metallicities, the blue loops for low-mass stars do not cross the Cepheid instability strip, whereas they do for low metallicities.

Alves et al. (1998) have used the MACHO database to discover 1,470 Cepheid variables in the LMC. About 70 of them turned out to be multimode (beat) Cepheids. Alves et al. also found that ~140 Cepheids lie within one arcminute of known Large Cloud clusters. Many of these will, presumably, turn out to be physical cluster members. In a complete subset of their sample Alves et al. found an excess of Cepheids with $M(\text{puls}) \approx 5\,M_{\odot}$, over the frequency distribution that would have been expected for a uniform rate of star formation. Although this conclusion is quite model dependent, it suggests that a short burst of star formation may have taken place in the LMC some 50–100 Myr ago. Alcock et al. (1999) date this burst as having occurred ~1.15×10^{8} years ago.

The slope of the Cepheid period–luminosity relation has been discussed by Gascoigne & Kron (1965) and more recently by Madore & Freeman (1991). These results are, however, superseded by the massive and homogeneous study of 290 LMC and 590 SMC Cepheids carried out by the EROS consortium (Bauer et al. 1998). Their most important result is that the Cepheid period–luminosity relation, for objects pulsating in their fundamental mode, appears to change slope at a period of ~2 days. From the very uniform

EROS database it is found that Cepheids with $P > 2.0$ days have:

$$\text{LMC}\langle V \rangle = 17.60 \pm 0.03 - (2.72 \pm 0.07) \log P, \tag{6.2a}$$

$$\text{SMC}\langle V \rangle = 18.20 \pm 0.02 - (2.80 \pm 0.06) \log P. \tag{6.2b}$$

However, Cepheids with $P < 2.0$ days yield:

$$\text{LMC}\langle V \rangle = 17.86 \pm 0.03 - (3.59 \pm 0.57) \log P, \tag{6.3a}$$

$$\text{SMC}\langle V \rangle = 18.40 \pm 0.02 - (3.47 \pm 0.18) \log P. \tag{6.3b}$$

Taken at face value the difference between the LMC and SMC period–luminosity relations would appear to indicate that the distance modulus of the SMC is 0.5 to 0.6 mag larger than that of the LMC. However, it should be noted that the zero-point of the period–luminosity relation may be sensitive to metallicity and might therefore differ somewhat between the Large Cloud and the Small Cloud.

Sandage & Tammann (1968) have shown that the period–luminosity relation for Cepheids has an intrinsic width $\Delta M_{\langle V \rangle} \approx 2.5 \Delta(B - V)$, in which $\Delta(B - V)$ is the width in color of the Cepheid instability strip. It is interesting to note that this slope is similar to that of the reddening line $A_V \approx 3.1 E(B - V)$. It follows (van den Bergh 1968a, p. 28) that the quantity $\langle W \rangle \equiv \langle V \rangle - 3(B - V)$ should be *almost* independent of both reddening and position of a Cepheid within the instability strip. That this is indeed the case for the Cepheids in the LMC and in the SMC is shown in Figure 6.16. Presently available samples of Cepheids suffer from a variety of sources of incompleteness. Future study of Cepheid variables in the MACHO database should provide an essentially complete sample of Cepheids variables near the center of the Large Cloud.

Imbert (1994) has found that the LMC Cepheid HV 883 is a member of a triple system formed by the Cepheid and a couple of B5 V stars. Using the MACHO database, Welch

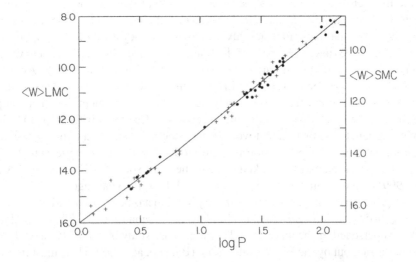

Fig. 6.16. Plot of $\langle W \rangle \equiv \langle V \rangle - 3(B - V)$ for Cepheids in the LMC (filled circles) and in the SMC (plus signs). Note that the LMC and SMC scales are shifted relative to each other by 0.55 mag.

Table 6.5. *RR Lyrae variables in LMC globulars*

Cluster	$\langle V \rangle$	n(var)	$E(B-V)$	[Fe/H]
NGC 1466	19.33	38	0.09	−1.85
NGC 1786	19.27	9	0.07	−1.87
NGC 1835	19.37	33	0.13	−1.79
NGC 1841	19.31	22	0.18	−2.11
NGC 2210	19.12	9	0.06	−1.97
NGC 2257	19.03	39	0.04	−1.8
Reticulum	19.07	32	0.03	−1.71

(1996; Welch et al. 1997) has discovered that the 17.5 day Cepheid HV 5756 is a member of an eclipsing binary system with a period of 416 days. The primary eclipse lasts between one and two weeks. Since this variable has $\langle V \rangle \approx 14.55$, it is bright enough out-of-eclipse to obtain accurate radial velocities of it. Such observations should provide valuable information on the mass of this Cepheid in the Large Cloud.

From high-dispersion spectra of 10 LMC Cepheids Luck et al. (1998) find a mean metallicity $\langle [\text{Fe/H}] \rangle = -0.30 \pm 0.04$.

6.6.2 RR Lyrae stars

RR Lyrae variables were discovered in the Large Cloud cluster NGC 1466 by Thackeray & Wesselink (1953). A compilation of observations of variables in clusters has been given by Walker (1992) and is partially reproduced in Table 6.5. The most surprising conclusion of this work (also see Section 6.2.2) is that the RR Lyrae stars turn out to be so bright. From Walker's data $\langle V(\text{RR}) \rangle_0 = 18.95$, which, with a Large Cloud distance modulus of $(m-M)_0 = 18.5$, yields $M_V = +0.45$ for [Fe/H] ≈ -1.9. This value appears significantly brighter than the value $M_V(\text{RR}) \sim +0.7$ that Fernley et al. (1998a) derived from the statistical parallaxes of Galactic RR Lyrae stars, which have proper motions derived from *HIPPARCOS* observations. This discrepancy might be caused by (1) an error in the adopted distance modulus of the LMC, (2) a systematic difference between RR Lyrae stars in the Galaxy and in the LMC, or by (3) a difference between the luminosities of RR Lyrae stars in clusters and the field. Resolution of this question is of considerable importance because the highest weight determinations of the age of the Universe depend critically on the adopted magnitude level of horizontal branches of globular clusters. It would be important to obtain CCD observations of RR Lyrae field stars in the LMC to see if they are systematically fainter than those that have already been observed in Large Cloud globular clusters. Because of the effects of possible (1) photometric errors, (2) background enhancement in clusters, and (3) incompleteness effects, existing photographic studies of differences between the luminosities of field and cluster RR Lyrae stars cannot yet answer this question in a conclusive fashion. Furthermore, a possible complication is that the RR Lyrae field stars in the Large Cloud might, on average, be younger than their counterparts in globular clusters.[7] However, Catelan

---

[7] From the metallicities derived for LMC RR Lyrae stars, Beaulieu, Lamers & de Wit (1999) reach the opposite conclusion, i.e., that the field population of cluster-type variables might be older than that of the RR Lyrae stars in Large Cloud globulars.

Table 6.6. *Novae in the Large Magellanic Cloud[a]*

Year	α (2000) δ	Year	α (2000) δ
1926	05^{h}14^m 54^s −66°48′ 44″	1977b	05^h 05^m 16^s −70° 09′ 04″
1936	05 07 27 −66 39 12	1978a	05 05 50 −65 53 05
1937	05 57 04 −68 54 48	1978b	05 01 00 −67 12 45
1948	05 38 15 −70 20 26	1981	05 32 42 −70 23 31
1951	05 12 52 −69 58 36	1987	05 23 50 −70 00 45
1968	05 10 05 −71 39 25^2	1988a	05 35 29 −70 21 29
1970a	05 33 13 −70 35 04	1988b	05 08 01 −68 37 38
1970b	05 35 29 −70 47 14	1990a	05 23 22 −69 29 48
1971a	04 58 23 −68 05 34	1990b	05 09 58 −71 39 51^b
1971b	05 40 35 −66 40 35	1991	05 03 45 −70 18 14
1972	05 28 25 −68 49 43	1992	05 19 20 −68 54 35
1973	05 15 19 −69 39 47	1995	05 26 50 −70 01 24
1977a	06 05 43 −68 38 26	1997	05 04 27 −67 38 38

[a]Data for novae after 1990 kindly provided by F. Younger.
[b]Recurrent nova.

(1998) has argued that the similarity of the period–temperature relations for field and cluster RR Lyrae stars militates against a luminosity difference.

Kinman et al. (1991) have fit observations of RR Lyrae field stars in the Large Cloud to an exponential Disk with a scale-length of 2.6 kpc. From this fit they estimate that the total number of field RRab Lyrae stars in the LMC is $\sim 10^4$ (although their data are expected to be very incomplete for small-amplitude RRc variables). Extrapolating from these RR Lyrae observations Kinman et al. estimate that the Population II component of the LMC has $M_V \sim -15.2$ and $M \sim 1.6 \times 10^8 M_\odot$, which would correspond to $\sim 1\%$ of the baryonic mass of the LMC. It is noted in passing that the Population II disk scale-length of 2.6 kpc is larger than the photometric disk scale-length of 1.5 kpc derived by Bothun & Thompson (1988).[8] This appears to confirm the notion that the star forming region of the LMC has shrunk with time. According to Udalski et al. (1998), blue horizontal branch stars are absent in the LMC field, even though they are present in Large Cloud globular clusters.

6.6.3   Novae

A detailed discussion of novae in the Large Magellanic Cloud is given in van den Bergh (1988b). An updated listing of LMC novae is provided in Table 6.6. The data in this table suggest that LMC 1968 = LMC 1990b is a recurrent nova. A plot of the positions of all known novae in the Large Cloud is shown in Figure 6.17. This figure shows what appears to be a concentration of novae just south of the LMC Bar but no evidence for novae associated with either the Bar itself or with the 30 Doradus region. The centroid of the novae listed in Table 6.3 is found to be located at $\alpha(2000) = 5^h 21^m_{.}4 \pm 3^m_{.}4$, $\delta(2000) = -69°_{.}2 \pm 0°_{.}3$. The data listed in Table 6.3 show that

---

[8] It would be very worthwhile to repeat the Bothun & Thompson observations with one of the much larger CCD chips that are now available.

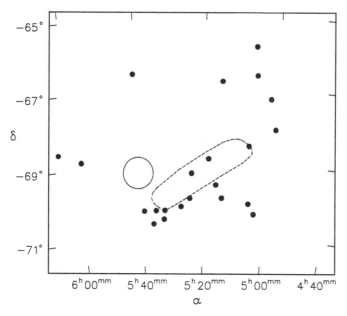

Fig. 6.17. Distribution of known novae over the face of the Large Cloud. The outline of the Bar is shown as a dashed ellipse, and the 30 Dor region is plotted as a circle. The observations suggest that there may be a concentration of novae just south of the LMC Bar.

this position is, within its errors, indistinguishable from the centroid of yellow light and the H I rotation center of the LMC. (If the probably recurrent nova 1968 = 1990b had only been counted once, the centroid would have shifted to $\alpha = 5^h21^m.9, \delta = -69°.1$.)

Data on the absolute magnitudes M_V (max) and the rates of decline for 15 novae in the LMC have been given by Della Valle & Livio (1995) and are plotted in Figure 6.18. In the figure t_2 is the time required for a nova to decline by 2.0 mag from V(max). The figure shows that the novae in M31 and in the LMC fall on approximately the same maximum magnitude versus rate-of-decline relationship if one adopts distance moduli of $(m - M)_0 = 18.5$ for the Large Cloud and $(m - M)_0 = 24.3$ for M31. It is of interest to note that the LMC appears to contain a larger fraction of rapidly declining novae than does M31. Since the mean age of stars in the Large Cloud is lower than that in the Andromeda galaxy, these observations might indicate that novae with small values of t_2 have young progenitors. Alternatively one might assume that the difference between the distribution of t_2 values in the LMC and in M31 is, in some way, related to the fact that the Large Cloud is more metal poor than M31. Della Valle & Livio (1998) find that the fast and bright novae in the Galaxy mainly have He/N spectra, whereas the slower fainter novae exhibit Fe II spectra.

The nova rate in the LMC is not well determined because it is not known how complete the searches for novae in the Large Cloud have been. Graham & Araya (1971) found two novae during a systematic search of the LMC that covered a 7-month interval. For the 28-year period between the beginning of 1970 and the start of 1998, a total of 19 novae have been discovered in the Large Cloud, yielding an apparent nova rate of about 0.7 ± 0.2 per year. This should be regarded as a lower limit to the true nova rate.

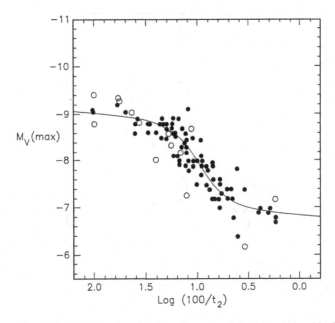

Fig. 6.18. Maximum magnitude versus rate-of-decline relationship for M31 (dots) and the LMC (open circles) for distance moduli $(m - M)_0 = 24.3$ and 18.5, respectively. The figure shows that the novae in the LMC, on average, decline faster in the Large Cloud than they do in the Andromeda Galaxy.

If $M_V = -18.5$ for the Large Cloud then the corresponding nova rate for a galaxy with $M_V = -21$ would be $\gtrsim 7$ per year. This is lower than the Galactic nova rate of ~ 20 per year estimated by Della Valle & Livio (1994) and much lower than the rate of 73 ± 24 Galactic novae per year found by Liller & Mayer (1987).

Three years after outburst, nova LMC 1995 was observed with the *ROSAT* X-ray satellite by Orio, Greiner & Della Valle (1998). These authors find that this object is one of only three known supersoft X-ray sources among classical novae. It is at present the only one that is still near its Eddington luminosity. Nova LMC 1995 was already bright in X rays six months after outburst. Its luminosity increased even more during the next three months and appears to have remained at a high, constant flux level since then.

Starrfield et al. (1998) have studied the evolution of the extraordinary nova LMC 1991, which was the most luminous [$V(\text{max}) \sim 9$], and fastest ($t_3 = 6 \pm 1$ days), yet observed. From *IUE* spectra the ejection velocity was found to exceed 3,000 km s^{-1}. There is no indication that the abundances of O, Ne, and Mg nuclei are enhanced. The outburst therefore most likely occurred on a CO white dwarf. Surprisingly, Starrfield et al. find that a model atmosphere abundance analysis yields $Z = 0.1 Z_\odot$, which is almost three times lower than that of the majority of stars in the LMC. Taken at face value this result suggests that the progenitor(s) of LMC 1991 might have belonged to Population II.

6.6.4   *Other types of variables*

W Virginis stars
Data on 17 W Virginis stars, sometimes called Cepheids of Population II, are listed in Table 6 of Payne-Gaposchkin (1971a). As expected the Cepheids that Payne-

Gaposchkin assigns to Population II on the basis of the shape and variability of their light curves (and on erratic variations in period) fall well below the period–luminosity relation for classical Cepheids of Population I. Modern period–luminosity (P–L) and period–luminosity–color (P–L–C) relations for W Virginis stars (and for their longer period counterparts the RV Tauri stars) in the LMC are given in Alcock et al. (1998a). From MACHO observations these authors find that the P–L and P–L–C relations for RV Tauri stars form an extension to longer periods of the corresponding relations for W Virginis stars. Adopting $(m - M)_0 = 18.5$ for the LMC these authors find the following period–luminosity relation for W Vir and RV Tau stars:

$$M_V = 1.34 \pm 0.45 - (3.07 \pm 0.35) \log P. \qquad (6.4)$$

Payne-Gaposchkin (1973) claimed that W Virginis stars are concentrated toward the Bar of the Large Cloud. However, it is not yet clear if this conclusion might have been affected by selection effects. The MACHO database should make it possible to reinvestigate this question in an unbiased fashion.

R Corona Borealis variables

Payne-Gaposchkin (1971a) lists four R CrB in the Large Cloud, of which W Mensae is the brightest. The discovery of two additional CrB stars is noted in Alcock et al. (1996). Clayton, Kilkenny & Welch (1999) report that MACHO observations have now increased the total known population of R CrB stars in the LMC to about a dozen. Since most of the newly discovered R CrB variables are fainter than those that were previously known, it follows that their mean luminosity is lower than was previously believed. The light curve of the Large Cloud R CrB star HV 5637 is shown in Figure 6.19. Element abundances in the LMC R CrB stars HV 966 (W Men) and HV 12842 have been studied by Pollard, Cottrell & Lawson (1994). These authors find an element abundance pattern similar to that observed in warm Galactic hydrogen-deficient C stars.

Eclipsing variables

A large number of eclipsing variables in the LMC are listed in Table 13 of Payne-Gaposchkin (1971a). A rich literature, which includes the last paper by Henry

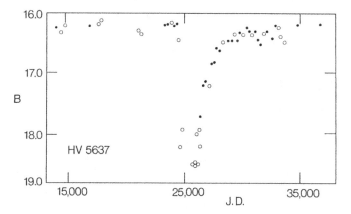

Fig. 6.19. Light curve of the R CrB variable HV 5637 adapted from Hodge & Wright (1969). Least certain data are plotted as open circles.

Norris Russell (1954), exists on the study of individual eclipsing variable stars in the LMC. Such observations can, for example, be used to derive the distance of the Large Cloud (Bell et al. 1993, Guinan et al. 1999).

Long-period variables (LPVs)

Using the MACHO database, Cook et al. (1997) find three sequences of long-period variables in the LMC. These represent the fundamental, first-, and second-overtone sequences for LPVs. As has already been mentioned in Section 6.2.2, van Leeuwen et al. (1997) have used a comparison between the period–luminosity relations of Galactic Mira stars studied by *HIPPARCOS* and Large Cloud Miras to obtain distance moduli in the range $18.47 \leq (m - M)_0 \leq 18.60$ for the LMC.

Hubble–Sandage variable

The Hubble–Sandage (H–S) variables are hot luminous objects that are surrounded by cool expanding envelopes. They exhibit light variations on timescale of years or decades. The Hubble–Sandage variables occupy an inclined instability strip in the Hertzsprung–Russell diagram that covers the range $-9 \gtrsim M_{bol} \gtrsim -11$ and $10,000 \lesssim T_{eff} \lesssim 35,000$ K. The best-known example of a Hubble–Sandage variable in the Large Cloud is S Doradus. The light variations of S Dor have been studied by Gaposchkin (1943), who finds its luminosity to have exhibited dips of about 0.7 mag in 1891, 1900, 1930, and 1940. The most recent dip in the light curve of S Dor lasted $\sim$10 yr and ended in 1973 (Thackeray 1973). Other examples of luminous blue H–S variables are R 71 = HDE 269006, R 110 = HDE 269662, R 127, and R 143 (Parker et al. 1993), which is located in the 30 Dor region. Observations by Clampin et al. (1993) show that R 127 is embedded in an equatorial Disk that may have been produced during a mass-loss episode. The stars HDE 269582 and HDE 269599, which are situated in Constellation II, are suspected H–S variables. Work by Sterken, de Groot & van Genderen (1998) suggests that R 49 (HD 6884), which has spectral type A Ia$^+$, may presently be developing into a luminous blue variable.

Voors et al. (1999) have detected the infrared signature of crystalline olivine at 23.5 μm in the Hubble–Sandage variable R 71. These authors also identify emissions from C-rich small grains composed of poly aromatic hydrocarbons (PAHs) at 6.2 μm and 7.7 μm. Voors et al. suggest that R 71 was probably a red supergiant, at the time when it produced these carbon-rich and olivine grains. Assuming a gas/dust ratio of 100, they find a time-averaged mass loss rate by the red supergiant of $\sim 7 \times 10^{-4}\ M_\odot\ \text{yr}^{-1}$.

6.7   Evolutionary history of the Large Cloud

6.7.1   *The rate of cluster formation*

Information on the history of star formation may be derived from (1) star clusters, (2) field stars, and from (3) the age–metallicity relation of stars or clusters. In the case of the LMC, detailed data on the age distribution of clusters are available from the study of Geisler et al. (1997). Figure 6.20 shows a plot of the age distribution that these authors obtain for clusters in the Large Cloud. The figure shows a few globular clusters with ages $T > 11.5$ Gyr, a gap extending from $T = 11.5$ Gyr to $T = 3.5$ Gyr, and an enormous burst of cluster formation beginning $\sim$3 Gyr ago. Only a single cluster

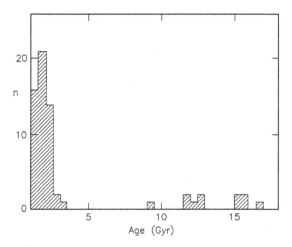

Fig. 6.20. Distribution of ages for LMC clusters with $T > 1.0$ Gyr according to Geisler et al. (1997). The figure shows that a great burst of cluster formation started $\sim$3 Gyr ago and continues to the present day. A number of globular clusters have $T > 11$ Gyr. Note the scarcity of clusters with ages in the range $3.5 < T$ (Gyr) < 11.5. The only outlying cluster ESO 121-S03, which has an age of $\sim$9 Gyr, falls in this interval that covers more than half of the age of the Large Cloud. Three clusters with ages of $\sim$4 Gyr have recently been discovered by Sarajedini (1998b).

(ESO 121-SC03) occurs during the 8-Gyr interval that extends from 3.5 Gyr to 11.5 Gyr (Mateo, Hodge & Schommer 1986). Possibly ESO 121-SC03 was tidally captured from the Small Magellanic Cloud, in which cluster formation appears to have proceeded at a relatively uniform rate. A possible argument against this suggestion is, however, the finding of Bica et al. (1998) that the stellar field population in the neighborhood of ESO 121-SC03 appears to have the same age of $\sim$9 Gyr as the cluster itself. However, Bica et al. also find a region in the northern part of the Large Cloud disk in which a secondary red giant clump appears to be located $\approx$0.45 mag below the dominant intermediate-age clump. Perhaps this fainter clump represents debris from a former LMC/SMC interaction that is presently located at the distance of the SMC.

It should be emphasized that the present ratio of young clusters to old clusters has been affected by the fact that some low-mass old clusters will not have survived to the present day. As a result the true ratio of young clusters to old clusters may originally have been smaller than it is presently observed to be. However, such differential destruction should not greatly affect the ratio of the number of 2-Gyr-old clusters to the number of 4-Gyr-old clusters, which (see Figure 6.20) is found to be 53 to 1. In other words the cluster data suggest that the rate of star formation might have increased by between one and two orders of magnitude $\sim$3 Gyr ago. This argument is, however, slightly weakened by Sarajedini's (1998b) recent discovery that the LMC clusters NGC 2155, SL 663, and NGC 2121 are relatively metal poor ([Fe/H] $\approx$ -1.0) with ages of $\sim$4 Gyr. Taken at face value these results suggest that the great starburst that took place 3 Gyr ago started to ramp up $\sim$4 Gyr ago. The results presented above appear to indicate that the rate of cluster formation increased much more rapidly in the LMC $\sim$2 Gyr ago than did the rate of star formation (see Section 6.7.2). In this connection (Hodge 1998a) it is of interest to note that the ratio of the present rate of *cluster* formation to the present rate of *star* formation is $\gtrsim 600$ times larger in the Large Magellanic Cloud than it is in the relatively quiescent Local Group dwarf IC 1613 (see Section 11.1). This suggests that the ratio of the rate of cluster formation to the rate of star formation may be greater in turbulent, strongly shocked regions than it is in more quiescent areas. In other words, cluster formation might be favored over star formation in galaxies that have undergone collisions or have recently been tidally harassed.

6.7.2   The rate of star formation

The idea that a violent burst of star formation in the LMC followed a rather lengthy period of quiescence was first introduced by Butcher (1977). He found a break in the slope of the main-sequence luminosity function of field stars in the LMC that was located about one magnitude above the main sequence turnoff point of Population II. From this observation Butcher concluded that a dramatic increase in the rate of star formation had taken place in the LMC 3–5 Gyr ago. This entirely unexpected conclusion was subsequently confirmed by Stryker (1983) and by Hardy et al. (1984). Recent discussions of the star forming history of field stars in the Large Cloud are given by Olszewski, Suntzeff & Mateo (1996), Gallagher et al. (1996), Elson, Gilmore & Santiago (1997), Holtzman et al. (1998), and by Olsen (1999). From an investigation of three star fields in the LMC, Bertelli et al. (1992) estimated that the mean rate of star formation in the Large Cloud was as much as ten times higher after the burst started than it had been previously. Using the *Hubble Space Telescope*, Geha et al. (1998) have studied the evolutionary history in three fields located in the outer disk of the Large Cloud. They conclude that the rate of star formation in the outer disk only increased by a factor of three about 2 Gyr ago. From their observations of a field located 1°3 from the center of the LMC Elson et al. (1997) found that only $\sim 5\%$ of the stars in this field belong to the old disk population that predated the great burst of star formation that started ~ 3 Gyr ago. These authors also claim to find evidence for a later burst of star formation that took place ~ 1 Gyr ago, which may correspond to the time of formation of the LMC Bar. Although the exact factor by which the rate of star and cluster formation increased ~ 3 Gyr ago is still highly uncertain, it seems possible that it might have resembled one of the milder examples of low surface brightness (LSB) galaxies during the "dark ages" that extended from ~ 11 Gyr to ~ 4 Gyr ago. In this connection it would be interesting to know if the specific cluster forming frequency is particularly low in LSB galaxies. Since the LMC is presently in a burst phase it might be the nearest example of the kind of objects that Babul & Ferguson (1996) have dubbed "boojums"[9] (i.e., blue objects observed just undergoing moderate starbursts).

6.7.3   Carbon stars and evolutionary history

The data in Table 6.7, which are plotted in Figure 6.21, show a strong correlation between the ratio of the number of carbon stars to the number of M 5–M10 stars. This correlation is due to the fact that: (1) Only a small number of C atoms need to be dredged

---

[9] From Lewis Carroll's (1876) *The Hunting of the Snark*
He had softly and suddenly vanished away–
For the Snark *was* a Boojum, you see.

Table 6.7. *Number ratio of C stars to M5–M10 stars*

Galaxy	M31	M33	LMC	SMC	NGC 6822
C/M[a]	0.08	1.0	0.8	4.3	33
M_V	−21.2	−18.9	−18.5	−17.1	−16.0

[a]From Cook et al. (1986).

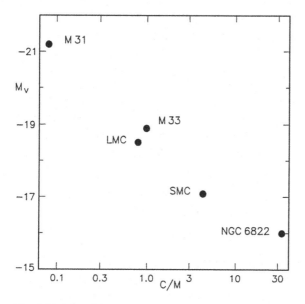

Fig. 6.21. Plot of the ratio of the number of carbon stars to the number of M5–M10 stars as a function of parent galaxy luminosity. The figure shows that the relative frequency of carbon stars increases with decreasing parent galaxy luminosity. The fact that C/M is slightly lower in the LMC than in M33 might be due to the fact that few C star progenitors were formed in the LMC during its "dark ages."

up from the interior of a metal-poor star to make the number of carbon atoms in its atmosphere larger than the number of oxygen atoms, and (2) the number of late M-type giants will be small in metal-poor dwarf galaxies because the abundance of TiO molecules is low in metal-poor stars. It is, however, surprising that the C/M ratio in the LMC is slightly *lower* than it is in M33, even though M33 ($M_V = -18.9$) is a bit more luminous than the LMC ($M_V = -18.5$). On the basis of the data on Local Group galaxies plotted in Figure 6.21 one might have expected the number of C stars in the LMC to have been about twice as large as is actually observed. Such a deficiency could, if real, be due to the paucity of progenitors of carbon stars that formed in the LMC during the "dark ages" between 4 Gyr and 11 Gyr ago.

Westerlund et al. (1991) have drawn attention to the remarkable fact that carbon stars with M_{bol} between −6 and −7 are present in the LMC field but appear to be absent from Large Cloud clusters. Marigo, Girardi & Chiosi (1996) note that this effect might be accounted for by assuming that the rate of cluster formation was depressed between

$\sim 2 \times 10^8$ and $\sim 6 \times 10^8$ years ago, while the rate of field star formation remained essentially constant. It is presently not clear why such a difference should have persisted throughout the LMC for a period of ~ 0.4 Gyr.

Additional information on the evolutionary history of the LMC may be obtained from the $C^{18}O/C^{17}O$ ratio in dense interstellar clouds. Heikkilä, Johansson & Olofsson (1998) estimate that the average gas-phase $C^{18}O/C^{17}O$ abundance ratio is 1.6 ± 0.3 in the LMC. This is a factor of two lower than the value found in Galactic molecular clouds. The observations of four different clouds in the LMC suggest that the $^{18}O/^{17}O$ ratio is globally low in the Large Cloud. Detailed calculations by Nomoto et al. (1997) show that ^{18}O is mostly produced by supernovae of Type II, with progenitors in the mass range 15 $M_\odot$ to 25 $M_\odot$, corresponding to main sequence spectral types O8 to B1. Since such stars are presently common in the Large Cloud, their apparent scarcity in the past is puzzling.

Observations of the rare hydrogen cyanide isotope $HC^{15}N$ toward the LMC have been reported by Chin et al. (1999). These authors find $^{14}N/^{15}N = 111 \pm 17$ in the massive star forming region N 113. This value is significantly smaller than the values $^{14}N/^{15}N = 270$, 300–400, and 500–1,000 in the solar system,[10] the local interstellar medium, and the Galactic center region, respectively. The observed trend in the $^{14}N/^{15}N$ ratios lends strong support to the notion that ^{15}N is predominantly ejected by supernovae of Type II.

6.7.4   *Globular clusters and evolution*

During its first active phase the LMC produced 13 globular clusters. In conjunction with $M_V = -15.0$ this yields a present specific globular cluster frequency of $S = 0.5$. If the LMC was two magnitudes fainter during the "dark ages" than it is now, then it would have had a specific frequency $S \approx 3$ during the period that it was a LSB galaxy. It would be interesting to check if known nearby LSB galaxies also have relatively high S values.

Da Costa (1991) and de Freitas Pacheco, Barbuy & Idiart (1998) have pointed out that there is a clear-cut dichotomy between the LMC globular clusters with ages >11 Gyr, which have metallicities $-2.1 < [Fe/H] < -1.4$, and open clusters with ages <4 Gyr that mostly have $-1.0 < [Fe/H] < 0.0$. According to Sarajedini (1998b) the three clusters with ages ~ 4 Gyr, which formed at the beginning of Butcher's great burst of star and cluster formation, have $[Fe/H] \sim -1.0$. This suggests that metal abundance in the Large Cloud increased by a factor of about two between the end of the globular cluster forming era and the beginning of Butcher's burst. This enrichment during the "dark ages" is discussed by Olszewski, Suntzeff & Mateo (1996).

It is not yet clear why the LMC suddenly started to form stars and clusters at such a high rate ~ 4 Gyr ago. Chappell & Scalo (1997) have recently suggested that galaxies may develop large-scale global oscillations, or bursts in their star formation rates, without the necessity for any organizing agent. In particular they find that such oscillations may require long (Gyr) "incubation times." Perhaps the "dark ages" prior to the LMC starburst ~ 4 Gyr ago might be regarded as such an incubation period. However, it is puzzling why no such incubation period appears to have occurred in the SMC.

As has already been discussed in Section 6.5.2 (see Figures 6.9–6.12), the distribution of clusters in the LMC shows that the active region of star and cluster formation in

---

[10] Presumably the observed solar system value represents an average for the region with $R_{GC} \sim 8.5$ kpc at ~ 4.5 Gyr ago.

the Large Cloud has shrunk with time. Furthermore, the distribution of Cepheids of various ages (see Figure 6.13) seems to show that star formation was concentrated in the Bar about 50 Myr ago. Ardeberg et al. (1997) find that a considerable fraction of the present Bar population is less than 0.5 Gyr old (i.e., the LMC Bar may be a relatively young feature). [But see Olsen (1999) for a conflicting conclusion.] If this conclusion is correct, then it indicates that Ir galaxies can transform themselves from unbarred to barred morphologies. Moreover, the fact that the LMC has spent more than half of its lifetime in the "dark ages" suggests the possibility that LSB galaxies can undergo a phase transition to normal Ir galaxies. Hunter (1997) has emphasized that irregular galaxies exhibit a much larger range in star formation rates than do spirals. The reason for this difference is not yet understood. Nor do we know why some irregulars maintain a star formation rate that is so much higher than that observed in others.

The structure of rich LMC clusters has been studied by Mateo (1987) and by Meylan & Djorgovski (1987). Meylan & Djorgovski studied 33 luminous clusters in the Large Cloud and found that one of them, NGC 2019, appeared to have a collapsed core. Their observations further suggest that NGC 1774 and NGC 1951 may also have collapsed cores. They note that the fraction of all LMC clusters that probably exhibits collapsed cores is small. This may be due to the fact that (1) many of the bright clusters in the LMC are younger than Galactic globulars and because (2) Cloud clusters typically have rather large radii. In a study of 10 Large Cloud clusters Mateo found that NGC 2005 and NGC 2019 probably also have collapsed cores. Mateo also notes that the most centrally concentrated clusters tend to occur near the core of the LMC, while the more open clusters are generally located in the outer regions of the Large Cloud.

6.8 Supernovae

According to van den Bergh & Tammann (1991) the supernova rates in late-type galaxies are 0.24 h^2 SNu for SNe Ia and 2.56 h^2 SNu for SNe Ib + II, in which one SNu is the number of supernovae per 10^{10} $L_\odot(B)$ per century. Adopting $L = 2.7 \times 10^9 L_\odot(B)$ for the LMC and $h^2 = 0.5$, one obtains an expected rates of 0.7 SNe Ia per millenium and of 7 SNe Ib + II per millennium for the Large Cloud. These values agree well with a rate of one supernova every (100 ± 20) yr that Filipović et al. (1998) derived from sources that are in common to radio and X-ray surveys of the LMC.

Abundances, and abundance ratios, of gas and stars in the LMC should eventually be able to place significant constraints on both the evolutionary history of the LMC and on the ratio of SN Ia to core-collapse supernovae of Types Ib + II. The discovery of a $\sim(1-2) \times 10^5$ yr old luminous white dwarf in the young LMC star cluster NGC 1818 by Elson et al. (1998a,b) strongly constrains the boundary mass M_c at which stars collapse catastrophically to become neutron stars, rather than evolving quietly into white dwarfs. From the white dwarf in NGC 1818 Elson et al. find $M_c \gtrsim 7.6$ $M_\odot$. This is consistent with the value ≥ 6.0 $M_\odot$ that Burleigh & Barstow (1998) find from the white dwarf companion of the Galactic B5 V star HR 2875.

6.8.1 SN 1987A

SN 1987A is the only supernova to have been observed in the Large Cloud in historical times. This object was of Type II and reached $V(\text{max}) \approx 3.0$. With $(m - M)_0 = 18.5$ and $A_V = 0.40$ mag, this yields the unusually faint value $M_V(\text{max}) = -15.9$.

It is probably no coincidence that this intrinsically faint SN II was discovered in a very nearby galaxy! Contrary to some theoretical speculations, no pulsar has been observed to be associated with SN 1987A. The progenitor of SN 1987A was the blue supergiant Sk $-69°$ 202. According to Walborn et al. (1989) this precursor had a spectral type in the range B0.7 to B3 and a luminosity class I. This progenitor star probably had a physical companion (van den Bergh 1987) with $V = 15.3$ at a distance of $3''$ and a second companion with $V = 15.7$ at a distance of $1''.5$. Since the components of this triple system are, presumably, coeval, the luminosity of the brightest companion (which would have a mass of ~ 10 $M_\odot$ if it is located on the main sequence) must be smaller than the mass of the supernova progenitor. This is consistent with the mass of ~ 20 $M_\odot$ that Woosley (1988) estimates for the progenitor of SN 1987A. Efremov (1991) points out that SN 1987A might have been located in the outer halo of the very sparse open cluster KMK 80, which is estimated to have an age of ~ 10 Myr.

Burrows (1998) has emphasized that SN 1987A is a case study in asphericity. This is so because: (1) X-ray and γ-ray fluxes, and optical light curves, require that blobs of matter containing the radioactive isotope ^{56}Ni were ejected from the core of SN 1987A; (2) the profiles of oxygen, iron, cobalt, and nickel lines are ragged in the infrared and exhibit red–blue asymmetries; (3) the light of the supernova was (slightly) polarized; and (4) the observations of hydrogen deep in the ejecta of SN 1987A strongly suggest that mantle instabilities mixed debris clouds and shattered initially spherical shells of ejecta. These results suggest that it might be wise to view geometrical distance determinations to SN 1987A with a certain amount of skepticism.

The expanding light echo produced of the optical outburst of SN 1987A has now been followed for a number of years (Xu, Crotts & Kunkel 1995). Investigation of the expansion of this light echo allows one to study the three-dimensional structure of the interstellar dust clouds in the LMC. In 1998 the expanding shell of SN 1987A reached the inner ring around SN 1987A, producing a small "hotspot" in Hα, [N II], and He I λ10830. It is expected that radio radiation from SN 1987A will start to increase again early in the 21st century, as more of the supernova ejecta start to interact with its ring.

For early summaries of observations and theoretical conclusions about the nature of this supernova, the reader is referred to the *ESO Workshop on SN 1987A* (Danziger 1987), to *Supernova 1987A in the Large Magellanic Cloud* (Kafatos & Michalitsianos 1988), and to *The Elizabeth and Frederick White Conference on Supernova 1987A* (Proust & Couch 1988). More recent results are summarized in *SN 1987A: Ten Years After* (Phillips & Suntzeff 1998).

Romaniello, Panagia & Scuderi (1998) have used the *Hubble Space Telescope* to study the stellar content of the area surrounding SN 1987A. From inspection of their color–magnitude diagram these authors conclude that several generations of star formation, with ages between 1 Myr and 150 Myr, have occurred in this region. These younger generations of stars are superposed on a much older field population. The youngest stars observed are T Tauri stars that are characterized by strong Hα emission. Gas along the line of sight toward SN 1987A has been studied in the ultraviolet with the *IUE* satellite by Welty et al. (1999). Groups of absorption lines with radial velocities between 193 and 225 km s^{-1}, which are likely due to relatively cool gas in the LMC, indicate that relative gas-phase abundance ratios are similar to those in Galactic halo clouds. Xu & Crotts (1998) have noticed insterstellar velocity components at $V = +269$ km s^{-1} and

$V = +301 \, \mathrm{km \, s^{-1}}$ that occur within $20''$ of SN 1987A. These structures are probably produced by the progenitor star's red supergiant wind.

6.8.2 Supernova remnants

Supernova remnants provide a record of supernovae that have exploded in gas-rich environments during the past few $\times 10^4$ years. The first SNRs in the LMC were discovered by Mathewson & Healey (1964), who noted that the H II regions Henize N49, 63A, and 132D were probably supernova remnants. Subsequently, Mathewson & Clark (1972) showed that the intensity ratio of [SII]/Hα in emission nebulae could be used to isolate SNRs. Finally, Long, Helfand & Grabelsky (1981) were able to use the *Einstein Observatory* to detect two dozen soft X-ray emitting SNRs in the Large Magellanic Cloud. Excellent reviews on supernova remnants in the Large Cloud have been published by Dopita (1984) and by Westerlund (1997, pp. 190–198).

The distribution of SNRs over the face of the Large Cloud has been discussed by van den Bergh (1988b). He finds that these objects are concentrated in (1) the 30 Doradus region, (2) the Bar of the LMC, and in (3) Constellation III, which (see Figure 6.8) is located in the northern part of the Large Cloud. From a statistical analysis of the distribution of 29 supernova remnants, Forest, Spenny & Johnson (1988) find that SNRs in the Large Cloud are, on average, closer to H II regions than would have been the case if these SNRs had been distributed at random over the face of the Large Cloud. From this Forest et al. conclude that the majority of supernova remnants in the LMC were produced by supernovae of Type II, which have massive young progenitors. It should, however, be emphasized that the effect observed by Forest et al. might also be attributed to the supernova remnants being rendered visible by their interaction with the interstellar medium. Such interstellar gas clouds are expected to be most frequent near active regions of star formation. Another method of estimating the relative number of SNe of types I and II is based on observation of the S/O ratio in the Large Cloud. From an oxygen-to-sulfur ratio S/O $= 0.021$ in Large Cloud planetary nebulae, Barbuy, de Freitas-Pacheco & Castro (1994) calculate that $\sim 70\%$ of all supernovae in the LMC were of Type II. Using coincidences between radio and X-ray source positions, Filipović et al. (1998) have been able to increase the number of SNRs and SNR candidates in the Large Cloud to 62, 74% of which are confirmed X-ray sources. From their data Filipović et al. estimate that the supernova rate in the LMC is one per 100 ± 20 years. Adopting this supernova rate they conclude that the star formation rate in the Large Cloud is $(0.7 \pm 0.2) \, M_\odot \, \mathrm{yr^{-1}}$. This estimate is somewhat higher than the value $0.26 \, M_\odot \, \mathrm{yr^{-1}}$ obtained by Kennicutt et al. (1995) from the integrated Hα flux of the LMC.[11] Chu (1997) has found high-velocity emission associated with a number of SNRs in LMC OB associations.

The oxygen-rich LMC remnant N132D is probably closely related to the Galactic SNR Cassiopeia A. Morse, Blair & Raymond (1998) have identified O, Ne, C, and Mg in the remnant of N132D. However, they find no evidence for S, Ca, Ar, etc., which are observed in Cas A. These elements are expected in models of supernovae of Type II. Therefore, Morse et al. suggest that the progenitor of N132D may have been an SN Ib.

---

[11] This discrepancy might be due to the fact that some Lyman continuum photons were able to escape from the LMC.

The Large Cloud remnant DEM 71 appears to be similar to the Galactic nonradiative shock remnants of Tycho (1572) and Lupus (1006). Since the Galactic prototypes of nonradiative supernovae are young, it appears likely that such objects in the Large Cloud are also youthful. Hughes, Hayashi & Koyama (1998) have used X-ray spectroscopy to determine the heavy element abundance in DEM 71. They find an iron abundance that is twice as large as that which they obtained in other LMC supernova remnants. This observation is consistent with the conclusion that DEM 71 was produced by a supernova of Type Ia. However, these authors find that other LMC SNRs have abundances of elements from oxygen to iron that are two to four times lower than the corresponding solar values. The remnant N 49 in the Large Cloud is an example of an old evolved radiative supernova shell, of which IC 443 and the Cygnus Loop are Galactic examples. Such radiative supernova shells are thought to have been produced by the explosion of massive supernovae of Type II. The supernova remnant N 49 is positionally coincident with the γ-ray burst of March 5, 1979. This was one of only four known soft γ-ray repeaters.

The supernova remnant N 63A, which was one of the first three SNRs discovered in the Large Cloud, appears to be a member of the small association NGC 2030 (van den Bergh & Dufour 1980). The currently most luminous star in this association is an Of star with a mass of $\sim$40 $M_\odot$ (Oey 1996). Since the progenitor of N 63A was, presumably, more evolved than this Of star, it must have had a main-sequence mass $>$40 $M_\odot$. The optical remnant N 63A has been studied in detail by Chu et al. (1998). She and her colleagues find that the two eastern lobes of this object, which is resolved into filamentary structures by the *Hubble Space Telescope*, exhibit high [S II]/Hα, indicating shock excitation. In contrast, the western lobe of N 63A exhibits diffuse emission and a spectrum that is characteristic of photoionization. A number of cloudlets, with diameters as small as 0.1 pc, are seen within the X-ray emitting region of this supernova remnant. The [S II]/Hα ratios observed in these cloudlets suggests that they are shocked objects that lag behind the shock front. Chu et al. believe that the morphology of these cloudlets shows them to be evaporating isotropically. Such evaporation will inject mass into the hot interior of the SNR, which will result in a high surface brightness of X-ray emission. N 63A provides the first clear example of cloudlets that are engulfed by a supernova remnant in the Large Cloud.

The SNR 0540 − 693 (Seward, Harnden & Helfand 1984) is, in many respects, similar to the Crab nebula, which contains a 33 ms pulsar. The pulsar in SNR 0540 − 693 has a period of 50 ms, has a mean optical pulsed magnitude of $\sim$23, and is embedded in a small synchrotron nebula with a half-power diameter of $\sim$4″ (1 pc). Manchester, Staveley-Smith & Kesteven (1993) use the spin-down rate of this pulsar to derive an age of $\sim$760 years, which is slightly younger than the 944 year age of the Crab. SNR 0540 − 693 may have been the last supernova to have occurred in the LMC before SN 1987A. Caraveo, Magnani & Bignami (1998) have published an image showing a peculiar spiral-shaped Hα emission nebulosity centered on this object. Spectra of this nebulosity exhibit a velocity-width of $\sim$1,300 km s^{-1}. Also seen is a ring of [N II] emission that appears to be expanding with a velocity $<$50 km s^{-1}. However, an expansion velocity of $\sim$200 km s^{-1} would have been required for this ring to grow to its present size, if its expansion had started $\sim$1,600 years ago. This ring must therefore have formed before the supernova explosion (i.e., it represents material ejected by the progenitor of the supernova). Ejected nebulosity that may be similar to pre-supernova ejecta has also been observed (Smith et al. 1998) near the luminous blue variables R 127 (= HDE 269858) and R 143 (= HDE 269929) and the Ofpe star S 119 (= HDE 269687). A clump of the material near R 143, which is observed to be enriched in nitrogen, appears to have been ejected from this object ~3,700 years ago.

The discovery of SN 1987A and SNR 0540-693 shows that at least two supernovae of Type II have occurred in the LMC during the past 800 years. This appears to conflict with the conclusion of de Freitas Pacheco (1998), who claims that the observed abundance ratios of heavy elements in the Large Cloud require that the frequency of SNe II in the LMC is one per 7,800 yr and that SNe Ia are almost four times as frequent in the Large Cloud as SNe II.

Using the *Rossi X-Ray Timer Explorer* satellite, Marshall et al. (1998) have discovered a 16 msec X-ray pulsar in the Crab-like (Wang & Gotthelf 1998) Large Cloud SNR N 157B = NGC 2060. The age of this pulsar is estimated to be ~5,000 yr. The interesting remnant DEM L 316 (Williams et al. 1997) appears to consist of two colliding supernova shells.

From X-Ray spectroscopy of supernova remnants in the LMC, Hughes, Hayashi & Koyama (1998) find a correlation between the diameters of SNRs and their oxygen abundance, with the smallest remnants having the highest average values of [O/H]. The most straightforward interpretation of this result is that the young remnants are more contaminated by supernova ejecta than are older ones, in which the shell has been diluted by swept-up interstellar gas. If the conclusion that SNRs are contaminated by oxygen from the exploding supernovae is correct, then one would expect SNRs to have a higher oxygen abundance than typical regions of the interstellar medium. It is therefore puzzling that the X-ray observations by Hughes et al. actually give lower O/H values than do abundance determinations in H II regions. Specifically, Hughes et al. find [O/H] = $12 + \log(O/H) = 8.21 \pm 0.07$ in SNRs, compared to [O/H] = 8.43 ± 0.08 in H II regions (Dufour 1984).

6.9 Planetary nebulae in the LMC

Both H II regions and planetary nebulae are emission objects. They can, however, be easily distinguished because the most luminous planetaries have diameters $\lesssim 1''$ (0.25 pc). The positions of 265 LMC planetaries have been plotted by Morgan (1994), who finds that the centroid of the entire planetary nebula system of the LMC is located at $\alpha = 5^h 23^m 6$, $\delta = -68° 58'$ (equinox 2000), which is north of the optical center of the Bar. The center of the outer envelope of the distribution of planetary nebulae, which has a radius of ~ $6°5$ (5.7 kpc), is located $0°7$ farther to the east, and $0°25$ farther to the north, at $\alpha = 5^h 31^m 1$, $\delta = -68°45'$. Spectroscopic studies (Barlow 1995) show that the mean oxygen abundance in the LMC planetaries is 0.3 dex lower than that of their Galactic counterparts but 0.28 dex higher that of the average planetary in the SMC. Dopita et al. (1997) have used the *Hubble Space Telescope* to determine both the ages and the metallicities of 10 planetary nebulae in the LMC. Their data suggest that the alpha process elements (O, Ne, Ar, and S) are systematically more abundant in planetaries with ages <1 Gyr than they are in planetaries with ages >3 Gyr. This is consistent with the notion that a burst of star formation (and associated SNe II) enriched the interstellar medium in the LMC a few billion years ago. Stanghellini et al. (1999) have used the *Hubble Space Telescope* to obtain high-resolution λ 5007 [O III] images of 27 planetary nebulae in the Clouds of Magellan. Morphologically the Cloud planetaries appear similar to those in the Galaxy, except that the Cloud sample does not contain multiple shell planetary nebulae. Of the sample 36% appear round, 32% are elliptical, and 32% have bipolar or quadrupolar shapes. A higher fraction of planetaries with a bipolar structure occurs among objects of high excitation. Using the MACHO database Alves et al. (1997) have discovered a variable planetary nebula in the Large Cloud. For a review of the literature on planetary nebulae in the LMC the reader is referred to Westerlund (1997, pp. 132–141).

6.10   Interstellar matter

The first 21-cm study of the neutral hydrogen gas in the Large Magellanic Cloud was carried out by Kerr, Hindman & Robinson (1954). These authors found that the gas in the LMC was less centrally concentrated than the stars and that it could be traced out to larger radii than the easily visible stellar population of the LMC. A much more detailed study (see Figure 6.3) of the gas in the LMC was undertaken by McGee & Milton (1964). The total amount of H I gas in the LMC is $7 \times 10^8 \, M_\odot$ (Westerlund 1997, p. 28), and the amount of H_2 is $\sim 1 \times 10^8 \, M_\odot$ (Israel 1998). Inspection of Figure 6.3 shows that the distribution of neutral hydrogen gas in the Large Cloud is very clumpy. The 30 Doradus complex appears to be a "blister" on the side of the most massive of these cloud complexes. Table 6.3 shows that the center of mass of the H I gas is displaced from the center of rotation of the LMC. Detailed studies of H I in the Large Cloud by Rohlfs et al. (1984) and by Luks & Rohlfs (1992) showed that most of the gas in the Large Cloud forms part of a rotating disk with a mass of $(2.2 \pm 0.4) \times 10^8 \, M_\odot$. An additional component with a mass of $(0.6 \pm 0.1) \times 10^8 \, M_\odot$ appears to be located in front of this disk. The 30 Doradus complex seems to be associated with this additional component. Luks & Rohlfs (1992) find that this component is situated 250–400 pc above (in front of) the LMC disk.

A much higher resolution study with the *Australia Telescope Compact Array* (Kim et al. 1998) emphasizes the turbulent and fractal structure of the interstellar medium in the LMC on small scales. The structure of this gas is dominated by H I filaments, shells, and holes (which are probably caused by dynamical feedback from processes associated with star and supernova formation). On a larger scale Kim et al. find that the rotating H I disk of the LMC is well organized and symmetric. The bulk of the H I resides in a disk having a diameter of 7.3 kpc. From these observations the total mass of the disk component of the Large Cloud is found to be $2.5 \times 10^9 \, M_\odot$ and the upper limit to all mass within a radius of 4 kpc is $\sim 3.5 \times 10^9 \, M_\odot$.

A number of quasars seen through the LMC have been listed by Crampton et al. (1997). In the future some of these objects might be used for high-resolution absorption line studies of the interstellar gas in the Large Cloud.

A detailed study of the distribution of H II regions in the LMC has been published by Kennicutt et al. (1995). Inspection of their Hα image demonstrates how very different the LMC morphology is in the continuum and in emission lines. It is particularly striking that the Hα image of the LMC does not show any evidence for the presence of large-scale spiral arm–like structures. This demonstrates that star formation does not need to be triggered by spiral density waves. Kennicutt et al. show that bright H II regions are usually embedded in more extended faint emission nebulosity. These authors find that such extended emission, including filamentary networks, accounts for 35% ± 5% of the

Table 6.8. *Element abundances in Magellanic Cloud H II regions*[a]

	He	C	N	O	Ne
LMC	10.91 ± 0.05	7.81 ± 0.22	6.92 ± 0.14	8.37 ± 0.09	7.55 ± 0.14
SMC	10.92 ± 0.02	7.20 ± 0.04	6.56 ± 0.07	8.02 ± 0.04	7.21 ± 0.05

[a]From Kurt & Dufour (1998).

integrated emission of the LMC, for which Kennicutt et al. obtain 2.7×10^{40} erg s^{-1}. This emission, which is equivalent to that of 540 O5 stars, is five or six times greater than that from the Small Magellanic Cloud. The integrated equivalent width of the Hα + [N II] emission from the LMC is 36 Å, which is roughly average for Magellanic irregular galaxies. From the integrated Hα flux of the LMC Kennicutt et al. estimate that the total star formation rate in the Large Cloud is 0.26 $M_\odot$ yr^{-1}. It should, however, be emphasized that this value is uncertain because it involves assumptions about the mass spectrum with which stars are formed in the Large Cloud.

Detailed abundance studies in eight LMC and six SMC H II regions have been published by Kurt & Dufour (1998). The results of this work are summarized in Table 6.8. The data in this table show that the mean oxygen abundance in the Large Cloud is 0.35 ± 0.10 dex larger in the LMC than it is in the SMC.

A nonlocal thermodynamic equilibrium analysis of the CO emission from the Large Cloud (Johansson et al. 1998) shows that individual clouds have kinetic temperatures in the range 10 to 50 K, with the highest temperatures found close to the 30 Doradus nebula. The bulk of the CO emission originates in clouds, or cloud complexes, with sizes no larger than 20 pc. Israel et al. (1993) observed that the strength of CO emission in the LMC is intermediate between that is seen in the Galaxy and that in the SMC. This result is not unexpected because Russell & Dopita (1992) find that the C and O abundances derived from H II regions and SNRs in the LMC are also intermediate between those in the Galaxy and the SMC. The velocity-integrated CO emission in the Large Cloud is at least a factor of three lower than it is in the Galaxy. The strongest CO emission is encountered in the 30 Doradus region. A CO map of the LMC by Cohen et al. (1988) shows that nine of the 41 CO clouds they identified are associated with the 30 Dor complex. Not unexpectedly, Figure 6.14 also shows that the most prominent dust clouds in the LMC occur in the vicinity of 30 Dor. Mochizuki et al. (1994) and Pak et al. (1998) find that molecular clouds in the LMC have a larger C$^+$ envelope, relative to their CO core, than do the molecular clouds in the Galaxy. This difference probably results from the lower dust abundance in the LMC, which allows UV photons (which convert CO molecules into C$^+$ ions) to penetrate deeper into gas clumps in the LMC than they do in our Galaxy.

De Boer et al. (1998b) have detected H$_2$ absorption in the LMC. They find an excitation temperature $T_{ex} \leq 50$ K for levels $J \leq 1$ and of $T_e \simeq 470$ K for levels $2 \leq J \leq 4$. This observation suggests that UV pumping influences the populations of even the lowest rotational states of the gas in the Large Cloud.

Chemical composition differences among the interstellar medium in the LMC, the Galaxy, and the SMC may also account for some of the differences among the interstellar absorption curves in these three galaxies. Absorption in the LMC rises more steeply toward shorter wavelengths than it does in the Galaxy. Furthermore, the λ 2175Å absorption bump in the LMC is less prominent in the Large Cloud than it is in the Galaxy. (Most regions in the SMC appear to exhibit no λ 2175 Å absorption bump.) Misselt, Clayton & Gordon (1998) find that the UV extinction curves inside and outside the supergiant shell LMC 2 (southeast of 30 Dor) exhibit significant differences in their 2175Å bump strengths, even though their far-UV extinctions are very similar. To date there appears to be no clear-cut correlation between environment and the properties of extinction curves. The observation that the wavelength of maximum polarization of starlight in the SMC is typically smaller than that in the Galaxy (Rodrigues et al. 1997) also shows that the size distribution of interstellar grains differs from galaxy to galaxy.

Wakker et al. (1998) have detected C IV absorption in five LMC stars using the high-resolution spectrograph on the *Hubble Space Telescope*. Four of these stars are too cool to photoionize C^{+3} on their own. One can therefore exclude the possibility that the C^{+3} is produced locally. This shows that at least some of the C^{+3} observed in the Large Cloud is distributed in a hot corona, similar to that enveloping the Milky Way system. A similar conclusion had previously been drawn by de Boer & Savage (1980) from less complete data. Assuming an exponential distribution of O^{+5} Widmann et al. (1998) find that the distribution of this hot gas may be approximated by an exponential with $n_o = (2.1 \pm 0.2) \times 10^{-8}$ cm^{-3} and a scale-height $h_o = 5.5 \pm 2.2$ kpc. More observations will be required to understand the properties of this $\sim 10^5$ K gas in the LMC in more detail.

6.11   X-ray and γ-ray emission

The C^{+3} observations of Wakker et al. (1998) suggest that the LMC is embedded in a 10^5 K halo of highly ionized gas. Evidence for even hotter gas at 10^6 K comes from soft X-ray observations. The hot gas that envelopes the Large Cloud was probably heated by the shocks originating in supernova remnants and stellar winds. The X rays from the Large Magellanic Cloud were discovered by Mark et al. (1969) during a brief rocket flight. These authors found the total emission from the LMC between 1.5 keV and 10.5 keV to be 4×10^{38} erg s^{-1}. A total of 105 discrete sources of X rays were detected by Wang et al. (1991). Of the sources that are probably physically associated with the LMC 28 are SNRs, six are binaries, and 20 appear to be related to OB associations. Chu & Mac Low (1990) find that seven H II complexes that are *not* associated with SNRs exhibit simple shell morphologies. Emission from the supergiant shell LMC 4 is discussed by Bomans, Dennerl & Kürster (1994). For a detailed discussion of individual stellar X-ray sources the reader is referred to Westerlund (1997, pp. 180–184), Cowley et al. (1993, 1997), and to Schmidtke et al. (1994, 1999). After eliminating background sources, Schmidtke et al. (1999) find four Be stars, one massive X-ray binary, one planetary nebula, one supersoft X-ray source, one pulsar, and six supernova remnants associated with individual X-ray sources, in the Magellanic Clouds. These authors note that the source population in the Clouds differs from that in the Galaxy, in which many identifications are with old sources. Such sources associated with an old stellar population appear to be almost entirely absent from the Clouds of Magellan. After subtracting the contribution by discrete sources, the spectrum of the remaining background radiation may be fit by emission from an optically thin plasma that has $T \sim 2 \times 10^6$ K in the western part of the LMC and $T \sim 10^7$ K near 30 Doradus. Wang et al. also believe that they have detected the shadowing effect of the cold gas in the LMC against the cosmic X-ray background.

On March 5, 1979 a very bright γ-ray burst was seen toward the SNR N49 in the Large Cloud. This burst was followed by a decaying oscillation (ringing) of softer γ-ray emission with a period of $\sim$8 s. Fainter, and much shorter, bursts were observed to emanate from this source over the next four years.

For the integrated emission of the LMC as a whole Allen et al. (1993) obtained an upper limit of 1.9×10^{-12} cm s^{-1} msr^{-1} for γ rays with energies $>$30 TeV. Hartmann, Brown & Schnepf (1993) used the upper limit on the integrated γ-ray emission from the Large Cloud to derive a limit to the LMC pulsar birth rate of one per 50 yr. Sreekumar et al. (1992) find that the integrated flux of the LMC above 100 MeV is $(1.9 \pm 0.4) \times 10^{-7}$ photons (cm^2 s)$^{-1}$. From this observation they conclude that the level of cosmic radiation in the Large Cloud is comparable to that in the Galaxy.

6.12   Interactions between the Magellanic Clouds

The centers of the LMC and of the SMC are separated in the sky by 20°7 ($\sim$19 kpc). *HIPPARCOS* proper motions (Kroupa & Bastian 1997) show that both of these galaxies are moving approximately parallel to each other on the sky,[12] with the Magellanic Stream (Mathewson, Cleary & Murray 1974) trailing behind. Orbital simulations by Byrd et al. (1994) suggest that the Magellanic Clouds may have left the neighborhood of M31 $\sim$10 Gyr ago and were captured by the Galaxy $\sim$6 Gyr ago. It should, however, be emphasized that such computations are uncertain because they contain a large number of free parameters. In none of the models considered by Byrd et al., does the capture of the Clouds occur $\sim$4 Gyr ago, when the LMC started the burst of star formation that persists to the present day. For all of the models considered by Byrd et al., the perigalactic distance of the LMC, during its previous encounter $\sim$6 Gyr ago, lies between 42 and 60 kpc. From an N-body simulation Gardiner & Noguchi (1996) find that the bridge between the LMC and SMC was created during a close encounter between the Magellanic Clouds $\sim$0.2 Gyr ago. They also conclude that the Magellanic Stream is a tidal plume that was created during the LMC–SMC–Galaxy encounter 1.5 Gyr ago. If this timescale is correct, then the great burst of star formation that started in the LMC $\sim$4 Gyr ago cannot have been triggered by this encounter.

It has been argued (e.g., Lynden-Bell & Lynden-Bell 1995) that some dwarf spheroidal galaxies are located in streams that fall along great circles on the sky. For example, the LMC, the SMC, and the Draco and UMi dwarf spheroidals all form part of Lynden-Bell's "Stream 2." However, it is of interest to note that the Draco system is metal poorer than all but the most metal-deficient LMC and SMC populations. In this connection Olszewski (1997) points out that "While the abundance arguments taken alone cannot rule out a scenario in which UMi and Dra are LMC or SMC tidal fragments, the age–metallicity relations for the Magellanic Clouds demand that this interaction happened 'in the beginning'." An additional argument against the hypothesis that the Draco and Ursa Minor systems represent tidal debris that was torn from the Magellanic Clouds is

---

[12] It is, however, of some concern that the proper motions by Anguita (1999), Kroupa and Bastian (1997), and by Jones, Klemola & Lin (1994) are only marginally consistent with each other.

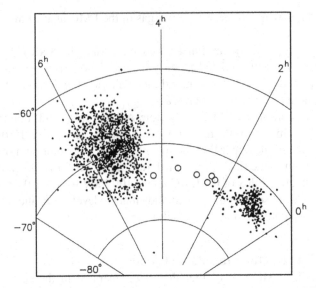

Fig. 6.22. Distribution of star clusters in and between the Magellanic Clouds adapted from Irwin (1991). Six associations discovered by Irwin, Demers & Kunkel (1990) are shown as open circles. The Wing of the SMC appears to be the brightest feature in the bridge between the Clouds.

that dark matter, which dominates the mass of these two dwarf spheroidals, would not have been entrained in a tidal arm torn from the LMC or SMC.[13] In summary it presently appears most likely that dSph galaxies are independently formed entities, rather than objects that condensed from tidal streams.

The results of interactions between the LMC and SMC are clearly shown by the distribution of OB stars between the Clouds (Irwin, Demers & Kunkel 1990) and by the distribution of star clusters (Irwin 1991), which is plotted in Figure 6.22. This figure appears to show that the "Wing" of the SMC at $\alpha \sim 2^h$, $\delta \sim -75°$ is the brightest part of a bridge between the Clouds that is highlighted by the six associations discovered by Irwin, Demers & Kunkel (1990). The Bridge and Wing form an extended structure with a length of $\sim 10°$ (10 kpc). Problems associated with star formation in a low-density gaseous bridge between the Clouds have been discussed by Christodoulou, Tohline & Keenan (1997).

The leading arm produced by the gravitational interaction between the Large Cloud and the Small Cloud, which also produced the trailing Magellanic Stream, was first detected (see their Figure 1) by Mathewson & Ford (1984). This feature was subsequently discussed in detail by Putman et al. (1998). In particular, the existence of both leading and trailing streams strongly militates against ram pressure models and supports scenarios in which such streams are torn from the Magellanic Clouds by tidal forces. The continuous leading arm, which extends for 25° from an area near the LMC toward the Galactic plane,

[13] The reason for this is that tides will stretch material, so that it has a lower density than it had in its progenitor. Furthermore, the density of dark matter in the LMC is expected to be much lower than it is in the more compact dSph galaxies.

is a natural prediction of tidal models (Gardiner & Noguchi 1996), which predict an ~1.5-Gyr-old leading arm that contains about 1/3 of the mass of the Magellanic Stream.

Hambly et al. (1994) have studied the chemical composition of two luminous B-type stars situated in the bridge between the Magellanic Clouds. One of these stars has a metallicity of at least −0.5 dex, while the other has a metallicity of −1.0 dex. Similarly low values for four B stars in the Bridge have been obtained by Rolleston & McKenna (1999). These latter low values suggest a physical association with the SMC, rather than with the LMC. In other words it appears that the Bridge may have been tidally extracted from the Small Cloud. Mebold (1991) has proposed that some of the high-velocity clouds observed near the Magellanic Stream may also be physically associated with it. In this connection it is of interest to note that Lu et al. (1998) have found that the high-velocity cloud $287.5 + 22.5 + 240$ has [S/H] = −0.6. They note that this low metallicity suggests an origin in the Magellanic Clouds; although an origin in the most metal-poor outer region of the Galactic Disk obviously cannot be excluded.

The motion of the LMC (Kroupa & Bastian 1997) is mainly directed toward the east. If the Large Cloud is moving through a circum-Galactic medium one would expect a bow shock to form along the eastern edge of the Large Cloud. The total velocity difference between the Galactic halo gas and the LMC gas is expected to be ~465 km s^{-1} (de Boer et al. 1998a). Such a velocity difference could produce compression of gas along the eastern edge of the LMC. Such compression *might* account for the fact (see Figure 6.3) that the edge of the LMC gas distribution is sharper toward the east than it is toward the west. However, this difference could equally well be due to the fact that the eastern edge of the LMC is dominated by the great gas complex in which 30 Dor is embedded. De Boer et al. have suggested that the bow shock produced by the interaction of LMC gas with that in the Galactic halo is responsible for the star formation associated with the 30 Dor complex and the supergiant shell LMC 4, which is located to the north of it. However, it appears more likely that these features are just a result of the presence of large amounts of interstellar gas. Moreover, inspection of Figures 6.9 and 6.10 shows little evidence for a predominance of star clusters along the eastern edge of the LMC. It is therefore concluded that there is still no conclusive evidence for star formation associated with a bow shock along the eastern edge of the Large Cloud.

6.13   Summary and conclusions

The first burst of activity in the Large Magellanic Cloud, which produced 13 globular clusters and a field population including stars that are now RR Lyrae variables, occurred more than 11 Gyr ago. This was followed by a period of quiescence that lasted for ~7 Gyr (i.e., for at least half of its lifetime). During these "dark ages" the LMC may have resembled a low surface brightness (LSB) galaxy. The observation that the ratio of carbon stars to late M-type giants in the LMC is smaller than it is in M33 might be due to the fact that few C star progenitors formed in the Large Cloud during its "dark ages." Butcher (1977) first showed that this low activity phase was terminated ~4 Gyr ago by a violent burst of star formation that continues to the present day. Observations of clusters suggest that the rate of cluster formation may have increased by one or two orders of magnitude. However, the rate of *star* formation appears to have increased by a smaller factor than the rate of cluster formation. Clusters formed during the first burst of star creation in the LMC occupied a much larger area than do those that formed during the last 4 Gyr. This shows that the region of active star formation in the LMC shrank with time.

The radial velocities of the 13 LMC globular clusters show that these objects exhibit only a small dispersion around the Large Cloud rotation curve. Taken at face value this result suggests that all of the LMC globular clusters are disk objects. This is puzzling because studies of their color–magnitude diagrams suggest that the Large Cloud globulars are all very old. In this respect the Large Cloud is quite different from M33, where the globular clusters appear to be relatively young, even though they have halo kinematics. It is presently a mystery why the early evolution of the LMC and M33 appears to have been so different, even though these two galaxies have rather similar luminosities.

Ardeberg et al. (1997) find that a considerable fraction of the stars in the Bar are younger than 0.5 Gyr. It is presently not yet clear if most of the older stars present in this region belong to the Bar, or whether they belong to the LMC disk. The isopleths of LMC carbon stars (Blanco & McCarthy 1983) show a strong concentration of C stars to the Bar of the Large Cloud. This shows that a significant fraction of the Bar population must consist of stars with ages >1 Gyr. Inspection of Figure 6.5 shows that a few globular clusters appear projected on the Bar. However, it is by no means certain that they are physically associated with it. In summary it is concluded that one cannot yet exclude the possibility that the LMC Bar is a morphological feature with an age of only a few billion years. If it is indeed relatively young, then the Large Cloud may have transformed itself from a low surface brightness nonbarred irregular, to a normal barred irregular, during the past few billion years. The absence of convincing evidence for star and cluster formation associated with a bow shock suggests that the Large Cloud is moving through a region of the outer Galactic halo that is presently essentially free of interstellar material that is kinematically tied to the Galaxy.

Proper motion observations show that the LMC and SMC are physical companions that are moving together through space. Orbital simulations suggest that they might have left the vicinity of M31 $\sim$10 Gyr ago and were subsequently captured into an orbit with a perigalactic distance of $\sim$50 kpc some 6 Gyr ago. Evidence for tidal interactions between the LMC and SMC is provided by (1) the Magellanic Stream, (2) some high-velocity clouds, at least one of which is known to have a metallicity similar to that of the SMC, and by (3) a $\sim$10 kpc long tidal feature that includes both the Wing of the SMC and half a dozen OB associations located between the Clouds. Metallicity observations of a luminous star suggest that the material in the Bridge was drawn from the Small Cloud.

Surprisingly, the mass spectrum of *star* formation in the compact cluster R 136 is found to be similar to that in the solar neighborhood, even though stars in these two regions were formed in environments that differed in density by over two orders of magnitude. However, the mass spectrum of *cluster* formation in the Clouds appears to differ significantly from that in the Galaxy and M31. Populous young clusters, such as NGC 1866 in the LMC, appear to have few (if any) Galactic counterparts. This suggests that the mass spectrum of cluster formation in the Large Cloud is more heavily weighted toward high-mass objects than it is in the Milky Way system. It may be speculated that the Galactic mass spectrum of cluster formation was similarly weighted toward high-mass clusters during the era of globular cluster formation.

One of the deepest enigmas associated with the Magellanic Clouds is why their RR Lyrae stars are brighter than expected. Adopting a distance modulus of $(m - M)_0 = 18.5$

it is found that the RR Lyrae stars in LMC globular clusters have $M_V = +0.45$ (Walker 1992). But the *HIPPARCOS* proper motions of Galactic RR Lyrae variables of similar metallicity yields $M_V \sim +0.7$. In addition, Catelan (1998) has shown that this discrepancy probably cannot be explained by assuming that M_V (RR) is systematically different for field and cluster variables. Resolution of this problem may have a profound impact on ideas about the age and size of the Universe.

It would be interesting to obtain radial velocities for the field RR Lyrae stars in the LMC to see if all of them share the disklike kinematics of the Large Cloud globulars, or if some of them have halo kinematics.

7

The Small Magellanic Cloud

7.1 Introduction

The Small Cloud is an irregular dwarf of DDO type Ir IV–V that has a low mean metallicity and a high mass fraction remaining in gaseous form. These characteristics suggest that the SMC is, from an evolutionary point of view, a more primitive and less evolved galaxy than the Large Cloud. The metallicity difference between the Galaxy and the LMC was discovered by Arp (1962), who wrote: "Taken together with the marked differences in the evolved giant branches and [C]epheid gaps in the SMC and [G]alactic clusters, there exists the inescapable implication that the chemical composition of the SMC stars is different from the chemical composition of the solar neighborhood."

The fact that the SMC contains only a single true globular cluster may indicate that star formation started off later, or more gradually, than it did in the Large Cloud.[1]

At the present time the SMC is forming stars less actively than is the LMC. Prima facie evidence for this is that the Small Cloud contains much smaller, and less spectacular, H II regions than does the Large Cloud. Furthermore, the LMC presently contains 110 massive Wolf–Rayet stars, whereas there are only 9 WR stars in the Small Cloud. Finally, CCD observations by Bothun & Thompson (1988) show that the SMC is redder than the LMC.[2] They find $B - V = 0.52 \pm 0.03$ for the Large Cloud, versus $B - V = 0.61 \pm 0.03$ for the integrated color of the Small Cloud. The observed color difference between the Clouds is reduced by the fact that the SMC is more metal poor than the LMC. As a result the Large Cloud contains more reddening dust grains than does the Small Cloud. Moreover, the lower metallicity of the SMC causes late-type stars of a given temperature in the Small Cloud to be bluer than similar stars in the Large Cloud. In other words the difference between the integrated colors of the LMC and SMC would have been even larger if they were to have had the same metallicity.

The most recent and complete reviews of observations of the SMC are provided in *The Magellanic Clouds* (Westerlund 1997), in *IAU Symposium No. 108* (van den Bergh & de Boer 1984), in *IAU Symposium No. 148* (Haynes & Milne 1991), and in *IAU Symposium No. 190* (Chu et al. 1999). A complete listing of recent reviews is given in Appendix 2 of Westerlund (1997). Excellent identification charts for star clusters, emission-line

---

[1] The LMC is more luminous by 1.4 mag than the SMC. For equal specific globular cluster frequencies one would therefore have expected the Small Cloud to contain $13 \times 0.275 = 3.6$ globulars.

[2] This result supersedes earlier work by Elsässer (1959), which appeared to show that the SMC was slightly bluer than the LMC.

Table 7.1. *Data on the Small Magellanic Cloud*

$\alpha(2000) = 0^h\ 52\overset{m}{.}6$		$\delta(2000) = -72°48'$
$\ell = 302\overset{°}{.}81$		$b = -44\overset{°}{.}33$
$V_r = +148.3 \pm 2.4$ km s^{-1} (1)		$i = 90°$ (2)
$V = 1.97$ (3)	$B - V = 0.52 \pm 0.03$ (3)	$R_h = 0\overset{°}{.}99 \pm 0\overset{°}{.}03$ (3)
$(m - M)_o = 18.85 \pm 0.10$ (4)	$E(B - V) = 0.06$ (4)	$A_V = 0.19$ mag (4)
$M_V = -17.07$	$D = 59$ kpc	$D_{LG} = 0.48$ Mpc
Type: Ir IV or Ir IV–V (14)		Disk scale-length
		$= 76\overset{°}{.}6$ (1.3 kpc) (5)
$M \gtrsim 8 \times 10^8 M_\odot$ (6)	$M_{HI} = 5 \times 10^8 M_\odot$ (7)	
$M_{H_2} = (7.5 \pm 2.5) \times 10^7 M_\odot$ (8)		
Total no. globulars $= 1$ (9)		
Total no. planetaries ≥ 50 (10)	Nova rate > 0.1 yr^{-1} (11)	
[Fe/H] $= -0.73 \pm 0.03$ (12)	$12 + \log$ (O/H)	
	$= 8.02 \pm 0.08$ (13)	

(1) Hardy, Suntzeff & Azzopardi (1989).
(2) Westerlund (1997, p. 28).
(3) Bothun & Thompson (1988).
(4) See Section 7.2.
(5) Fit to outer region only by Bothun & Thompson (1988).
(6) Hardy et al. (1989).
(7) Hindman (1967).
(8) Israel (1998).
(9) NGC 121.
(10) Olszewski et al. (1996).
(11) See Section 7.5.2.
(12) Luck et al. (1998).
(13) Kurt & Dufour (1998).
(14) Luminosity class difficult to determine because of its proximity.

objects, and supergiant stars, together with tables giving detailed information on these objects, have been published in the monograph *The Small Magellanic Cloud* (Hodge & Wright 1977). A compilation of data on the SMC is given in Table 7.1.

7.2 Distance and reddening

Information on the reddening of the SMC has been briefly reviewed by Bessel (1991). Bessell estimates that $\langle E(B - V)\rangle \approx 0.06$ for the SMC, with reddening values for individual regions reaching as high as 0.3 mag. From the RR Lyrae stars in the SMC globular cluster NGC 121, Walker & Mack (1988) obtain $A_V = 0.13 \pm 0.03$ mag.

Estimates of the distance modulus of the Small Cloud have been summarized by Westerlund (1997, p. 14). From four determinations via the Cepheid period–luminosity relation one obtains $\langle(m - M)_o\rangle = 18.94 \pm 0.05$. However, this value may have been affected in a systematic way by the fact that the metallicity of SMC Cepheids is significantly lower than that of those in the LMC and the Galaxy (Sasselov et al. 1997). Laney & Stobie (1994) find values of $(m - M)_o$ ranging from 18.87 in V to 19.02 in the K passband. Using their visual surface brightness technique, Barnes, Moffett & Gieren (1993) get $(m - M)_o = 18.9 \pm 0.2$ for the Cepheid HV 829 in the SMC.

From four RR Lyrae variables in NGC 121, Walker & Mack (1988) obtain $\langle V(RR)_o \rangle = 19.46 \pm 0.07$. Substituting this into the RR Lyrae luminosity calibration

$$M(RR)_V = (0.20 \pm 0.04)[Fe/H] + 1.03 \pm 0.14 \qquad (7.1)$$

obtained from statistical parallaxes using *HIPPARCOS* proper motions (Fernley et al. 1998ab), and assuming [Fe/H] $= -1.5$, one obtains a distance modulus of $(m - M)_o = 18.73 \pm 0.17$. The quoted uncertainty of this value does not take into account the possibility that the globular cluster NGC 121 might be slightly in front of, or behind, the main body of the SMC. This problem may be avoided by using field RR Lyrae variables as distance indicators. Reid & Stugnell (1986) have used Graham's (1975) photoelectrically calibrated photographic observations of RR Lyrae field stars in the SMC, in conjunction with the assumption that $M_V(RR) = +0.75$, to derive a distance modulus of $(m - M)_o = 18.78 \pm 0.2$. More recently Smith et al. (1992) have found $\langle B \rangle = 19.88 \pm 0.04$ and $\langle B \rangle = 19.95 \pm 0.02$ for field RR Lyrae stars in the vicinity of NGC 361 and NGC 121, respectively. Adopting $\langle B - V \rangle_o \approx 0.26$ (Hawley et al. 1986) for RR Lyrae stars and $E(B - V) = 0.06$, this yields $V_o(RR) = 19.38$ and $V_o(RR) = 19.45$, respectively. For $M_V(RR) = +0.75$ the corresponding distance moduli of SMC field stars are 18.63 and 18.70, respectively. Kaluzny et al. (1998) have observed twelve background SMC RR Lyrae variables in the field of 47 Tucanae. For these objects $\langle V \rangle = 19.70 \pm 0.05$. Adopting $A_V = 0.19$ (see Table 7.1) and $M_V(RR) = +0.75$, this yields a distance modulus of $(m - M)_o = 18.76$. The data discussed above are summarized in Table 7.2. It is of interest to note that the distance moduli for the Small Cloud obtained from RR Lyrae stars are ~ 0.2 mag smaller than those found from the Cepheid period–luminosity relation. Van den Bergh (1995a) has previously noted a similar discrepancy between the Cepheid distance modulus of the LMC and the Large Cloud distance modulus derived using statistical parallaxes and Walker's (1992) observations of the RR Lyrae stars in LMC

Table 7.2. *Distance modulus of the Small Magellanic Cloud*

Method	$(m - M)_o$	Reference
Cepheid P–L relations	18.94 ± 0.05	Westerlund (1997, p. 14)
P–L relation in V	18.87	Laney & Stobie (1994)
P–L relation in K	19.02	Laney & Stobie (1994)
Surface brightness	18.9 ± 0.2	Barnes et al. (1993)
RR Lyrae in NGC 121	18.73 ± 0.17	Walker & Mack (1988)
Field RR Lyrae	18.78 ± 0.2	Reid & Stugnell (1986)
NGC 121 field RR	18.73	Smith et al. (1992)
NGC 361 field RR	18.66	Smith et al. (1992)
47 Tuc field	18.76	Kaluzny et al. (1998)
LMC vs. SMC RR[a]	18.93	See Section 7.2.
Red clump giants	18.56 ± 0.07	Udalski et al. (1998)
Red clump giants	18.82 ± 0.20	Cole (1998)
Planetary nebulae	19.09 ± 0.29	Jacoby (1997)

[a] Assuming an LMC modulus of 18.5 and a metallicity difference of 0.4 dex between the LMC and SMC.

globular clusters. It is of particular interest that the data in Table 7.2 show similar distance moduli for SMC field RR stars and for the variables in the globular cluster NGC 121. This shows that the discrepancy between the Cepheid and RR Lyrae distance moduli cannot be due to a systematic difference between field and cluster RR Lyrae variables. This conclusion is consistent with that of Catelan (1998), who finds that the period–temperature distributions of Galactic field and cluster RR Lyrae stars are essentially indistinguishable, which suggests that there is no luminosity difference between them. A distance that is even smaller than that obtained from RR Lyrae stars is derived by Udalski et al. (1998). Comparing the *HIPPARCOS* calibration of red giant clump stars with the apparent magnitudes of clump stars in the SMC, these authors find $(m - M)_0 = 18.56 \pm 0.03$, with an estimated systematic uncertainty of ± 0.06 mag. However, Cole (1998) finds that $(m - M)_0 = 18.82 \pm 0.20$ after population effects on the luminosities of red clump stars are taken into account. Twarog et al. (1999) have compared the mean apparent magnitude of red clump stars in the SMC with the absolute magnitudes of red clump stars in the Galactic outer Disk clusters NGC 2420 and NGC 2506. After taking metallicity differences into account, these authors find $(m - M)_0 = 18.91 \pm 0.17$ for the SMC. Excluding the distance modulus of Udalski et al., and the indirect distance estimate from the difference between the LMC and SMC distance moduli, the data listed in Table 7.2 yield a formal mean $\langle (m - M)_0 \rangle = 18.85 \pm 0.04$. It is, however, a source of concern that the four Cepheid determinations yield $(m - M)_0 = 18.93 \pm 0.03$, which is inconsistent with the value of 18.73 ± 0.02 obtained from five distance determinations based on RR Lyrae variables. A factor that might contribute to this difference (W. E. Kunkel 1998, private communication) is that many of the SMC Cepheids are located in the young tidal tail behind (see Section 7.3) the main body of the Small Cloud. However, one would expect the old RR Lyrae stars to be centered on the main body of the SMC. This suspicion is confirmed by the work of Olszewski, Suntzeff & Mateo (1996), who found that the luminosity of SMC RR Lyrae stars has only a small dispersion in regions of the Small Cloud in which Cepheid variables indicate a significant depth along the line of sight.

Storm, Carney & Fry (1999) have used the Baade–Wesselink technique to determine the distances to four Cepheids in the SMC. From these observations they find that $(m - M)_0 = 18.93 \pm 0.09$, which is in good agreement with other results cited above. From their data Storm et al. find that $\Delta M_V / \Delta [\text{Fe/H}] = -0.45 \pm 0.15$, in the sense that metal-rich Cepheids are brightest.

An alternative approach to the determination of the distance to the SMC is to compare the brightness of RR Lyrae variables in NGC 121 in the SMC to those of the RR Lyrae stars in LMC globulars. From Walker (1992) $\langle V(\text{RR})_0 \rangle = 18.95 \pm 0.04$ for seven globular clusters in the Large Cloud, compared to $\langle V(\text{RR})_0 \rangle = 19.46 \pm 0.07$ from the observations of four RR Lyrae stars in NGC 121 by Walker & Mack (1988). This yields a formal difference in distance modulus between the SMC and LMC of 0.51 ± 0.08 mag. According to Eq. 7.1 this value decreases to 0.43 mag when the difference in metallicity between NGC 121 and the LMC globular clusters is taken into account. If we adopt $(m - M)_0 = 18.5$ for the LMC, the distance modulus of the SMC then becomes 18.93.

On the basis of the data listed in Table 7.2 it will tentatively be assumed that the true distance modulus of the SMC is 18.85, with an uncertainty that is probably $\sim$0.1 mag. With assumed distances of 50.1 kpc and 58.9 kpc to the LMC and SMC, respectively, the linear separation between the centers of the Clouds is 21 kpc. Adopting a distance of 8.5 kpc to the Galactic center, the Galactocentric distances of the LMC and SMC are 49.6 kpc and 56.1 kpc, respectively.

Observations of SMC Cepheids by Caldwell & Coulson (1986) appear to indicate that the Wing of the SMC has a distance modulus that is ~0.3 mag smaller than that of the centroid of the SMC. Taken at face value this result supports the suggestion that the Wing is part of a (tidal) bridge linking the Large and Small Clouds.

7.3   Global properties

The main features of the Small Magellanic Cloud are clearly revealed by the counts of stars with $B < 16$ (de Vaucouleurs 1955b), which are shown in Figure 7.1. This figure shows that the main body (Bar)[3] of the SMC has a position angle of ~45°. Furthermore, the Wing of the Small Cloud, which is probably a young tidal feature, is situated almost 2° east of the Bar. More detailed isopleths of the Wing and other regions of the SMC are given by Gardiner & Hatzidimitriou (1992). In Figure 7.2 the distribution of the brightest individual Small Cloud stars (Florsch 1972) is compared with that of neutral hydrogen gas (Hindman 1967). A 1.4-GHz radio map of the SMC by Haynes et al. (1986) exhibits a distribution similar to that of the young stars and gas. However, the 45-MHz radio emission from the SMC (Alvarez, Aparicio & May 1989) shows emission

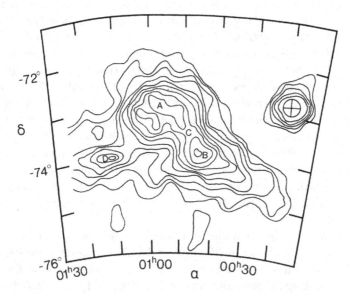

Fig. 7.1. Equidensity contours for counts of stars with $B < 16$ in the SMC. The position of the optical centroid of the Small Cloud "Bar" is marked C. Maxima of the star counts in this bar are denoted by A and B, and the position of the "Wing" by D. The position of the foreground globular cluster 47 Tucanae is shown by ⊕. Figure adapted from de Vaucouleurs (1955b).

---

[3] The widely used term "Bar" of the SMC is somewhat misleading because it refers to the brightest portion of the major axis of the rather chaotic main body of the Small Cloud. It should be emphasized that the SMC is *not* an irregular galaxy of the barred subtype, of which the Large Cloud is the prototype.

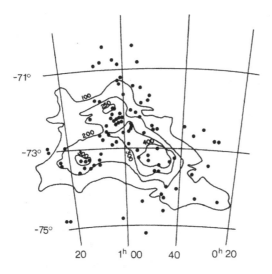

Fig. 7.2. Comparison between the distribution of the brightest stars (Florsch 1972) and of neutral hydrogen gas (Hindman 1967) in the SMC. The figure shows that gas and the brightest stars in the Small Cloud have broadly similar distributions.

from a large halo with (deconvolved) dimensions of $1°3 \times 3°1$ (1.3 kpc × 3.2 kpc). The most detailed optical surface photometry of the Small Cloud has been published by Bothun & Thompson (1988), who used a wide-field CCD camera to observe this galaxy in B, V, and R. They find that the isophote with $B = 25$ mag arcsec^{-2} has a diameter of $4°6 \pm 0°2$ (4.7 ± 0.2 kpc) and that the half-light radius of the SMC is $r_h = 0°99 \pm 0°03$ (1.0 kpc). Bothun & Thompson conclude that the choice of an exponential disk scale-length is somewhat ambiguous with possible values ranging from 51′ to 76′.

The intrinsic shape of the SMC remains a subject of lively controversy (cf. Welch et al. 1987). Observations of the distances to individual Cepheids by Mathewson, Ford & Visvanathan (1986, 1988) and by Caldwell & Laney (1991) indicated that the Small Cloud has an unexpectedly large depth along the line of sight. This conclusion is supported by observations of the magnitudes of stars in the red giant clump by Gardiner & Hawkins (1991). Some uncertainty in their results is, however, introduced by the fact (Cole 1998) that the luminosities of individual red giant clump stars may depend on their population characteristics. Taken at face value the results of Gardiner & Hawkins suggest that the Small Cloud has a tidal tail that, as seen from the Milky Way, appears projected on the main body of the SMC. A three-dimensional model of the Small Cloud, and the tidal arm emanating from it, has been published by Caldwell & Maeder (1992).

The positions of the centroids of various population components in the SMC have been tabulated by Westerlund (1997, p. 34). The locations of most of these centroids probably differ by no more than their uncertainties. The position of the centroid for six novae given by Westerlund lies somewhat to the west of that of most of the other components. However, the addition of the recent SMC nova 1992 (see Section 7.5.2) moves the centroid to $\alpha = 00^h 45^m 5$, $\delta = -72°18'$ (equinox 2000), which is close to that of most of the other centroids tabulated by Westerlund. The position of the centroid of the 408-MHz radiation of the SMC lies well to the east of the Bar. Probably its position has been affected by a contribution from the young population component associated with the Wing of the Small Cloud.

7.4   Star clusters

Early catalogs of clusters in the SMC have been published by Kron (1956 – K numbers) and by Lindsay (1958 – L numbers). The locations of these clusters may be found on the Small Cloud charts published by Hodge & Wright (1977). A more recent, and more complete, catalog of 554 SMC clusters (of which 284 are new) is given by Bica & Schmitt (1995). Their catalog also includes clusters in the bridge between the LMC and SMC as far east as $\alpha = 3^h40^m$. Additional clusters are listed by Pietrzyński et al. (1998). Battinelli & Demers (1992) have found 78 associations in the bridge between the Magellanic Clouds. They note that these associations have dimensions similar to those of associations in the central part of the SMC, even though their total stellar population is up to 80 times smaller. Such low-density associations probably also exist in the main body of the SMC, but they would be undetectable against the rich stellar background that prevails there. Demers & Battinelli (1998) have studied the color–magnitude diagrams of some of these associations and find that distances decrease from west (SMC side) to east (LMC side) and are compatible with the hypothesis that they form a bridge between the LMC and SMC. The main sequences of these associations indicate that a burst of star formation occurred in the Bridge between 10 Myr and 25 Myr ago. However, star formation seems to have started earlier in the Wing of the Small Cloud where evolved main-sequence stars are found to have ages of ~60 Myr.

A compilation of published UBV colors of SMC clusters is given by van den Bergh (1981b). He finds that a color–magnitude diagram for the integrated light of clusters in the Small Magellanic Cloud exhibits a less pronounced color gap between red and blue clusters, than does a similar diagram for clusters in the Large Cloud. The most straightforward explanation for this difference is that the SMC did not experience a major hiatus in its star cluster formation rate, like that which has occurred in the LMC (Butcher 1977). This conjecture is supported by rather sparse data on the ages of individual Small Cloud clusters (Sagar & Pandey 1989; Olszewski, Suntzeff & Mateo 1996; Da Costa & Hatzidimitriou 1998). Alternatively, it might be assumed that the color gap for the integrated colors of SMC clusters is smaller than it is in the LMC because the asymptotic giant branch phase of evolution is more extended in the metal-poor Small Cloud stars.

Maeder (1999) has drawn attention to the fact that the fraction of all B stars that exhibit emission lines seems to be larger in metal-poor galaxies than it is in metal-rich ones. In particular he finds that the fraction of all early B stars that are Be stars is significantly greater in the luminous young SMC cluster NGC 330, than it is among early B-type stars in the Milky Way.[4] The rotational velocities of stars in metal-poor galaxies might therefore be systematically higher than they are among metal-rich stellar populations. This suggests that the binary frequency might depend on metallicity. In addition, rapidly rotating metal-poor stars might mix, which could enhance their surface nitrogen abundance. Support for the hypothesis that partial mixing of CN-cycled gas from the interiors has reached the surfaces of some rotating stars is provided by the

---

[4] Keller & Bessell (1998) point out that diffuse H II emission near NGC 330 might account for some of the observed excess of "emission line" stars.

observations of Venn (1999). Venn finds that the star-to-star variations in N among SMC A-type supergiants are much larger than they are for other elements. Maeder's work suggests that the frequency distribution of stars in the color–magnitude diagram (Hess diagram) of galaxies might depend on their rotational velocity, as well as on their age distribution.

NGC 121, with an age of 12 ± 2 Gyr, is the only real globular cluster associated with the Small Cloud. Some other old clusters (Da Costa & Hatzidimitriou 1998; Mighell, Sarajedini & French 1998; de Freitas Pacheco, Barbuy & Idiart 1998) are L 1 (10 ± 2 Gyr), K 3 (9 ± 2 Gyr), NGC 416 (7 ± 1 Gyr), and L 113 (6 ± 1 Gyr). From the metallicities and ages given by Da Costa & Hatzidimitriou, it is found that [over the range $0 < T$ (Gyr) < 10] the metallicity of individual SMC clusters scatters around the relation

$$[\text{Fe/H}] = -0.6 - 0.05 \, \text{T}. \tag{7.2}$$

The SMC field RR Lyrae stars at $\langle[\text{Fe/H}]\rangle = -1.8$ and $T \sim 14$ Gyr (Butler, Demarque & Smith 1982) lie 0.5 dex below the extrapolation of Eq. 7.2. This suggests that [Fe/H] initially increased faster than it has during the past 10 Gyr.

It is noted in passing that NGC 121, with an ellipticity $\epsilon \approx 0.25$, is one of the most highly flattened globular clusters known. Since it is the only (and hence most luminous!) globular in the SMC it conforms to the curious relation (van den Bergh 1996a) that the brightest globular cluster in each galaxy also appears to be the most highly flattened one. Shara et al. (1998) have discovered 23 blue straggler stars in NGC 121. They find that these objects are more concentrated to the cluster core (and hence presumably more massive) than cluster subgiants, red giants, and horizontal branch stars.

7.5   Variable stars

Early work on the variable stars in the Small Magellanic Cloud has been discussed in detail by Gaposchkin & Gaposchkin (1966). Useful identification charts for SMC variables are given in the atlas published by Hodge & Wright (1977). During the EROS survey (Palanque-Delabrouille et al. 1998) 5.3 million stars were examined for possible variability. In future, such massive surveys should provide us with valuable complete data sets of various types of variable stars. Of particular interest are the EROS observations of the binary microlensing event MACHO SMC-98-1 (Afonso et al. 1998), which allows one to calculate the proper motion of this object. These observations show that SMC-98-1 probably does not lie in the Galactic halo or disk; that is, it is likely to be a member of the SMC itself.

7.5.1   Cepheids

Periods and amplitudes for large numbers of SMC variables are tabulated by Hodge & Wright (1977). Such data show (van den Bergh 1968a) that (1) short-period Cepheids are relatively more common in the Small Cloud than they are in the Galaxy and that (2) the pulsation amplitudes of short-period Cepheids in the SMC are, on average, larger than those of Galactic Cepheids of the same period. The fact that the characteristics of LMC Cepheids are intermediate between those of Cepheids in the Small Cloud and in the Galaxy suggests (van den Bergh 1958, Arp 1962) attributing these differences to the fact that $[\text{Fe/H}]_{\text{SMC}} < [\text{Fe/H}]_{\text{LMC}} < [\text{Fe/H}]_{\text{Galaxy}}$. The observed dependence of Cepheid periods on metallicity is caused by the metallicity dependence of the blue loops of evolutionary tracks in the color–magnitude diagram, in the sense that metal-poor evolved stars exhibit larger excursions to the blue than do metal-rich ones. Smith et al. (1992) find that, at periods shorter than three days, the Cepheid period–luminosity relation splits into two sequences; the brighter one is due to Cepheids pulsating in the first overtone mode, while the fainter sequence consists of variables pulsating in the fundamental mode. This conclusion has been confirmed from Fourier analysis of SMC Cepheid light curves by Buchler & Moskalik (1994), and by the EROS observations of Bauer et al. (1998), which are discussed in Section 6.6.1. The period–luminosity relations for fundamental mode Cepheids with $P > 2.0$ days and $P < 2.0$ days are given in Equations 6.2b and 6.3b, respectively. Imbert (1994) has found that the SMC Cepheids HV 837 and HV 1157 are spectroscopic binaries. Data on the distance to the SMC determined from Small Cloud Cepheids have been summarized in Table 7.2.

Luck et al. (1998) have obtained high-dispersion spectra of six Cepheids in the Small Cloud. From these observations they find that $\langle[Fe/H]\rangle = -0.74 \pm 0.03$. This is 2.75 times smaller than the value $\langle[Fe/H]\rangle = -0.30 \pm 0.04$ that these same authors find for Cepheids in the Large Cloud. Unexpectedly Luck et al. find that $\langle[O/Fe]\rangle = -0.30 \pm 0.05$ in the SMC, which is quite similar to the value $\langle[O/Fe]\rangle \sim -0.3$ obtained in metal-rich Galactic stars with $[Fe/H] \sim 0.0$. A similar result has also been obtained by Hill, Barbuy & Spite (1997). Taken at face value this result suggests that the fractional contribution of SNe II to the enrichment of the interstellar medium in the SMC is similar in the Galaxy and in the LMC. The [O/Fe] values obtained by Luck et al. (1998) for individual Cepheids in the Large Cloud exhibit too much scatter to draw any conclusions about the evolutionary history of the LMC. However, Hill et al. find that the oxygen-to-iron ratio for iron-poor stars in the LMC is similar to that of iron-rich stars in the solar neighborhood.

Gilmore & Wyse (1991) have shown that the present value of [O/H] in the interstellar medium of a galaxy will depend on its star formation history. This is so because oxygen enrichment by short-lived SNe of Type II is essentially instantaneous. In contrast, super-novae of Type Ia may explode and form iron up to a few billion years after the formation of their stellar progenitors. The iron abundance will therefore increase gradually after a burst of star formation, whereas the oxygen abundance remains constant. As a result the oxygen-to-iron ratio will decrease with increasing time after a burst of star formation. Gilmore & Wyse show that evolutionary histories that encompass multiple bursts of star formation can account for a wide range of [O/Fe] values. These results suggest that the observed oxygen-to-iron ratios in the LMC and SMC can be explained by scenarios involving a strong initial starburst followed by one or more secondary bursts.

7.5.2   *Novae*

The characteristics of SMC novae that occurred prior to 1987 have been discussed by van den Bergh (1988b). Since then an additional nova (de Laverny et al. 1998) has been discovered in the Wing of the Small Cloud. A compilation of the positions of all novae known to have occurred in the SMC is given in Table 7.3 and is plotted in Figure 7.3. From these data the centroid of the distribution of Small Cloud novae is found to be at $\alpha = 00^h45^m5$, $\delta = -72°18'$ (equinox 2000). This position does not differ significantly from that of most of the SMC population centroids listed by Westerlund (1997, p. 34). However, the rather small number of novae that have so far been discovered in the SMC

Table 7.3. *Novae in the Small Magellanic Cloud*

Year	α(2000)	δ(2000)
1897	$01^h00\overset{m}{.}0$	$-70°$ 15′
1927	00 34.1	-73 15
1951	00 35.3	-72 57
1952	00 48.5	-73 31
1974	00 26.0	-74 01
1986	00 36.9	-72 05
1992	01 17.7	-73 31
1994	00 51.5	-73 20

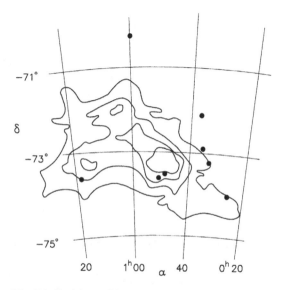

Fig. 7.3. Positions of SMC novae superimposed on neutral hydrogen isophotes of the Small Cloud. The distribution of novae may be broader than that of the neutral hydrogen and of the brightest stars shown in Figure 7.2.

suggests, but does not yet prove, that the distribution of novae might exhibit less central concentration than that of the brightest stars and hydrogen gas.

From very incomplete searches, six SMC novae have been discovered during the past half century. The nova rate in the Small Cloud must therefore be >0.12 yr^{-1}.

7.5.3 *Other types of variable stars*

The star R 40 is the only luminous Hubble–Sandage variable known in the Small Magellanic Cloud (Szeifert et al. 1993). During a recent brightening of 0.5 mag the spectrum of R 40 was observed to change from B8 Ia to A3 Ia–O. Barbá et al. (1995) have reported that the binary WR star HD 5980 began to exhibit some of the characteristics of a luminous blue variable after it had brightened by 2.3 mag in V light. These observations show that even quite metal-poor stars, such as those in the SMC, can become variables of the Hubble–Sandage type. During its eruption HD 5980 was, for about five months, the most luminous star in the Small Cloud. The spectral type of this object changed from H-poor WN3 some 20 years before outburst, to H-rich WN11 during outburst. Moffat et al. (1998) speculate that component A of this system is normally a quiescent luminous O7 star, but that it became a WN11 star during outburst. They believe that component B remained an H-poor WR star.

It is believed that opacity provided by metals drives the pulsation of β Cep variables. If this is so then one might expect such objects to be absent from the metal-poor SMC. The apparent lack of β Cep variables in the Small Cloud (Balona 1992) appears to support this hypothesis.

The semiregular variables in the Small Cloud have been discussed by Lloyd-Evans (1992). He finds that most of the semiregular variables in the SMC are carbon stars, whereas semiregular M star variables outnumber C star variables in the LMC. On average the semiregular C-type variables are found to have longer periods than do the M-type variables. The main reason why the ratio of C stars to M giants is so much greater in the SMC, than it is in the LMC, is that only a small number of carbon atoms need to be dredged up from the interior of a metal-poor star to make the number of C atoms in its photosphere larger than the number of O atoms. An additional factor contributing to the scarcity of late M-type giants in the SMC is that the abundances of titanium and oxygen (and hence of TiO molecules) are low in cool, metal-poor, Small Cloud stars.

7.6   Evolutionary history

Excellent reviews on the evolution and spatial distributions of stellar populations in the Small Magellanic Cloud have been given by Gardiner & Hatzidimitriou (1992) and by Olszewski, Suntzeff & Mateo (1996). The picture that emerges from these studies is that the stars and clusters with ages $\lesssim 1$ Gyr are mainly concentrated in the Wing and in the main body (Bar) of the SMC, whereas the older populations appear to be distributed throughout a larger volume of space. On the whole, the distributions of luminous young stars (see Figure 7.2), of emission nebulosity, and of young clusters (Bica & Schmitt 1995) appear rather similar to that of neutral hydrogen gas (Hindman 1967). Perhaps the best insight into the distribution of intermediate-age stars in the SMC is provided by the survey of carbon stars carried out with the UK Schmidt telescope by Morgan & Hatzidimitriou (1995). These objects are generally believed to have ages that lie in the range $1 \text{ Gyr} \lesssim T \lesssim 10 \text{ Gyr}$. The data by Morgan & Hatzidimitriou, which are plotted in Figure 7.4, show a smooth distribution of C stars. The core region of this distribution is flattened and resembles an E5 galaxy. However, the outer halo of C stars appears more nearly circular. The carbon star halo of the SMC is seen to extend out to $\sim 6°$ (6 kpc). The observation that only seven out of 1,634 C stars were found between $\alpha = 2^{h}30^{m}$ and $\alpha = 4^{h}00^{m}$ is consistent with the view that the bridge between the LMC and SMC is a relatively young feature. The fact that the Wing of the SMC does *not* contain a concentration of C stars confirms that the Wing, like the bridge between the LMC and SMC, is younger than ~ 1 Gyr. The distribution of 1,707 carbon stars in the main body of the SMC has been plotted by Rebeirot, Azzopardi & Westerlund (1993).

Cook, Aaronson & Norris (1986) have compared the ratio of the number of carbon stars to the number of M3–M10 stars in various Local Group galaxies. They find that the C/M number ratio increases with decreasing parent galaxy luminosity (i.e., faint

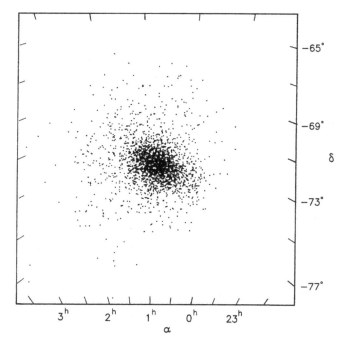

Fig. 7.4. The core of the distribution of carbon stars in the Small Cloud is seen to resemble an E4 galaxy, whereas the isopleths of the carbon star halo appear be more circular. The figure is adapted from Morgan & Hatzidimitriou (1995).

metal-poor galaxies contain more C stars than do luminous metal-rich ones). This effect is, no doubt, partly due to the dredging up of carbon from the stellar interior, which increases the number of C atoms in the atmospheres of metal-poor stars so much that the number ratio C/O becomes larger than unity. However, this is not the whole story because most metal-poor globular cluster stars do not contain carbon stars. The reason for this is that globulars (in contrast to dwarf spheroidal galaxies that mostly do contain C stars) do not have an intermediate-age population.

From spectroscopic observation of a relatively small sample of C stars Hardy, Suntzeff & Azzopardi (1989) obtained a heliocentric velocity of $+148.3 \pm 2.4$ km s^{-1} for the Small Could. The data give no evidence for systematic rotation of the C star component of the SMC. This suggests that the carbon stars in the SMC form a nonrotating halo. Hardy et al. (1989) find that the velocity dispersion of the C stars within the main body of the SMC is 27 km s^{-1}. Dopita et al. (1985) have obtained a very similar velocity dispersion of 25.3 ± 2.8 km s^{-1} for the planetary nebulae within the main body of the SMC. The kinematics of various SMC population components are compared in Table 7.4. These data (Suntzeff et al. 1998) suggest that field stars, planetary nebulae, and carbon stars all have similar velocity dispersions but that the velocity dispersion of relatively young Cepheids and star clusters may be smaller. Furthermore, Dopita et al. also find no evidence for any rotation of the system of planetary nebulae in the Small Cloud. This suggests that the majority of planetary nebulae belong to the same population as do the SMC carbon stars.

Table 7.4. *Velocity dispersions*

Sample	σ(km s^{-1})	N
NGC 121 field[a]	25.7 $\pm$ 2	36
PNe	25 $\pm$ 3	44
C stars	27 $\pm$ 2	131
Cepheids	22 $\pm$ 3	61
Clusters	16 $\pm$ 4	7

[a]From Suntzeff et al. (1998).

Hardy et al. (1989) have used the velocity dispersion of the carbon stars in the SMC to estimate the mass of the Small Cloud. Such estimates are, of course, uncertain because they depend strongly on the assumed mass distribution. Using the projected mass distribution method (see Eq. 6.1) of Bahcall & Tremaine (1981), they obtain a mass of 0.8 $\mu \times 10^9 M_\odot$, where $\mu = 2/3$ for circular orbits, 1 for an isotropic velocity distribution, and 2 for linear orbits.

The clump on the red side of the horizontal branch is one of the most prominent features in the color–magnitude diagram of the Small Cloud. This clump is due to the fact that core helium burning red giants have (almost) the same mean absolute magnitude over a wide range of ages and metallicities. The ages of such red clump stars lie in the range 1 Gyr $\lesssim T \lesssim 10$ Gyr. Gardiner & Hatzidimitriou (1992) have used UK Schmidt plates to study the distribution of red clump stars over the face of the SMC. Such objects are seen to have a distribution that is similar to that which Morgan & Hatzidimitriou (1994) found for carbon stars. The outermost isopleths of the carbon star distribution at $r \approx 5°$ are seen to be almost circular. Gardiner & Hatzidimitriou have used the color difference between clump stars and red giant branch stars, at the magnitude level of the horizontal branch, to estimate the mean ages of stellar populations in various regions of the Small Cloud. They find a median clump age of 3–4 Gyr for the innermost region of the SMC [$r \sim 1°5$ (1.5 kpc)] near NGC 411 and NGC 152. The mean age of clump stars increases rapidly with radius to $\sim$10 Gyr for stars with $r > 2.3°$ (2.3 kpc). Gardiner & Hatzidimitriou estimate that the oldest population component of the SMC, which includes RR Lyrae stars and the cluster NGC 121, comprises $\sim$7% of the stellar population in the outer regions of the Small Cloud. The color–magnitude diagram of the SMC reveals that the Small Cloud Population II is mainly composed of stars of the red horizontal branch variety. Blue horizontal branch stars, if they exist at all, are an order of magnitude less abundant than are old red horizontal branch members.

Hatzidimitriou & Hawkins (1989) have taken advantage of the fact that the average magnitude of red clump stars is expected to be almost independent of age and metallicity, to study the depth of the line of sight in various parts of the SMC.[5] They find that the northeastern regions of the Small Cloud have a depth that is $\sim$10 kpc greater than that

---

[5] The real situation might, however, be more complicated. Cole (1998) has argued convincingly that the detailed physics of mass loss, and the relations between heavy element abundance, helium abundance, and core mass of evolved stars, may introduce variations of a few tenths of a magnitude in the luminosities of red clump giants.

Table 7.5. *Ages and metallicities of SMC clusters*[a]

Cluster	[Fe/H]	Age (Gyr)
NGC 121	−1.19	12 ± 2
L 1	−1.01	10 ± 2
K 3	−0.98	9 ± 2
L 113	−1.17	6 ± 1
NGC 339	−1.19	4 ± 1.5
L 11	−0.81	3.5 ± 1
Recent[b]	−0.74	0

[a]Da Costa & Hatzidimitriou (1998).
[b]Luck et al. (1998).

observed in the southwestern regions. They conclude that this difference is due to the presence of a tidal tail resulting from a gravitational interaction with the Large Cloud. Since red clump stars are older than 0.5 Gyr this result suggests that the tidal interaction, which formed the SMC tail, took place at least 0.5 Gyr ago.

Da Costa & Hatzidimitriou (1998) have attempted to derive an age–metallicity relation for the Small Magellanic Cloud by determining [Fe/H] for giant stars in clusters of known age. Their data (see Table 7.5) show that the oldest SMC clusters (NGC 121, L 1, and K 3) have metallicities that are about half of that of recently formed objects in the Small Cloud. However, the scatter in the relation between [Fe/H] and age is still too large to permit any strong conclusions about the shape of the age–metallicity relation in the Small Cloud. All that can presently be said with certainty is that the SMC iron abundance has increased by a factor of ~2 during the last ~10 Gyr.

As has already been pointed out in Section 7.5.1 the [O/Fe] ratio in the SMC is similar to that in the solar neighborhood, even though the Small Cloud metallicity is much lower than that of interstellar material near the Sun. Following Gilmore & Wyse (1991) it will be assumed that this can be accounted for by the hypothesis that star formation in the Small Cloud has experienced a number of bursts.

7.7 Interstellar matter

The hydrogen gas in the Small Magellanic Cloud was first detected by Kerr, Hindman & Robinson (1954). More detailed observations of the SMC, which were obtained with the Parkes 64-m reflector, are discussed by Hindman (1967) and are shown in Figure 7.5. This figure shows that the neutral hydrogen gas in the Small Cloud has a rather smooth distribution. In this respect it differs from that in the Large Cloud (see Figure 6.3), in which the gas exhibits a much clumpier distribution. This is a valid comparison because the LMC and SMC were observed with the same telescope, and hence with the same angular resolution. Possibly the clumpier distribution of LMC gas is due to the fact that the higher metallicity in the Large Cloud produced more dust, which in turn resulted in more intense gas cooling.

Stavely-Smith et al. (1997) have used H I aperture synthesis to study the small-scale structure of the neutral hydrogen gas in the SMC. Their beautiful map, which does not show structural features with size ≥0.6°, reveals the presence of numerous expanding

Table 7.6. *Types of shells in the Small Cloud*

Type	Parent Population	Log E(ergs)	Log R(pc)
Bubble, SNR	single O, WR star	51	0–1
Superbubble	OB association	52–53	1–2
Supergiant shell	starburst	≥ 54	2–3

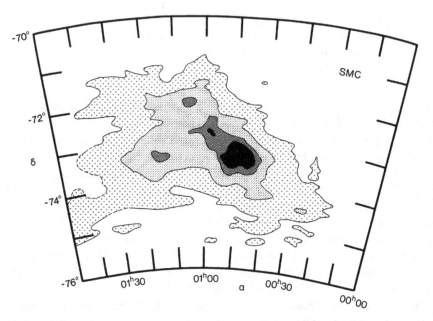

Fig. 7.5. Distribution of neutral hydrogen gas in the SMC, adapted from Hindman (1967). Note that the distribution of gas in the SMC is less clumpy than that in the LMC, which is plotted in Figure 6.3.

young shells and supershells. These authors find that the total kinetic energy of these expanding shells is a significant fraction of the binding energy of the Small Cloud (unless the SMC is embedded in a massive dark halo).

Table 7.6, which is taken from Oey (1999), lists the various types of shell-like structures that are observed in the SMC and their probable parent populations. An atlas showing the SMC in the light of Hα + [N II], which shows numerous shells, has been published by Davies, Elliot & Meaburn (1976).

Hindman (1967) found a total hydrogen mass of $\sim 5 \times 10^8 M_\odot$ for the SMC. Since the total mass of the SMC is estimated to be $\geq 8 \times 10^8 M_\odot$ (Hardy, Suntzeff & Azzopardi 1989) it follows that a large fraction of the total Small Cloud mass is in the form of gas. This observation shows that the SMC is presently still in a rather unevolved, and therefore primitive, state.

Inspection of Figure 7.5 shows that most of the Small Cloud gas is aligned with the major axis of the Small Cloud Bar. A secondary gas clumping is centered on the Wing of the Small Cloud. The major concentration of gas is found to be in the southern half of

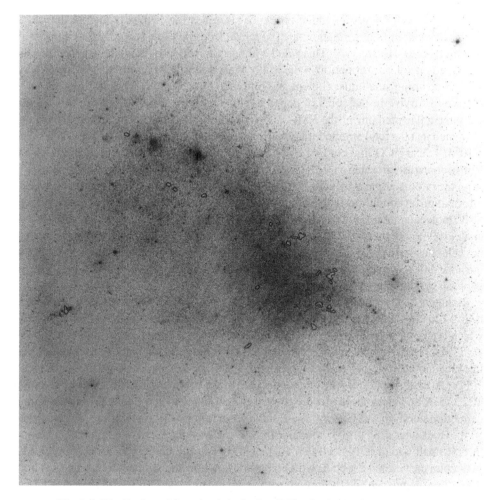

Fig. 7.6. Distribution of dust clouds in the Small Cloud. The SMC (van den Bergh 1974a) is seen to contain fewer, and less conspicuous, dust clouds that the LMC, which was shown in Figure 6.12.

the Bar. This is also the region (see Figure 7.6) in which the most conspicuous SMC dust clouds occur. Owing to its lower metallicity the dust in the SMC is much less in evidence than it is in the Large Cloud (see Figure 6.14). Note that both the LMC and SMC dust clouds were outlined on similar plate material obtained with the Curtis Schmidt telescope at the Cerro Tololo Observatory.

In the Magellanic Clouds it is possible to study the distribution of H II on scales that range from ~3 pc to ~3 kpc. In the SMC the large-scale distribution of emission nebulosity (and therefore that of star formation) roughly follows that of H I (le Coarer et al. 1993, Kennicutt et al. 1995). Bright emission structures are embedded in fainter diffuse clouds of emission. Kennicutt et al. find that $41\% \pm 5\%$ of all SMC emission originates in diffuse clouds and extended networks of filaments. The integrated Hα + [N II] equivalent width for the SMC is 24 Å, which is somewhat lower than that for typical irregular galaxies. Kennicutt et al. find that the total integrated Hα flux from the

SMC is 4.8×10^{39} erg s^{-1}, which is five or six times lower than that from the LMC. The corresponding star formation rate (which depends on the assumed stellar luminosity function) is $\sim$0.046 $M_\odot$ yr^{-1}. This value is almost two orders of magnitude lower than the estimated rate of star formation in the Milky Way system. Because of its low metallicity, and the resulting small dust content and high UV flux density, CO molecules will only be able to survive in the cores of dense gas clouds in the Small Cloud. In this respect the interstellar medium in the SMC differs from that in the Galaxy. Israel (1998) has estimated that the total amount of H$_2$ gas in these cloud cores in the SMC is (0.75 ± 0.25) $\times 10^8 M_\odot$. The ^{12}CO line emitted by SMC molecular clouds is often optically thick, so that it contains information on the surface layers of such clouds. However, emission from the rarer isotopic molecules ^{13}CO and C^{18}O is optically thin and therefore provides information on conditions in the cloud cores. Such observations show that the interiors of SMC molecular clouds are warmer than those in typical Galactic molecular clouds. Chin et al. (1998) note that, in both the LMC and in the SMC, I(^{13}CO)/I(C^{18}O) is larger than the values usually encountered in the Galactic disk. It is not yet clear if this represents an intrinsic difference between these galaxies, or whether it is due to either isotope-selective photodissociation or chemical fractionation.

Richter et al. (1998) have detected H$_2$ absorption in SMC gas. In a velocity component near $+120$ km s^{-1} they detect a cool component with an excitation temperature of $\simeq$70 K, which is comparable to the kinetic gas temperature. For another component at $+160$ km s^{-1} Richter et al. find an excitation temperature $>$2,300 K, from which they conclude that this cloud must be highly excited by strong UV radiation from its energetic environment.

For the majority of stars the interstellar extinction curve in the SMC exhibits no bump at 2,175 Å and rises more steeply toward the ultraviolet than does the corresponding extinction curve in the Galaxy. From their observations of polarization of stars in the SMC, Rodrigues et al. (1998) conclude that (1) the wavelength of maximum polarization in the Small Cloud is typically smaller than it is in the Galaxy, (2) the typical SMC extinction curve can best be fit with amorphous carbon and silicate grains, and (3) both the carbon abundance and the grain size distribution must be different in the Small Cloud from that which prevails in the Milky Way system. Using *IUE* observations, Gordon & Clayton (1998) have, however, found a single star in the Wing of the SMC that has a 2,175-Å bump and that has far weaker UV extinction than do typical stars in the main body of the SMC. These authors also note that the UV extinction in the Small Cloud resembles that observed in starburst galaxies. A similar conclusion has been drawn by Gordon, Calzetti & Witt (1997). These authors note that both starburst galaxies at $z > 2.5$ and stars in the main body of the SMC lack a 2,175-Å bump. Zubko (1999) finds that he can reproduce the properties of the dust absorption in the Small Cloud with a model in which the major grain constituents are silicates, organic refractories, and nanosized silicon particles. Because of the strong UV radiation field, the icy mantles of SMC grains (and of grains in starburst galaxies at $z > 2.5$) are probably converted into organic refractories. However, such processing does not proceed to the point where the organic refractory materials are transformed into amorphous carbon.

Observations in the ultraviolet (Prinja & Crowther 1998) also show that the terminal velocities of the stellar winds of O-type dwarfs in the SMC are systematically lower than the mean Galactic values by $\sim$600–1,000 km s^{-1}. This difference (Kudritzki, Pauldrach & Puls 1987) is due to the lower metallicity of stars in the Small Magellanic Cloud.

Sreekumar et al. (1993) have observed the Small Magellanic Cloud with the *Energetic Gamma Ray Experimental Telescope* (*EGRET*) on the *Compton Observatory*. These observations showed that the cosmic-ray energy density in the SMC is much lower than it is locally in the Galaxy. This proves that the energy density of cosmic rays cannot be uniform throughout space. However, it should be emphasized that this discovery does not preclude the possibility that the very highest energy ($>10^{19}$ eV) cosmic rays have an extragalactic origin.

7.8 Supernova remnants

The first SNR to be discovered in the Small Magellanic Cloud is associated with the radio source 0045−73.4 (Mathewson & Clark 1972), which is located in the emission nebula N19. These authors took advantage of the fact that the ratio ([S II] 6731 + 6731)/(Hα + [N II]) is typically an order of magnitude larger in collisionally ionized remnants of SNe II, than it is in radiatively excited H II regions. However, this technique cannot be used to discover the nonradiative shocks, which are believed to be associated with the remnants of SNe Ia. Such remnants, of which the SMC supernova remnant 1E0102-7219 is an example, only exhibit Balmer emission lines. The remnant 1E0102-7219 is surrounded by a remarkable high-excitation halo that exhibits strong He II lines, which are probably excited by the remnant itself. A nice review of information on the supernova remnants in the SMC is given by Dopita (1984).

Filipović et al. (1998) have combined radio and X-ray surveys of the Small Cloud to produce a listing of 12 probable SNRs and two SNR suspects. From these data Filipović et al. derive a rate of one supernova per 350 ± 70 years. The corresponding integrated star formation rate in the SMC is (0.15 ± 0.05) $M_\odot$ yr^{-1}. This is higher than the value 0.046 $M_\odot$ yr^{-1} that Kennicutt et al. (1995) estimated from the integrated Hα flux of the SMC. This difference might be due to the escape of Lyman continuum photons from the Small Cloud. Figure 7.7 shows a plot of the distribution of the Small Cloud supernova remnants listed by Filipović et al. The figure shows that the SMC supernovae are strongly concentrated in regions that presently contain luminous stars (see Figure 7.2) and in areas of high gas density. This is, no doubt, due to the fact that massive core-collapse SNe are young and will therefore preferentially occur in regions of active star formation. However, a second reason is that the presence of a relatively dense ambient interstellar medium is required to render the shock waves generated by an exploding supernova visible.

Rosado, Le Coarer & Georgelin (1994) have used a Fabry–Perot interferometer to study the kinematics of a number of SMC supernova remnants. They find that the H II complex N 19 exhibits violent internal motions that suggest the presence of three SNRs. They also investigated the SNRs 0046-73.5 in N 24 and 0050-72.8 in N 50. The shell-type SNR 0101-7226 has been studied by Ye et al. (1995). These authors suggest that the low X-ray brightness of this SNR might result from a nearby OB association that has punched a hole in the supernova shell through which the hot gas that was once contained in the shell was able to leak out. A variable pointlike X-ray source, which is spatially coincident with SNR 0101-7226, is found to be in a binary containing a star of type Be. Hughes & Smith (1994) find that the source SNR 0104-72.3 also contains a pointlike X-ray source. Amy & Ball (1993) have published a high-resolution radio map of the oxygen-rich SMC supernova remnant 1E 0102.2-7219, which is found to have a shell-like structure with a diameter of $\sim 40''$ (11 pc). Morse, Blair & Raymond

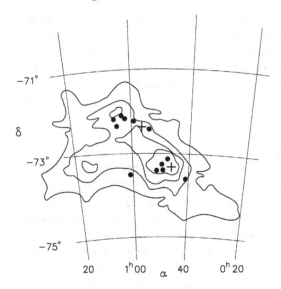

Fig. 7.7. Distribution of SMC supernova remnants listed by Filipović et al. (1998). Probable SNRs are shown as dots, possible SNRs as plus signs. The figure shows that supernova remnants occur preferentially in regions with a high gas density and active star formation.

(1998) have obtained images and ultraviolet spectra with the *Hubble Space Telescope* of 1E0102.2 and N132D, which are both oxygen-rich supernova remnants that also exhibit O, Ne, C, and Mg emission. However, these objects differ from their Galactic counterpart Cassiopeia A in that their spectra provide no evidence for the presence of S, Ca, and Ar.

Using *ROSAT*, Israel et al. (1998) have found that the source 2E 0053.2-7242 is a pulsar with a period of 59 s and $\dot{P} = -0.016$ sec yr^{-1}, yielding a spindown age of $\sim$2,000 yr. Another pulsar with a period of 75 s has recently been reported by Yokugawa & Koyama (1998). Israel et al. (1998ab) report that the SNR 0101-724 contains an X-ray pulsar with a sinusoidal light curve and has a period of 345 s. Yokogawa & Koyama have also found a pulsar with a period of 3.3 s in the SNR DEM S128.

Tsujimoto et al. (1995) conclude that the abundance pattern in the SMC, in particular the ratio of oxygen to iron, suggests that $n(\text{SNIa})/n(\text{SNII}) = 0.15$. This conclusion should, however, be treated with some reserve because the O/Fe ratio will depend on the detailed history of supernova production. In particular O/Fe will be higher just after a burst of star formation than it will be a few billion years after the termination of such a burst. This is so (Gilmore & Wyse 1991) because most of the oxygen produced by massive SNe II will be ejected just after a starburst, whereas the abundance of ejected iron will build up gradually as more and more SNe Ia explode. The fact that Venn (1999) finds very small star-to-star abundance variations (except for N) in SMC A-type supergiants suggests that the gas in the Small Cloud is presently quite well mixed.

7.9   Interactions of SMC with LMC

Evidence for tidal interactions between the Large and the Small Magellanic Clouds has already been discussed in Section 6.12. There it was noted there that two B stars in the bridge between the Clouds have metallicities that appear to indicate that they were formed from metal-poor SMC gas (Hambly et al. 1994), rather than from more metal-rich LMC gas.

Rolleston & McKenna (1999) have obtained high-resolution spectra of four B-type stars that are associated with the Bridge between the LMC and SMC. Since these stars are relatively unevolved, their photospheric abundances should be similar to that of the interstellar material from which they formed recently. Rolleston & McKenna find that these inter-Cloud stars have [Fe/H] = −1.1, which is ∼0.6 dex lower than that of similar stars in the SMC. Taken at face value, this result suggests that gas in the Bridge has not been enriched by the supernovae of Type Ia that have occurred in the Magellanic Clouds during the past ∼10 Gyr.

Evidence for the tidal distortion of the SMC is provided by the great depth of the Small Cloud along the line of sight (see Section 7.3), which probably results from a tidal arm behind the SMC being projected onto the main body of the Small Cloud. Optically the most striking evidence for a bridge between the LMC and the SMC is provided by the Wing of the Small Cloud. This region of recent active star formation was discovered by Shapley (1940) and first discussed in detail by de Vaucouleurs & Freeman (1972). From far-UV imaging Courtès et al. (1995) have been able to show that the Wing of the SMC has a faint extension that projects 3°7 to 5°8 (3.5 kpc to 5.5 kpc) from the center of the SMC toward the LMC. The young inter-Cloud population has been studied in detail by Demers & Battinelli (1998). They find that distances increase from east to west, as would be expected for a bridge linking the LMC ($D = 50$ kpc) and SMC ($D = 59$ kpc). Main sequence ages of associations in the bridge between the Clouds are found to be 10 to 25 Myr. Demers & Battinelli find that star formation started earlier, perhaps 60 Myr ago, in the Wing of the Small Cloud.

Buonanno et al. (1994) have suggested that young globular clusters, such as Terzan 7, Arp 2, Ruprecht 106, and Palomar 12, might have been tidally stripped from the Magellanic Clouds. This idea now seems improbable for Terzan 7 and Arp 2, because both of these objects appear to be associated with the Sagittarius dwarf. Furthermore, Irwin (1999) notes that Pal 12 may form part of the tidal tail of the Sagittarius dwarf galaxy. Tidal stresses would have destroyed this dwarf spheroidal galaxy if it had been pulled from one of the Magellanic Clouds. The isolated young cluster Ruprecht 106 has [Fe/H] = −1.8. This low metallicity implies that it could only have been torn from the LMC or SMC very long ago, (i.e., before these galaxies started to increase their metallicity). The young cluster Palomar 12 has [Fe/H] = −0.9. This either implies that (1) it was torn from the LMC very long ago, before the metallicity of the Large Cloud reached its present level of [Fe/H] ≈ −0.3, or (2) that it was tidally stripped from the Small Cloud ([Fe/H] ≈ −0.7) more recently. The outlying LMC cluster ESO 121-SC03 is the only Large Cloud cluster with an age that falls within the "dark ages" that extended from ∼3.5 Gyr to 11.5 Gyr ago. The position, the age of ESO 121-SC03, and its metallicity ([Fe/H] ≈ −0.9) are all consistent with the hypothesis that this cluster *might* have been tidally stripped from the Small Magellanic Cloud. An orbital model, in which the Sagittarius dwarf was tidally deflected by a gravitational interaction with the Magellanic Clouds, has been proposed by Zhao (1998). On the basis of its radial velocity and position on the sky Lu et al. (1998) have suggested that the metal-poor ([Fe/H] = −1.5) high-velocity cloud 287.5 + 22.5 + 240 may also represent material that was tidally detached from the Magellanic Clouds before it reached its present higher metallicity.

7.10   Summary and conclusions

The Small Cloud is a low luminosity galaxy of DDO type Ir IV or Ir
IV–V. No evidence for spiral arms is seen. This shows that spiral density waves are
not required to trigger star formation. Both the low metallicity of SMC gas and the high
fraction of its total mass remaining in gaseous form show that the Small Magellanic
Cloud is still a relatively primitive and unevolved galaxy. The existence of the Magellanic Stream suggests that the LMC and SMC have suffered a violent tidal interaction
in the past. A number of lines of evidence also point to a more recent tidal interaction, which has profoundly affected the morphology of the outer regions of the SMC:
(1) observations of young Cepheids, and of old red giant clump stars, appear to show that
the Small Cloud has a considerable depth along the line of sight. This depth is probably
caused by a tidal arm behind the main body of the SMC. (2) A bridge containing gas
and associations of young luminous stars links the LMC and SMC. The fact that some of
the young stars in this bridge have SMC-like metallicities suggests that this bridge was
tidally drawn out of the Small Cloud. The cluster ESO 121-SC03, which is located in the
outermost part of the LMC, might have been tidally detached from the SMC, although
some difficulties remain for this hypothesis.

The rate of cluster formation in the SMC appears to have remained relatively constant
over time, whereas that in the LMC exhibited a giant burst that started 3–5 Gyr ago. The
fact that the Large Cloud exhibits such a burst, but that the Small Cloud does not, militates
against the suggestion that the starburst in the LMC was triggered by a close encounter
with the SMC. The ages of star clusters, and the color–magnitude diagrams of field stars,
do not appear to show any evidence for major bursts of star and cluster formation in the
history of the SMC. It is therefore puzzling that the low [O/Fe] abundance in the SMC
indicates that star formation, and hence the formation of oxygen, took place in bursts.

The best present estimate for the distance modulus of the SMC is $(m - M)_0 = 18.85$
$\pm$ 0.10, corresponding to a distance of 59 kpc, with the Wing being about 3 kpc closer
than the main body of the Small Cloud. It is, however, a source of some concern that the
SMC distance derived from RR Lyrae variables is $\sim$0.2 mag smaller than that obtained
from Cepheids.

The LMC is more massive than the SMC. Tidal interactions between these two objects
will therefore affect the Small Cloud more than the Large Cloud (M. Mateo 1999, private
communication). As a result, the SMC has probably suffered from greater tidal stress
than the LMC. This may account for some of the differences in the evolutionary histories
of these two galaxies. The neutral hydrogen gas in the SMC is seen to have a much
smoother distribution than does the gas in the LMC. The reasons for this may be that
the higher metallicity in the Large Cloud produces more cooling and hence a clumpier
morphology of the interstellar medium. This greater clumpiness will also contribute to
a higher specific rate of star formation in the LMC, which will in its turn result in the
formation of giant gas shells. The rate of supernova formation in the SMC is estimated
to be one per (350 $\pm$ 70) years, which is substantially lower than that in the LMC.

Unfortunately the number of SMC clusters with well-determined ages and metallicities
is presently too small to place significant constraints on the history of metal enrichment
in the Small Cloud. The existence of a single globular cluster (NGC 121) shows that star
and cluster formation in the Small Cloud must have started $\gtrsim$10 Gyr ago. It is presently
not understood why star clusters of all ages are more flattened in the SMC (and to a lesser
extent in the LMC) than they are in the Galaxy.

8

The elliptical galaxy M32 (= NGC 221)

8.1   Introduction

The compact E2 galaxy M32 is the closest companion to the Andromeda galaxy. The projected separation of these two objects on the sky is only 24′ (5.3 kpc). It was first suggested by Schwarzschild (1954) that the tides induced by M32 were responsible for the distortion of the spiral structure of M31 and the warping of its disk. Later Faber (1973) noted that CN and Mg absorption in M32 was stronger than might have been expected from its luminosity (i.e., it has the metallicity usually seen in significantly more luminous objects). She therefore proposed that M32 was initially a much more luminous galaxy had suffered severe tidal truncation by M31. The very compact object NGC 4486B, which is a companion to M87, appears to be another example of a similar type of galaxy that has suffered tidal truncation. The absence of globular clusters in M32 probably also results from its outer swarm of globulars being stripped off by tidal forces. [The innermost M32 globulars might have been sucked into its massive semi-stellar nucleus by tidal friction (Tremaine, Ostriker & Spitzer 1975).] It would be interesting (M. Mateo 1999, private communication) to study the distribution of globular clusters in NGC 4486B, which, like M32, is thought to have suffered severe tidal truncation.

Kormendy (1985) and Ziegler & Bender (1998) have pointed out that M32 is quite a unique object and that there are very few other ellipticals like it. Compared to other dwarf galaxies having similar values of M_V, its central surface brightness is four orders of magnitude higher, and its core radius r_c is three orders of magnitude smaller. No other known member of the family of elliptical galaxies is as faint as M32 (i.e., this object probably lies close to the lower cutoff of the luminosity function of true elliptical galaxies).

M32 is of particular interest because it is the only true elliptical galaxy in the Local Group. The luminosity profiles of such elliptical galaxies can be represented by an $R^{1/4}$ law, whereas those of most (dwarf) spheroidals are best represented by an exponential disk (de Vaucouleurs 1959b). The dE galaxies represent the true low-luminosity extension of the classical E galaxy family. For detailed discussions of the differences between dE and dSph galaxies the reader is referred to Wirth & Gallagher (1984) and to Kormendy (1985). The structure of the nuclear bulges of spirals resembles elliptical galaxies. One therefore cannot entirely rule out the possibility that M32 was once an early-type (Sa) galaxy that had its outer disk torn off during tidal interactions with M31. The rotation of the central region of M32 was discovered by Walker (1962) and has been studied in more detail by, among others, Carter & Jenkins (1993) and van der Marel, de Zeeuw & Rix (1997). These observations show that the (projected) rotational velocity of M32 drops from $\sim$45 km s^{-1} at $r = 4''$ (15 pc) to $\sim$30 km s^{-1} at $r = 30''$ (110 pc). The large velocity dispersion observed within the nucleus of this galaxy is diagnostic for the presence of a massive central black hole (Kormendy & Richstone 1995).

The luminosity profile of M32 has been studied by Kent (1987), who finds that it is well represented by a de Vaucouleurs profile with $r_e = 32''$ (120 kpc) over the range $15'' < r < 100''$. Kent finds that the main body of M32 has an ellipticity $\epsilon = 0.17$. For $r < 15''$ the surface brightness of M32 falls below that predicted by a de Vaucouleurs profile. According to Kent, M32 exhibits an excess of light, relative to a de Vaucouleurs profile, at $r > 100''$ (370 kpc). There is no evidence for an outer truncation of the M32 light distribution, although such measurements are difficult to make because this object appears projected on the outer spiral structure of the Andromeda galaxy. Possibly this outermost light is the vestige of an ancestral disk. Integrating over the brightness profile of M32 Kent finds V(total) $\approx$ 8.06. From its radial velocity and presumed tidal limit Cepa & Beckman (1988) derive a (very uncertain) orbit. They conclude that the orbital period of M32 around the Andromeda galaxy is $\sim$8 $\times$ 10^8 yr, that the orbit is retrograde, and that it has an orbital radius of $\sim$12 kpc. If the orbit of M32 is retrograde then it must have been captured by M31 (i.e., it was not formed within the proto-Andromeda galaxy). According to Cepa & Beckman, M32 is presently above the fundamental plane of the Andromeda galaxy, and about to pass through it. This conclusion is consistent with the observation that no dust lanes associated with the disk of M31 are seen in projection on the image of M32 (van Dokkum & Franx 1995). Emerson (1974) has placed an upper limit of 1.5 $\times$ 10^6 $M_\odot$ on the amount of H I in M32. More recently Sage, Welch & Mighell (1998) have used CO observations to place an upper limit of 5,100 $M_\odot$ on the amount of H$_2$ in M32. Sofue (1994) has suggested that the gas in M32 might have been stripped from this object by ram pressure during a previous passage of this galaxy through the fundamental plane of the Andromeda galaxy.

8.2   The nucleus of M32

When viewed through a small telescope M32 is seen to have a bright nucleus. This nucleus is listed as BD + 40° 147 in the *Bonner Durchmusterung* (Argelander 1903). Michard and Nieto (1991) find that the intensity profile of M32 starts to rise above the de Vaucouleurs profile inside $r = 3''$. For the core region they obtain colors of $B - V = 0.96$ and $U - B = 0.56$. Moreover, their data show no evidence for a color gradient over the central 10$''$ of this galaxy. The fact that the nucleus of M32 is not bluer than its envelope militates against the suggestion that captured globulars, which are expected to be metal poor and relatively blue, provided a significant contribution to the integrated light of the nucleus (T. R. Lauer, private communication). From observations with the *Hubble Space Telescope* Lauer et al. (1992) conclude that the central brightness profile may either be described as a single cusp with an $R^{-1/2}$ surface brightness distribution or as a core with a half-power radius $r_c = 0''.11$ (0.4 pc). More recent observations by Lauer et al. (1998) show that the image of M32 exhibits a central cusp with a power-law slope of $\approx$0.5 down to the resolution limit of the *Hubble Space Telescope*. The corresponding central space density is $>$10^7$M_\odot$ pc^{-3}. Lauer et al. also find that the $V - I$ and $U - V$ profiles of the core of M32 are essentially flat and that there is no sign of an inner disk, dust, or any type of substructure. Furthermore, Cole et al. (1998) detect no color gradient between the far-UV (1,600 Å) and the V-band (5,500 Å). They also conclude that population synthesis models require no intermediate-age stellar population to account for the observed 1,600–5,500 Å color of the central region of M32.

From spectroscopic observations, Bender, Kormendy and Dehnen (1996) derive a maximum rotational velocity of $V(\text{max}) = 55 \pm 3$ km s^{-1} and a central velocity dispersion of $\sigma \approx 92 \pm 5$ km s^{-1} for M32. By combining photometric and radial velocity observations of the central region of M32, Bender et al. find that the nucleus of M32 contains a black hole with a mass of $(3.0 \pm 0.5) \times 10^6\ M_\odot$. Using more sophisticated dynamical models, van der Marel et al. (1998) find a black hole mass of $(3.4 \pm 0.7) \times 10^6\ M_\odot$. From X-ray observations Loewenstein et al. (1998) conclude that the X radiation from M32 is dominated by a single, possibly variable, source that is offset to the east of the nucleus by $\sim 10''$ (37 pc). This source is probably a single super-Eddington X-ray binary. The nucleus itself is not an observable X-ray source, indicating that it is fuel starved. This is not surprising since M32 appears to be an almost dust-free, and hence presumably gas-free, galaxy.

Magorrian & Tremaine (1999) have pointed out that the tidal disruption of stars on nearly radial orbits will produce fuel that can feed the central black hole in M32. Much of this debris will get ejected, but a portion of it remains bound to the central black hole and emits a "flare" that will last for between a few months and a year. Such a flare is expected to emit mainly thermal radiation that peaks in the extreme UV or soft X-ray region. Such flares are expected to occur more frequently in the nuclei of low-luminosity galaxies like M32, in which the central density is high, than they will in high-luminosity galaxies, such as M87, which have low-density cores. Furthermore, the massive black holes in M87-like galaxies will swallow stars whole, whereas flare-producing gas can be liberated during encounters with less-massive black holes. Magorrian & Tremaine calculate that the flare frequency will be as high as $\sim 1 \times 10^{-4}$ yr^{-1} in M32, compared to a much lower rate of $\sim 10^{-8}$ yr^{-1} in M87. For black holes with masses of $10^6\ M_\odot$ and $10^7\ M_\odot$, they find V-band flare luminosities of $2 \times 10^7\ L_\odot$ and $8 \times 10^8\ L_\odot$, respectively. The corresponding luminosities are $M_V = -13.4$ and $M_V = -17.4$.

8.3   Stellar populations

The strength of the λ 4216 band of CN provides a powerful discriminant for the segregation of giants and dwarfs with late G-and early K-type spectra (Lindblad 1922, 1923; Lindblad & Stenquist 1934). Using this criterion it is easy to see from spectrum scans (van den Bergh & Henry 1962) that cyanogen giants provide the dominant contribution to the integrated light of M32. Figure 8.1 shows a comparison between the spectral energy distributions of M31 and M32. This plot shows that (1) M32 is bluer than M31 and that (2) the blue cyanogen break is weaker in M32 than it is in M31. These observations suggest that the dominant stellar population in M32 is metal deficient compared to that in M31 and (or) that the stars in M32 might, on average, be somewhat younger than those in the nuclear bulge of the Andromeda galaxy. There is conflicting evidence on the existence of a metallicity gradient in M32. Hardy et al. (1994) claim to find a well-defined gradient for $r \gtrsim 15''$ from integrated spectra, whereas Jones & Rose (1994) and Peletier (1993) find no such gradient out to $r = 30''$. From B, V, and I photometry of individual stars in a field, at $r = 100''$, Davidge & Jones

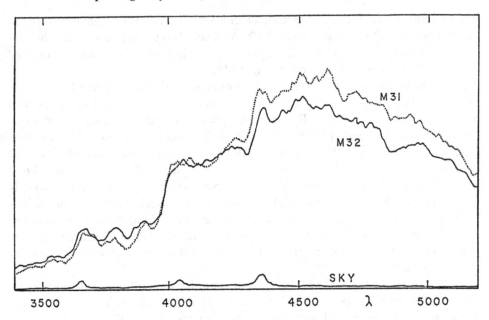

Fig. 8.1. Comparison between the spectral energy distributions of M31 and M32 (van den Bergh & Henry 1962). The figure shows that (1) the integrated light of M32 is bluer than that of the nuclear bulge of the Andromeda galaxy and (2) that the λ 4216 CN absorption in M32 is weaker than that in M31. These observations suggest that the dominant stellar population in M32 is more metal poor, or perhaps younger, than that in the central bulge of M31.

(1992) conclude that there is a real difference in metallicity between the central and outer regions of M32. Their finding that the brightest stars in the center of M32 are half a magnitude brighter than they are at $r = 100''$ is, of course, susceptible to the effects of image crowding (Grillmair et al. 1996, 1997; Sodemann & Thomsen 1998). From their observations Davidge & Jones find that the M32 giant branch is quite broad, indicating the presence of stars as metal poor as [Fe/H] ~ -2.2. The conclusion that M32 has a broad red giant branch is strongly confirmed by the *Hubble Space Telescope* observations of Grillmair et al. (1996). However, these authors note that the observed metallicity distribution of M32 stars needed to account for the observed width of the giant branch is considerably narrower than would be predicted by a "closed-box" model of chemical evolution. Grillmair et al. conclude that their (luminosity-weighted) data are best represented by a model in which $\langle [Fe/H] \rangle \approx -0.25$, and in which the average stellar age is 8.5 Gyr. On the basis of near-infrared (JHK) photometry Freedman (1992) has argued that some very luminous red asymptotic giant branch stars in M32 were produced during a star forming episode that took place less than about 5 Gyr ago. However, Renzini (1998) has used the luminosity distribution per pixel in images of M32 to argue that no solid evidence exists yet for an intermediate-age stellar population in M32. This conclusion appears to be slightly at variance with earlier population modeling by Barbuy, de Freitas Pacheco & Borges (1994). These authors had concluded that available data were well represented by a history with continuous star formation from 15 Gyr to 11 Gyr ago, when a strong wind blew out the remaining gas. It is of interest to note that the references cited above still exhibit a wide range of opinions on the true nature and evolutionary history of M32. If we do not even understand the nearest elliptical, then some surprises may still be in store for us about the evolution of distant ellipticals!

Using *Hubble Space Telescope* imaging in the ultraviolet, Brown et al. (1998) were able to detect 183 stars that appear to be evolving along post–asymptotic giant branch and asymptotic giant branch manqué paths. A total of 27 planetary nebulae have been found in M32 (Ciardullo et al. 1989). Owing to the high central surface brightness of this galaxy the search for planetaries is only considered complete for $r > 7''$. The number of planetaries is not large enough to use the planetary nebula luminosity function to obtain an independent distance estimate for M32. However, Ciardullo & Jacoby (1992) have compared the luminosity function of M31 with the combined luminosity functions of its companions NGC 185, NGC 205, and M32. These authors find that the break in this combined luminosity function is $0.28^{+0.11}_{-0.17}$ mag fainter than that for the luminosity function of M31 planetaries. This result shows that these luminosity functions are only marginally different, even though the parent galaxies with which these planetary nebulae are associated have significantly different metallicities.

Nine planetary nebulae in M32 have been studied spectroscopically by Richer, McCall & Stasińska (1998). Their O/H observations of these planetaries, and those in other Local Group galaxies, are compared in Table 8.1. These data show that the O/H distributions are fairly similar in M31 and in the bulge of the Galaxy. However, in both of these systems they are, on average, higher than they are in the Magellanic Clouds. The O/H ratio in M32 appears to be intermediate between that in the Clouds and in M31.

Nolthenius & Ford (1986) have observed the radial velocities of 15 of the planetary nebulae in M32 and obtain a velocity dispersion $\sigma = 42$ km s^{-1}, from which $M = (1.1 \pm 0.3) \times 10^9 \, M_\odot$. With this value the mass of the central black hole in M32 is 0.3% of the total mass of this galaxy. This value is close to the logarithmic mean value of 0.6% of the total mass that Merritt (1998) obtains for eight black holes in early-type (E, S0, Sa) galaxies. This result suggests that tidal forces cannot have stripped a significant fraction of the main body of M32, but it places no constraints on the amount of disk material that might have been removed from it.

Table 8.1. *Abundances in planetary nebulae*[a]

Galaxy	M_V	[O/H]
M31 (bulge)	−21.2	−0.19 ± 0.21
Galaxy (bulge)	−20:	−0.30 ± 0.27
M32	−16.5	−0.58 ± 0.18
NGC 205	−16.4	−0.40
NGC 185	−15.6	−0.78
Sagittarius	−13.8:	−0.93
Fornax	−13.1	−0.95

[a] Most data are from Richer et al. (1998).

Table 8.2. *Data on M32 (= NGC 221)*

$\alpha(2000) = 00^h\ 42^m\ 41\overset{s}{.}9$		$\delta(2000) = +40°51'55''$
$l = 121\overset{\circ}{.}15$		$b = -21\overset{\circ}{.}98$
$V_r = -205 \pm 8\,\mathrm{km\,s}^{-1}$ (1)		
$V = 8.06$ (2)	$B - V = 0.95$ (1)	$U - B = 0.48$ (1)
$(m - M)_0 = 24.4^a$	$E(B - V) = 0.062^a$	$A_V = 0.15\,\mathrm{mag}^a$
$M_v = -16.5$	$D = 760\,\mathrm{kpc}$	$D_{LG} = 0.30\,\mathrm{Mpc}$
Type = E2		size = $8' \times 12'$ (1.8 kpc
		$\times 2.7\,\mathrm{kpc}$) (3)
$M = (1.1 \pm 0.3) \times 10^9\,M_\odot$ (4)	$M_{\mathrm{H\,I}} < 1.5 \times 10^6\,M_\odot$ (8)	$M_{\mathrm{H}_2} < 5.1 \times 10^3\,M_\odot$ (9)
No. globular clusters = 0	No. planetary nebulae ≥ 27 (5)	
M(nucleus) = (3.4 ± 0.7)		$[\mathrm{Fe/H}] \approx -0.25$ (7)
$\times 10^7 M_\odot$ (6)		

^{a} Assumed equal to that for M31.
(1) de Vaucouleurs et al. (1991).
(2) Kent (1987).
(3) Holmberg (1958).
(4) Nolthenius & Ford (1986).
(5) Ciardullo et al. (1989).
(6) van der Marel et al. (1998).
(7) Grillmair et al. (1997).
(8) Emerson (1974).
(9) Sage et al. (1998).

8.4   Summary

The E2 galaxy M32 (= NGC 221) is an unusual object near the lower edge of the luminosity function for normal ellipticals. The fact that its metallicity is higher than that of other dwarf galaxies of comparable present luminosity strongly suggests that it has been severely truncated by tidal interactions with the much more massive Andromeda galaxy. This suspicion is strengthened by the observation that the tidal arms of M31 are severely warped by the tides of M32. It is, however, puzzling that the photometry by Kent (1987) shows no evidence for tidal truncation of M32, even though the metallicity–luminosity relation strongly suggests significant mass loss. Perhaps this discrepancy is due to the great difficulty of measuring the outer parts of M32, which are superposed on the outer spiral arms of M31. The fact that the central surface brightness of M32 is four orders of magnitude higher than that of typical spheroidals of comparable total luminosity suggests that dissipative processes played a mayor role in the early history of this galaxy.

From its present luminosity M32 would be expected to have 10–20 globular clusters. If its original luminosity was higher than it is presently then M32 might originally have contained >20 globular clusters. None are presently seen. This suggests that the majority of M32 globulars were tidally stripped off, while some of the innermost globulars might have been drawn into the M32 nucleus by tidal friction. Since M32 is metal deficient compared to M31 one would expect that a significant fraction of the most metal-poor clusters in the M31 globular cluster system were originally formed in the outer regions of M32. Such clusters might be expected to have smaller radii than those that were born in the outer halo of the Andromeda galaxy.

The history of star formation in M32 remains somewhat controversial. It was once thought that this galaxy contained a significant intermediate-age population. However, most recent authors have argued that there is little evidence for any star formation during the last ~11 Gyr. The possibility that M32 once contained a younger outer disk that was subsequently stripped off by M31 cannot yet be excluded. It is very humbling to realize that the structure, stellar content, and evolutionary history of the nearest elliptical galaxy still remain so uncertain and controversial.

A summary of data on M32 is given in Table 8.2.

9

The irregular dwarf galaxy NGC 6822

9.1 Introduction

NGC 6822 is a barred dwarf galaxy of type Ir IV–V that is located at relatively low Galactic latitude. The first detailed study of this galaxy was by Hubble (1925c) who discovered 11 Cepheid variables. Hubble wrote "Cepheid variables, diffuse nebulae, dimensions, density, and distribution of stellar luminosities agree in defining the system as a curiously faithful copy of the [C]louds, but removed to a vastly greater distance. N.G.C. 6822 lies far outside the limits of the [G]alactic system, even as outlined by the globular clusters, and hence may serve as a stepping-stone for speculation concerning the habitants of space beyond." From these observations Hubble (1936, p. 124) was able to establish that NGC 6822 is a relatively nearby member of the Local Group. Hubble also discovered nine nonstellar objects in this galaxy. Some of these are young star clusters embedded in nebulosity, whereas others appear to be old star clusters. Hubble's cluster VII has been studied by Cohen & Blakeslee (1998), who find that it is a typical metal-poor globular cluster with a metallicity [Fe/H] $= -1.95 \pm 0.15$ and an age of $\sim$11 Gyr. It would be interesting to determine the morphology of the horizontal branch of this cluster. Hodge (1977) gives $V = 16.28$ for the cluster Hubble VII, from which $M_V \approx -8.0$. Cluster Hubble VI is a much younger object with an age of $\sim$2 Gyr and a metallicity of [Fe/H] ≈ -1.0, a value similar to the present abundance of the interstellar medium in this galaxy. Perhaps the most surprising discovery (Roberts 1972) made since Hubble's early work is that the relatively small optical core (bar) of NGC 6822, which has dimensions of $6' \times 11'$ (0.9 kpc $\times$ 1.6 kpc), is embedded in an enormous flattened neutral hydrogen envelope. Figure 9.1 shows that this envelope has dimensions of $42' \times 89'$ (6.1 kpc $\times$ 12.9 kpc). Observations by Gottesman & Weliachew (1977) show that the center of the radio rotation curve of NGC 6822 lies within a fraction of a kiloparsec from the optical centroid of this galaxy. This suggests that most star formation in NGC 6822 took place near the bottom of its potential well.

9.2 Distance and reddening of NGC 6822

Hubble (1925c) published his observations of Cepheids in NGC 6822 before he published his detailed observations of the Cepheids in M33 (Hubble 1926) and in M31 (Hubble 1929). His observations of NGC 6822 therefore have an important place in the history of modern astronomy. Hubble (1925c) was well aware of the significance of his new results and wrote "Of especial importance is the conclusion that the Cepheid

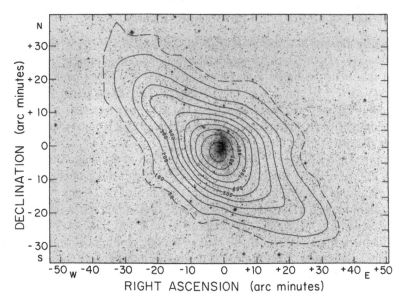

Fig. 9.1. Photograph of NGC 6822 with hydrogen isophotes superposed. The figure shows that this galaxy is embedded in a huge flattened gaseous envelope. Reproduced with permission from Roberts (1972).

criterion functions normally at this great distance. The Cepheid variables have recently been found in the two largest spiral nebulae, and the period–luminosity relation places them at distances even more remote than N.G.C. 6822. This criterion seems to offer the means of exploring extra-galactic space; N.G.C. 6822 furnishes a critical test of its value for so ambitious an undertaking."

Observations of the distance to, and reddening of, NGC 6822 have been reviewed by Gallart, Aparicio & Vílchez (1996). They conclude that $E(B - V) = 0.24 \pm 0.03$. A rather more complex situation emerges from the UBV photometry of OB stars by Massey et al. (1995). These authors find that the smallest reddening values $E(B - V) \approx 0.26$ occur at the eastern and western ends of the galaxy, whereas the higher value $E(B - V) \approx 0.45$ appears to be appropriate near the center of NGC 6822. The higher values found by Massey et al. from photometry of OB stars may, at least in part, result from these young objects still being associated with the dusty gas clouds from which they formed only $\sim 10^7$ years ago. In the subsequent discussion a global value $E(B - V) = 0.25$, from which $A_V \sim 0.8$ mag, will be adopted.

The most recent distance determinations to NGC 6822 are by Lee, Freedman & Madore (1993b) and by Gallart et al. (1996). Lee et al. use the magnitude level of the tip of the giant branch of old stars to derive $(m - M)_0 = 23.46 \pm 0.10$. From infrared observations of three Cepheids Visvanathan (1989) found $(m - M)_0 \sim 23.2$. More recently Gallart et al. have used BVRI photometry of eight Cepheids to derive a distance modulus of $(m - M)_0 = 23.49 \pm 0.08$. The weighted mean of these values yields $(m - M)_0 = 23.48 \pm 0.06$, corresponding to a distance of 497 kpc. In the subsequent discussion a distance of 500 kpc will be adopted.

9.3   Stellar populations and star forming history

9.3.1   *Young objects*

Hodge et al. (1991b) have studied the distribution of young stars in NGC 6822. They find that both the bright young blue stars and red supergiants have a clumpy distribution with typical clump sizes of 150–200 pc. The fainter stars are observed to be distributed more widely and more smoothly. These authors have also used their data to derive the luminosity distribution of stars in NGC 6822. They find a total of 363 stars in this galaxy that are brighter than $M_V = -6.5$.

Images of NGC 6822 (see, for example, panel 330 of Sandage & Bedke 1994, or Volders & Högbom 1961) show that this galaxy is not forming stars as vigorously as M33 and the Large Cloud. This conclusion is confirmed by Hodge (1993), who has estimated the present star formation rate in NGC 6822 to be 0.021 $M_\odot$ yr^{-1} from its global Hα emission. The brightest H II regions in NGC 6822 are those associated with clusters Hubble I, II, and III, which are located near the northern edge of NGC 6822. A catalog and identification charts for 157 H II regions in this galaxy are given by Hodge, Kennicutt & Lee (1988).

According to Massey et al. (1995) the most luminous stars in NGC 6822 have $M_{\rm bol} = -10$, from which a mass of $\sim 60 M_\odot$ is inferred. These authors also find that lines in the spectra of stars in NGC 6822 are considerably weaker than those in comparable stars in the LMC and SMC. This result is in line with expectations because NGC 6822 ($M_V = -16.0$) is significantly less luminous than the SMC ($M_V = -17.1$). Marconi et al. (1995) conclude that star formation in NGC 6822 has been proceeding more or less continuously (but at a slow rate) for the last ~ 1 Gyr, with a mass spectrum that has a slope similar to that of the Salpeter (1955) function. However, Gallart et al. (1996) suggest that there may have been an enhancement in the rate of star formation during the past 100–200 Myr.

Hodge (1980) was able to identify 31 star clusters in NGC 6822. Hodge (1977) found that this galaxy also contains 16 OB associations, from which he obtained a formation rate of 0.7 associations per million years. Using objective computer-based techniques Wilson (1991) was able to identify 13 associations with a mean mass $\langle M \rangle = 500\ M_\odot$ and a mean age $\langle T \rangle = 8 \times 10^6$ yr.

From a UBV survey of M31, M33, and NGC 6822 Massey (1998b) finds a clear progression with color, metallicity, and luminosity in the relative numbers of the most luminous red supergiants. This correlation between metallicity and spectral type (Spinrad, Taylor & van den Bergh 1969) occurs, no doubt, because the classification of late M-type stars depends on the strength of the TiO bands, which is very sensitive to metal abundance. The fact that both M_V and $M_{\rm bol}$ of red supergiants correlate with the metallicities of their parent galaxies indicates that considerable caution should be exercised when the luminosities of the brightest M-type supergiants are used as distance indicators. Massey suggests that the observed dependence of red supergiant numbers on metallicity is due to the fact that high metallicities will result in high mass-loss rates. Consequently, a metal rich star of a given luminosity will spend a large fraction of its helium burning lifetime as a Wolf–Rayet star, rather than as a red giant. According to Massey stars more luminous that $M_{\rm bol} = -7.5$ will become WR stars in M31 [$12 + \log({\rm O/H}) = 9.0$]. The corresponding limits are $M_{\rm bol} = -8.5$ in M33 [$12 + \log({\rm O/H}) = 8.4$] and $M_{\rm bol} = -9.0$ in NGC 6822 [$12 + \log({\rm O/H}) = 8.2$].

9.3.2 Old and intermediate-age objects

A detailed five-color CCD study of the stellar population in NGC 6822 has been published by Gallart et al. (1996). From their photometry these authors conclude that NGC 6822 probably started forming stars from metal poor gas 12–15 Gyr ago. (The globular cluster Hubble VII was a member of this earliest population.) During the past few billion years the rate of star formation in NGC 6822 was either constant or declined slightly, with a possible slight increase of star forming activity 400 Myr ago, and a further enhancement 100–200 Myr ago. At its present rate of star formation the total number of stars now observed in NGC 6822 could have been formed in $\sim$16 Gyr (Hodge 1993).

9.4   Interstellar matter

A catalog with identification charts for 157 H II regions in NGC 6822 has been published by Hodge, Kennicutt & Lee (1988). The morphology of these H II regions was studied by Collier, Hodge & Kennicutt (1995). These authors conclude that the most luminous H II regions are also the largest. No other significant correlations were found between physical parameters such as mass, gas density and distribution, exciting star(s), etc. Patel & Wilson (1995) detected four distinct components (bright, halo, diffuse, and field) to the Hα emission of NGC 6822. These authors noted that about half of all OB stars in NGC 6822 are located in the field, while only about a quarter of them are found in bright and halo regions. They concluded that the OB stars in NGC 6822 spend about three quarters of their lifetimes outside of classical H II regions. It should, however, be emphasized that the fraction of their lifetime that O stars spend inside H II regions depends critically on spectral type. Van den Bergh (1988c) finds that 75% of Galactic O5 stars are located in H II regions, compared to only 29% of O9.5 stars. There are two reasons for this effect: (1) Early O-type stars are able to ionize larger volumes of space and (2) old H II regions around some late O-type stars may expand themselves out of existence.

A single supernova remnant in NGC 6822 has been found by D'Odorico, Dopita & Benvenuti (1980) from its high [S II]/Hα emission line ratio. From inspection of images of NGC 6822 Hodge (1977) has been able to identify 11 discrete dust clouds. In the mean these clouds have smaller sites and exhibit lower opacities than do the dark clouds in the LMC and SMC. This difference is, no doubt, due to the low metallicity of the interstellar gas in NGC 6822, which was first noted by Peimbert & Spinrad (1970). Skillman, Kennicutt & Hodge (1989) find an oxygen abundance of $12 + \log(O/H) = 8.14$ for emission nebulae in NGC 6822. A compilation of values for the nitrogen-to-oxygen abundance ratios in low-luminosity Local Group galaxies by Mateo (1998), which is plotted in Figure 9.2, appears to show that [N/O] is low in NGC 6822. Taken at face value this might indicate (Kobulnicky & Skillman 1998) that nitrogen, which is mainly provided by intermediate-mass stars, has not yet had a chance to build up after a recent oxygen producing starburst.

The low heavy element abundance in NGC 6822 is, no doubt, responsible for the small optical depth found for the molecular gas in this galaxy. The optical thinness of CO emission from NGC 6822 allows one to calculate the mass of molecular gas in this galaxy directly. Petitpas & Wilson (1998) find molecular masses of $(1.2-2.1) \times 10^5 \, M_\odot$. This value agrees with the virial masses obtained for these clouds to within a factor of $\sim$3. Three molecular clouds in NGC 6822 have been observed in CO by Wilson (1994).

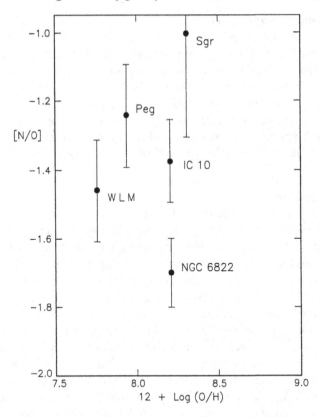

Fig. 9.2. Logarithmic nitrogen-to-oxygen ratio versus oxygen abundance for faint Local Group members from a compilation by Mateo (1998). Note the low value of [N/O] for NGC 6822.

She finds that these clouds are similar to clouds in the SMC but are somewhat less massive than those in Local Group giant spirals. Petitpas & Wilson (1998) also note that their molecular cloud No. 2 is located within $1''$ of Hubble V, which is the most luminous H II region in NGC 6822. The Hα luminosity of Hubble V is an order of magnitude lower than that of NGC 604, which is the brightest H II region in M33.

Killen and Dufour (1982) have made a search for emission objects in NGC 6822. They found 31 diffuse objects and 36 apparently stellar ones. Of these "stellar" objects some are likely to be very compact H II regions, whereas the others are probably emission-line stars and planetary nebulae. Spectra of these planetaries (Dufour & Talent 1980, Richer & McCall 1995) yield $12 + \log(\text{O/H}) = 8.1 \pm 0.1$. This value is close to that obtained for the H II regions in NGC 6822 (Skillman, Kennicutt & Hodge 1989).

9.5   Radio radiation and X rays

From a 21-cm study Volders & Högbom (1961) found NGC 6822 to have a total hydrogen mass of $1.5 \times 10^8 M_\odot$. The very extended $40' \times 80'$ (5.8 kpc $\times$ 11.6 kpc) distribution of H I (Roberts 1972) is shown in Figure 9.1. From the rotation curve of

Table 9.1. *Data on NGC 6822*

$\alpha(2000) = 19^{\text{h}}44^{\text{m}}56\overset{\text{s}}{.}0$		$\delta(2000) = -14°48'\,06''$ (1)
$\ell = 25\overset{\circ}{.}34$		$b = -18\overset{\circ}{.}39$
$V_{\text{r}} = -56 \pm 4\ \text{km s}^{-1}$ (2)		
$V = 8.52 \pm 0.08$ (2,3)	$B - V = 0.79 \pm 0.06$ (3)	$U - B = 0.04 \pm 0.20$ (3)
$(m - M)_{\text{o}} = 23.48 \pm 0.06$ (4)	$E(B - V) = 0.25$ (4)	$A_V = 0.8\ \text{mag}$ (4)
$M_V = -15.96$	$D = 500\ \text{kpc}$ (4)	$D_{\text{LG}} = 0.67\ \text{Mpc}$
Type = Ir IV–V Disk scale-	Size = $20' \times 20'$(2.9 kpc	
length $\sim120''$ (290 pc) (5)	$\times$ 2.9 kpc) (6)	
$M = 1.9 \times 10^9\,M_\odot$ (5)	$M_{\text{HI}} = 1.5 \times 10^8\,M_\odot$ (7)	$M_{\text{H}_2} = (1.2 - 2.1) \times 10^5\,M_\odot$ (8)
No. globulars ≥ 1	No. planetaries >2	
	$12 + \log(\text{O/H}) = 8.14$ (9)	

(1) Gallouet et al. (1975).
(2) de Vaucouleurs et al. (1991).
(3) Hodge (1977).
(4) See Section 9.2.
(5) Hodge et al. (1991b).
(6) Holmberg (1958).
(7) Volders & Högbom (1961).
(8) Petitpas & Wilson (1998).
(9) Skillman et al. (1989).

this galaxy Volders & Högbom derived a total mass of $1.5 \times 10^9\,M_\odot$, from which $M(\text{H I})/M(\text{total}) = 0.1$. Higher resolution $(2\overset{.}{.}3)$ mapping of the central part of NGC 6822 has been carried out by Gottesman & Weliachew (1977). From a comparison of their H I distribution, with counts of young blue stars, they concluded that the rate of star formation in this galaxy is proportional to the surface density of H I to the power 1.5. The 21-cm centroid of NGC 6822 determined by Gottesman & Weliachew (1977) lies $61'' \pm 20''$ (150 pc) south of the optical centroid determined by Gallouet, Heidmann & Dampierre (1975).

A 21-cm continuum source is found to be associated with the supernova remnant in the H II region Hubble V. Two other nonthermal sources seen in this field may *not* be physically associated with NGC 6822. Eskridge & White (1997) have discussed an apparently variable X-ray source. They note that this source appears to be associated with Hodge's H II region Ho 12, which was listed as a supernova remnant by D'Odorico, Dopita & Benvenuti (1980).

9.6   Summary and conclusions

NGC 6822 is a barred irregular dwarf of DDO type Ir IV–V that has $M_V = -16.0$. Study of this object is rendered difficult by the fact that it is located relatively close to the Galactic plane, in a region with a moderately high density of foreground stars and absorbing dust. A detailed discussion of the variable stars, star clusters, etc. in this galaxy was published by Hubble (1925c). This study had a significant impact on ideas about the nature of galaxies because it was published before similar investigations of M33 (Hubble 1926) and M31 (Hubble 1929) appeared in print. Perhaps the most surprising observation (Volders & Högbom 1961) is that NGC 6822, which appears quite small on optical images, is embedded in an enormous neutral hydrogen envelope. The radio rotation center and the centroid of the optical image of NGC 6822 are found to coincide to within a small fraction of a kiloparsec. This suggests that star formation in this object mainly took place near the bottom of its potential well. The high luminosity of the brightest M-type red supergiants in NGC 6822 can probably be attributed to the fact that mass-loss rates are lower in its metal-poor red supergiants than they are in the more metal-rich supergiants in M31 and M33. The presence of a metal-poor globular cluster and of old field stars shows that star formation started early in NGC 6822. Since these early days stars have continued to form at a low rate that has either remained constant or declined slowly with time.

A summary of available data on NGC 6822 is given in Table 9.1.

10

The starburst galaxy IC 10

10.1 Introduction

Mayall (1935) was the first person to draw attention to this highly obscured galaxy, which is located at $b = -3°3$. He pointed out that "the original negatives show, almost conclusively, that IC 10 is an extra-galactic object." Hubble (1936, p. 147), who called it "one of the most curious objects in the sky," first raised the possibility that IC 10 might be a member of the Local Group. This suspicion was subsequently confirmed by the Cepheid distances of Saha et al. (1996) and Wilson et al. (1996). Photographs in blue and red light of this galaxy have been published by Roberts (1962) and by de Vaucouleurs & Ables (1965). From an inspection of these images the DDO classification of this object appears to be Ir V:, the luminosity class is uncertain because of the large Galactic foreground obscuration. Roberts (1962) notes that the optical and 21-cm radio centers of IC 10 agree within their errors.

10.2 Distance and reddening

The integrated colors of IC 10 have been measured by de Vaucouleurs & Ables (1965). From these colors they estimated the reddening of this galaxy to be $\langle E(B-V) \rangle = 0.87$. Using the colors of four Cepheids, Wilson et al. (1996) find a range in reddening of $0.6 \lesssim E(B-V) \lesssim 1.1$ mag. From their data they adopt a mean value $E(B-V) \approx 0.80$. Using this value, Wilson et al. obtain a true distance modulus of 24.57 ± 0.32, corresponding to a distance of 820 ± 80 kpc. A larger reddening value $E(B-V) = 0.98$, and hence a smaller distance $D = 580$ kpc, was derived by Tikhonov (1999) from R versus $R - I$ observations of the red giant branch of IC 10. Multiwavelength observations of Cepheids in IC 10 (Sakai, Madore, & Freedman 1998) yield a reddening $E(B-V) = 1.16 \pm 0.08$ and a true Population I distance modulus of $(m-M)_0 = 24.10 \pm 0.19$, corresponding to a distance of 660 kpc. Sakai et al. find a significantly smaller distance to IC 10, if they adopt the Population I reddening for the stars at the tip of the (Population II) red giant branch. These authors therefore assume that the IC 10 halo stars suffer only a foreground reddening of $E(B-V) \simeq 0.85$, whereas the Cepheids (which may still be associated with the gas and dust clouds from which they formed) have a mean reddening that is ~ 0.3 mag larger. These values were adopted in the compilation given in Table 10.1. It should, however, be emphasized that detailed studies of the patchy interstellar absorption and line-of-sight effects will be required to establish a definitive value for the distance to IC 10.

Table 10.1. *Data on IC 10*

$\alpha(2000) = 0^h\,20^m\,24^s$		$\delta(2000) = +59°17'30''$ (1)
$\ell = 118°97$		$b = -03°34$
$V_r = -344 \pm 3$ km s^{-1} (1)		$i = 60° \pm 10°$ (2)
$V = 10.4 \pm 0.2$ (2)	$B - V = 1.30 \pm 0.03$ (2)	$U - B = 0.25 \pm 0.03$ (2)
$(m - M)_o = 24.10 \pm 0.19$ (3)	$E(B - V) \simeq 0.85$ (3)	$A_V \approx 2.6$ mag
$M_V = -16.3$	$D = 660 \pm 60$ kpc (3)	$D_{LG} = 0.27$ Mpc
Type = Ir IV:	Optical size = 5.5×7.0 (1.1 kpc	
	$\times$ 1.3 kpc) (4)	$r_e = 2.0 \pm 0.2$ (0.5 kpc) (2)
$M \sim 6 \times 10^8\,M_\odot$ (3,6)	$M_{HI} = 1.2 \times 10^8\,M_\odot$ (6)	$M_{H_2} = 1.0 \times 10^7\,M_\odot$ (3,8)
		$L_B/L_{H\alpha} = 27$ (7)
	$12 + \log(O/H) = 8.2$ (5)	

(1) de Vaucouleurs et al. (1991).
(2) de Vaucouleurs & Ables (1965).
(3) Sakai et al. (1998).
(4) Massey & Armandroff (1995).
(5) Skillman et al. (1989).
(6) Huchtmeier (1979).
(7) Hunter et al. (1993).
(8) Wilson et al. (1996).

The angular separation of IC 10 and M31 on the sky is 18°4, which is only slightly larger than the 14°8 distance between M31 and M33. This observation suggests that IC 10 might be an outlying member (see Figure 2.1) of the M31 subgroup of the Local Group. The observed radial velocities of -344 km s^{-1} for IC 10 and -301 km s^{-1} for M31 are consistent with this suggestion. However, the new distance determination by Saki, Madore & Freedman (1999) militates against this idea.

10.3   Stellar population

According to Massey & Armandroff (1995), IC 10 contains 5.1 WR stars kpc^{-2}, compared to only 0.9 WR stars kpc^{-2} in the SMC and 2.0 WR stars kpc^{-2} in the LMC. In fact, the global star formation rate per kpc^{-2} in IC 10 is comparable to that found in active star forming regions in the disk of M33. This suggests that IC 10 is a galaxy that is presently undergoing a burst of star formation. If this conclusion is correct then IC 10 is the nearest starburst galaxy[1] and the only such object in the Local Group.

IC 10 is a rather small galaxy. Its effective radius (i.e., the radius that encloses half of its luminosity) is $r_e = 0.5$ kpc (de Vaucouleurs & Ables 1965). This value is only half of that of the Small Magellanic Cloud for which $r_e = 1.0$ kpc, even though IC 10 and the SMC have comparable present luminosities. The higher oxygen abundance of IC 10 [$12 + \log(O/H) = 8.20$] compared to that for the SMC[$12 + \log(O/H) = 7.98$] (Skillman, Kennicutt & Hodge 1989) suggests that IC 10 may have had a slightly

---

[1] The second nearest starburst galaxy is NGC 1569 (Israel 1988), which is located at a distance of 2.2 Mpc. Hunter, Howley & Gallagher (1993) make the interesting comment that "This galaxy [IC 10] resembles NGC 1569 in being surrounded by an amazingly complex sea of filaments and diffuse gas that extends to at least 1 kpc from the galaxy's center in the plane of the sky." IC 10 is X-ray dim, whereas NGC 1569 is bright in X rays.

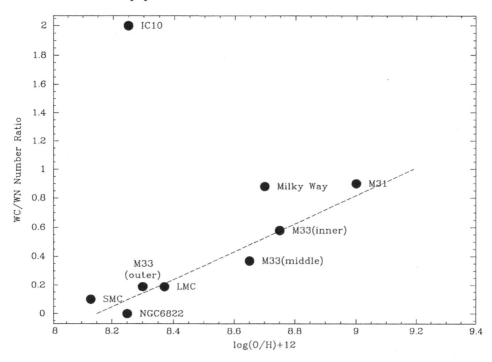

Fig. 10.1. Dependence on oxygen abundance of the ratio of the number of WC stars to the number of WN stars. The figure shows that the WC/WN ratio increases with increasing metallicity. The anomalously high WC/WN ratio in IC 10 is attributed to an excess of massive stars in starbursts. (Reproduced with the kind permission of Phil Massey.)

higher past mean rate of star formation than the Small Cloud. With the distance to IC 10 adopted in Section 10.2, the observed Hα flux from IC 10 (Thronson et al. 1990) yields a total rate of star formation of (0.04–0.08) $M_\odot$ yr^{-1}. At this rate of consumption the gas supply in IC 10 would last only a few billion years. The presence of numerous H II regions (Hodge & Lee 1990) attests to the vigor of massive star formation in this galaxy at the present time.

Stars of type WC (which show the products of He burning) are more highly evolved than WN stars (which reveal products of H burning via the CNO cycle). Mass loss from the envelopes of evolved stars will proceed most rapidly in those giants that have the highest metallicities. *Other factors being equal*, one would therefore expect the ratio of WC stars to WN stars to be highest in metal rich galaxies. This expectation is confirmed by the WC/WN versus log(O/H) plot shown in Figure 10.1 (Massey & Johnson 1998). However, the figure shows that IC 10 exhibits a WC/WN ratio that is an order of magnitude larger than what one might have expected for its metallicity. Massey & Armandroff (1995) propose that this high WC/WN ratio in IC 10 is due to the formation of an excess of massive stars (which have high luminosity, and hence a high mass-loss rate) during a starburst. However, the observation that the slope of the mass spectrum of star formation in the supercluster R 136 in the LMC appears to have the same slope as that in the solar neighborhood (Massey & Hunter 1998) suggests that this excess of massive stars is probably not due to a change in the mass spectrum of star formation during violent star forming events. A more likely explanation is that the excess of WC stars is an age effect produced by the rapid variation in the rate of star formation during a starburst. Such an explanation is only viable if a significant fraction of all of the WC stars in IC 10 were formed almost simultaneously in a relatively small volume of space. This does indeed appear to be the case. More than half of all known WC stars in IC 10 are located in a region with a diameter $<1\!\!\stackrel{.}{}0$ (<240 pc). These objects might therefore have been formed almost simultaneously in a single supershell.

10.4   Interstellar matter

In the case of IC 10 the term interstellar material is somewhat inappropriate because the hydrogen gas associated with this object extends far beyond its optical limits. From 21-cm line observations with the 100-m Effelsberg radio telescope Huchtmeier (1979) finds that IC 10 is embedded in a centrally concentrated gaseous envelope with dimensions of $62' \times 80'$, which is an order of magnitude larger than the optical galaxy, whose dimensions are $5\!\!\stackrel{.}{}5 \times 7\!\!\stackrel{.}{}0$ (Massey & Armandroff (1995). IC 10 therefore belongs to the same class as the Local Group irregular galaxies NGC 6822 and IC 1613, which are also embedded in enormous hydrogen envelopes. Shostak & Skillman (1993) found that the outer hydrogen envelope of IC 10 has a well-defined velocity gradient, with velocities that range from ~ -300 km s^{-1} in the south to ~ -360 km s^{-1} in the north. However, the core of this galaxy shows a velocity gradient with the same position angle, but with the opposite sign (i.e., IC 10 has a counter-rotating core). Assuming an inclination of $45°$ and a distance of 660 kpc yields a rotational mass of $\sim 5 \times 10^8 \, M_\odot$ for the central disk in IC 10. Wilcots & Miller (1998) find that the gas in this galaxy can be modeled as a regularly rotating disk that is embedded within an extended and complex distribution of hydrogen. Within this disk, the distribution of gas is dominated by holes and shells, which continue to be shaped by the stellar winds produced by numerous luminous WR and O-type stars. In the disk the hydrogen density is high ($\sim 5 \times 10^{21}$ cm^{-2}), whereas it is much lower ($\sim 1 \times 10^{20}$ cm^{-2}) in the streamers, spurs, and plumes that surround it.

Madden et al. (1997) have used the *Kuiper Airborne Observatory* to measure the 158-μm line of [C II] in IC 10. This radiation is observed to cover an area of about 2.0 kpc $\times$ 2.6 kpc. Madden et al. find that standard H I clouds and H II regions can each account for only $\sim 10\%$ of the observed emission. These authors therefore propose that 80% of the observed [C II] emission is radiated by dense photodissociation regions. These regions are speculated to contain a column density of H_2 that is five times that present in H I. In this model, small CO cores are embedded in relatively large [C II]-emitting envelopes, where H_2 is self-shielded. The mass of the molecular gas in these envelopes may exceed that in their cores by two orders of magnitude. Two of the molecular clouds in IC 10 have been studied in detail by Petitpas & Wilson (1998).

The radio continuum radiation from IC 10 has been measured at three wavelengths by Yang & Skillman (1993). They find that most of the point sources of continuum radiation can be identified with H II regions, which have been catalogued by Hodge & Lee (1990), or with nonthermal background sources. Furthermore, a shell centered at $\alpha = 00^h \, 20^m \, 28^s$, $\delta = +59° \, 16' \, 49''$ (equinox 2000), which has a diameter of $\approx 48''$ (150 pc), is, no doubt, physically associated with IC 10. Yang & Skillman tentatively identify the center of this super bubble with two WR stars. They speculate that the bubble was formed by multiple supernova explosions. Brandt et al. (1997) have identified a variable X-ray source (which

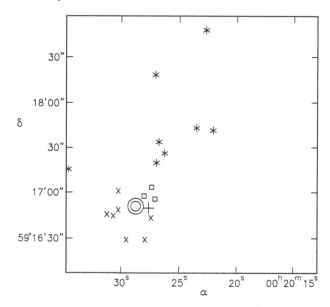

Fig. 10.2. Relative positions of the X-ray source (concentric circles), the center of H II region No. 113 (large plus sign), molecular clouds (squares), possible supernova remnants (x marks), and Wolf–Rayet stars (asterisks). Adapted from Brandt et al. (1997).

is probably an X-ray binary) that is located at the center of the nonthermal bubble in IC 10. The relative positions of the X-ray source (which coincides with the Wolf–Rayet star WR 17), the center of the H II region No. 113 of Hodge & Lee (1990), and of molecular clouds and possible supernova remnants are shown in Figure 10.2.

10.5   Summary

At the present time, IC 10 is the only starburst galaxy in the Local Group. Because it is located at low Galactic latitude, and therefore highly obscured, it has not yet been studied in much detail. Its location and distance suggest that it may be an outlying member of the M31 subgroup of the Local Group. Radio observations show that IC 10 is embedded in an enormous rotating, centrally condensed, H I envelope. Surprisingly, the core of this gas mass (which coincides with this small optical galaxy) rotates in the opposite sense to the outer envelope. The reason for this is not yet understood. It is also not clear how the gas in the outer envelope of IC 10 has been able to maintain itself over a period of $\sim$10 Gyr.

IC 10 resembles its fellow Local Group irregulars NGC 6822 and IC 1613, which are also embedded in enormous H I envelopes. Surprisingly, IC 10 does not obey the same relation between oxygen abundance and the ratio of the number of WC to WN stars, as is observed for other Local Group galaxies. It is tempting to assume that this peculiarity is, in some way, related to the fact that IC 10 is the only Local Group starburst galaxy. Since most of the WC stars in IC 10 occur in a small region with a diameter $<1'$ (<190 pc), it is possible that they were formed, more or less simultaneously, during a single short burst.

11

Faint dwarf irregular galaxies

11.1 IC 1613

11.1.1 Introduction

The dwarf irregular galaxy IC 1613 was discovered by Wolf (1906) and is described as "F,eeL" (i.e., faint and most extremely large) in the *Second Index Catalogue of Nebulae* (Dreyer 1908). The true nature of IC 1613 was first recognized by Baade (1935), who determined its distance by using the period–luminosity relation for Cepheids having periods ranging from 14 days to 42 days. Baade concluded that "it is without doubt a system of low luminosity." Subsequently it was included as a bona fide member of the Local Group by Hubble (1936, p. 145). IC 1613 is a nonbarred irregular that serves as the prototype for DDO type Ir V. The fact that IC 1613 was already known almost a century ago, even though it is quite faint ($M_V = -15.3$), suggests that the inventory of all but the dimmest Local Group members is (excepting objects at low Galactic latitude) probably reasonably complete. A blue image taken with the Palomar 1.2-m Schmidt telescope is shown in Figure 11.1. A higher resolution photograph of IC 1613, which was obtained with the Palomar 5-m reflector, is shown in Volders & Högbom (1961). A beautiful photograph in the light of Hα is reproduced in Sandage (1971).

11.1.2 Distance and reddening

The discovery of Cepheids in IC 1613 by Baade (1935) was followed up by a more detailed investigation (Baade 1963, pp. 218–226) that resulted in the identification of 25 Cepheids with periods ranging from 2.4 days to 146 days. Using photoelectric calibrations of the photographic observations of Baade's faintest Cepheids, Carlson & Sandage (1990) were able to show that the slopes of the period–luminosity relations for IC 1613 and the Large Magellanic Cloud are indistinguishable. Sandage (1971) obtained photoelectrically calibrated B and V observations of 24 Cepheids. From his photometry he found $E(B - V) \simeq 0.03$ and $(m - M)_0 = 24.43$. Subsequently Freedman (1988c) employed CCD photometry in the B, V, R, and I bands to obtain $(m - M)_0 = 24.3 \pm 0.1$ and $E(B - V) = 0.04 \pm 0.04$. Baade (1963, p. 225) and Sandage (1971) noted the presence of a sheet of old Population II stars in IC 1613. These old red stars are (as was also the case in the LMC) more widely dispersed than the bright blue ones that have formed recently. From the apparent magnitude of the tip of the red giant branch of Population II, Lee, Freedman & Madore (1993b) find $(m - M)_0 = 24.27$. Using RR Lyrae stars in IC 1613, Saha et al. (1992) derive $(m - M)_0 = 24.10 \pm 0.27$. Presently available data on the

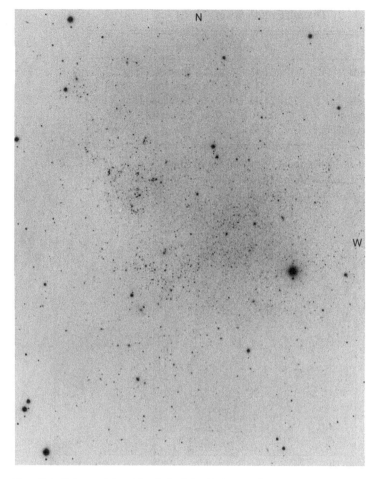

Fig. 11.1. Palomar 1.2-m blue light Schmidt image of IC 1613. The rich star cloud in the NE part of the galaxy contains a number of beautiful ring-shaped H II regions (Sandage 1971).

distance modulus of IC 1613, are summarized in Table 11.1. On the basis of these data a distance modulus of $(m - M)_0 = 24.3 \pm 0.1$, corresponding to a distance of 725 kpc, will be adopted in the subsequent discussion. With this distance modulus the absolute magnitude of IC 1613 becomes $M_V = -15.3$. It is noted that the distance to IC 1613 derived from RR Lyrae stars is, although at a marginal level of significance, smaller than that obtained from Cepheids. A similar phenomenon has been observed in the LMC (van den Bergh 1995a), in which RR Lyrae stars also yield a smaller distance than do the Cepheids.

Hutchinson (1973) has discussed what appears to be a W Virginis star with a period of 14.36 days in IC 1613. This object lies 0.9 mag above the Cepheid period–luminosity relation for this galaxy. However, Baade & Swope (1963) show that W Vir stars in M31 lie 1.7 mag *below* the Cepheid period–luminosity relation. This suggests that IC 1613 V39 (if it is indeed a W Vir variable) has a distance modulus that is 2.6 mag smaller than that of IC 1613 (i.e., that it is located at a distance of 220 kpc). If this interpretation is correct then V39 is the most distant known isolated field star.

Table 11.1. *Distance modulus of IC 1613*

Method	$(m - M)_o$	References
Cepheids	24.43	Sandage (1971)
Cepheids	24.3 ± 0.1	Freedman (1988b)
Cepheids[a]	24.42 ± 0.13	Kennicutt et al. (1998)
RR Lyrae	24.10 ± 0.27	Saha et al. (1992)
Tip red giant branch	24.27	Lee et al. (1993b)

[a]Based on a rediscussion of the data by Freedman (1988b), which takes into account the possible dependence of the Cepheid period–luminosity relation on metallicity.

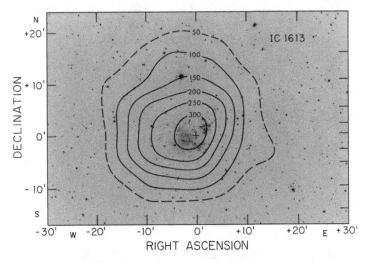

Fig. 11.2. Distribution of neutral hydrogen gas in IC 1613. The core of the optical image of this galaxy is seen to be much smaller than its hydrogen envelope. Reproduced with permission from Roberts (1962).

11.1.3 *Interstellar material*

Figure 11.2 (Roberts 1962) shows that IC 1613 is embedded in a large neutral hydrogen envelope. The ratio of the size of this galaxy in 21-cm radiation to that at optical wavelengths is, however, not as extreme as it is in for NGC 6822 and for IC 10. The hydrogen gas in this galaxy has been studied in great detail with the *Very Large Array* by Lake and Skillman (1989), who find that the rotation curve of IC 1613 is very regular and does not exhibit major noncircular motions. As a result, the analysis of the rotation of this galaxy is quite straightforward. The gas distribution is found to be lumpy in the inner parts of IC 1613, but the contours of 21-cm emission are more regular at larger radii. Hodge et al. (1991a) find that hydrogen clumps with densities $>10^{21}$ at cm^{-2} are well correlated with known H II regions. Lake & Skillman derive an inclination of $38° \pm 5°$, which lies within the range of 36° to 50° previously found by Ables (1971) and Hodge

Table 11.2. *Data on IC 1613*

$\alpha(2000) = 01^{\rm h}\, 04^{\rm m}\, 47\overset{\rm s}{.}3$ (1)		$\delta(2000) = +02°\, 08'\, 14''^a$ (1)
$\ell = 129\overset{\circ}{.}73$		$b = -60\overset{\circ}{.}56$
$V_{\rm r} = -232$ km s^{-1} (2)		$i = 38° \pm 5°$ (2)
$V = 9.09$ (3)	$B - V = 0.71 \pm 0.05$ (3)	
$(m - M)_{\rm o} = 24.3 \pm 0.1$ (4)	$E(B - V) \simeq 0.03$ (5)	$A_V \approx 0.09$
$M_V = -15.3$	$D = 725 \pm 35$ kpc (4)	$D_{\rm LG} = 0.47$ Mpc
Type = Ir V	Optical size $16' \times 20'$ (3.4 kpc $\times$ 4.2 kpc) (1)	
	H I size $31' \times 32'$ (6.5 kpc $\times$ 6.7 kpc) (6)	
$M \approx 1 \times 10^8 M_\odot$ (2)	$M_{\rm HI} = 6.5 \times 10^7 M_\odot$ (7)	$L_{\rm B}/L_{\rm H\alpha} = 130$ (8)
No. globulars = 0	Nova rate >0.02 yr^{-1} (9)	
[Fe/H] = -1.3 (10)	$12 + \log({\rm O/H}) = 7.86$ (11)	

^{a}Photometric center.
(1) Ables (1971).
(2) Lake & Skillman (1989).
(3) Hodge (1978).
(4) See Section 11.1.2.
(5) Sandage (1971).
(6) Roberts (1962).
(7) Volders & Högbom (1961).
(8) Kennicutt et al. (1998).
(9) Baade (1963, p. 222).
(10) Lee et al. (1993b).
(11) Skillman et al. (1989).

(1978). Lake & Skillman measure the center of the rotation for IC 1613 to be located at $\alpha = 01^{\rm h}\, 04^{\rm m}\, 46^{\rm s}$, $\delta = +2°\, 08'\, 46''$ (J 2000), which is close to the optical centroid given in Table 11.2. Adopting a velocity dispersion of 7.5 km s^{-1}, Lake & Skillman find a rotation curve that keeps rising and reaches $V({\rm max}) = 21$ km s^{-1} at $R = 2.6$ kpc. This rise shows that IC 1613 is embedded in a massive dark halo. It should, however, be noted that the maximum dark matter *density* allowed by the rotation curve of IC 1613 is an order of magnitude lower than that required to explain the observed velocity dispersion in dwarf spheroidal galaxies, such as Draco and Ursa Minor. From star counts Hodge et al. (1991a) find a (very uncertain) tidal cutoff at a galactocentric distance of $1420'' \pm 25''$ (5.0 kpc). A reasonable fit to the inner part of the rotation curve of IC 1613 is obtained with a total gas mass of $\sim 6 \times 10^7 M_\odot$ and a stellar mass of $\sim 4 \times 10^7 M_\odot$. If this fit is correct, then less than half of all of the gas in this system has yet been transformed into stars. In other words IC 1613 is, from an evolutionary point of view, a very primitive galaxy. From the strength of Hα emission Hunter, Elmegreen & Baker (1998) estimate the present star formation rate in the core of IC 1613 to be only $\sim 300–400 M_\odot$ kpc^{-2} per million years. Hodge (1978) lists 11 possible dust clouds in IC 1613. He comments that the dust clouds in this object are not very conspicuous. This suggests that the dust content of IC 1613 may be even lower than it is in the Small Magellanic Cloud. This, no doubt, results from the metallicity of the gas in IC 1613 being lower $[12 + \log({\rm O/H}) = 7.86$ (Skillman et al. 1989)] than it is in the SMC $[12 + \log({\rm O/H}) = 8.1$ (Massey 1998b)].

Images showing the Hα emission in IC 1613 have been published by Sandage (1971) and by Meaburn, Clayton & Whitehead (1988). This emission is mainly concentrated in a number of shells that are located in the most active area of star formation. Meaburn et al. suggest that these interlocking shells, which are expanding with velocities $\sim$40 km s^{-1}, will eventually merge into a single supershell. From Hα/[S$\textsc{ii}$] measurements D'Odorico, Dopita & Benvenuti (1980) suspected Sandage's H $\textsc{ii}$ region No. 8 to be a supernova remnant. This suspicion is confirmed by the spectrophotometry of Peimbert, Bohigas & Torres-Peimbert (1988), who find this to be the remnant of a supernova of Type II. The shell of swept-up material in this SNR is calculated to have a mass of $\sim$2,000 $M_\odot$. Both the helium abundance (Y = 0.230) and metallicity (Y = 0.0014) are found to be low. Lozinskaya et al. (1998) find that a luminous X-ray source coincides with the supernova remnant Sandage No. 8. Most of the remaining X radiation from the direction of IC 1613 appears to originate in a background cluster. Over the range 0.1–2.4 keV, S 8 has a luminosity of 4×10^{36} erg s^{-1}, which makes it one of the most luminous X-ray sources in the Local Group.

The largest H $\textsc{ii}$ ring in IC 1613 has a diameter of 44.″6(157 pc), which is much smaller than the largest H $\textsc{ii}$ shells in more luminous galaxies. Hodge, Lee & Gurwell (1990) have studied 77 H $\textsc{ii}$ regions in IC 1613. They find that these objects have a power-law luminosity function with exponent -1.6. Furthermore, these authors find that the size distribution of the H $\textsc{ii}$ regions in IC 1613 is exponential with a scale size of 56 pc. This value is similar to the values of 56 pc and 44 pc that van den Bergh (1981c) obtained for the LMC and SMC, respectively. Goss & Lozinskaya (1995) find that Sandage's H $\textsc{ii}$ region No. 3, which is centered on a WO star, has a mass of $3 \times 10^4 M_\odot$. This object is situated near the southern edge of a giant dust complex that has been detected in the infrared by the *IRAS* satellite. Armandroff & Massey (1985) have listed eight WR candidates in IC 1613, seven of which have known types. Of these, six are of type WN and only one of type WC. Such an excess of WN stars is expected (see Figure 10.1) in stars that have formed recently from metal poor interstellar gas.

Eskridge (1995) has discovered a cluster of galaxies at $z \approx 0.2$ that is projected only 6′ from the center of IC 1613. Blue objects in this cluster might enable one to undertake absorption-line studies of the interstellar medium in IC 1613.

11.1.4 *Stars and clusters*

Sandage & Katem (1976) have published a color–magnitude diagram for the brightest stars in IC 1613. Their data show that this galaxy exhibits a well-populated main sequence that extends to $B = 17.0(M_B = -7.5)$, $B - V = -0.15$, a half a dozen variable red supergiants with $\langle V \rangle \approx 17.0(M_V \approx -7.4)$, and a sprinkling of bright supergiants with intermediate colors. According to Humphreys (1980) the most luminous of these has a spectral type A0 Ia and $V = 16.38(M_V = -8.0)$, $B - V = 0.10$. Hodge (1978, 1980) has been able to identify 20 OB associations in IC 1613. Seven additional associations were identified by Georgiev et al. (1999). These associations are mainly concentrated in the NE and NW quadrants of this galaxy. Assuming that the 20 associations in IC 1613 were formed at a uniform rate, and that the oldest of them has an age of 3.6×10^7 yr, Hodge obtains a formation rate of 1.8 associations per million years.

Baade (1963, p. 231) remarks that "not a single star cluster is found in IC 1613, although there is a large super association." It might, perhaps, be argued that the apparent absence of clusters in IC 1613 is due to its great distance. It will, therefore, be more difficult to discover clusters in IC 1613 ($D = 725$ kpc) than in the SMC ($D = 59$ kpc). To test this hypothesis van den Bergh (1979) compared a Hale 5-m reflector plate of IC 1613 taken on Palomar Mountain in ∼1″ seeing with a Curtis Schmidt plate of the Small Magellanic Cloud taken on Cerro Tololo. Both plates were taken on 103aD emulsion behind a yellow (GG 14) filter, which suppresses emission nebulosity. The Curtis Schmidt image was printed out of focus to give the same *linear* scale for stellar images in the SMC and IC 1613. Intercomparison of these two images strongly supports Baade's claim that the Magellanic Clouds are forming clusters much more vigorously than IC 1613. A total of 15 obvious clusters are visible on the out of focus image of the SMC, whereas not a single cluster is seen on the comparable print of IC 1613. However, a few faint objects, which might be open clusters, are visible on my best Hale 5-m plates. A caveat is that IC 1613 is located in front of a distant cluster of galaxies (Eskridge 1995, Georgiev et al. 1999), so that some of the apparent clusters might, in fact, be background galaxies at $z \approx 0.2$. Hodge (1978) has listed 43 cluster candidates in IC 1613. More recently, Hodge & Magnier (Hodge 1998a) have used the *Hubble Space Telescope* to search for clusters in IC 1613. They find that the specific cluster frequency is at least 600 times greater in the LMC than it is in IC 1613. This result strongly suggests that the rate of cluster formation in galaxies does not always track their rate of star formation.

11.1.5 Summary

IC 1613 is a faint ($M_V = -15.3$), unobscured dwarf of type Ir V, which is embedded in a large hydrogen envelope. The rotation curve of this object demonstrates that it is embedded in a halo of dark matter. However, the density of this invisible component is an order of magnitude lower than that of the dark matter in some dwarf spheroidal galaxies. This demonstrates that not all dwarf spheroidals can have formed by sweeping gas from a dwarf irregular. The observation that the stellar mass in IC 1613 is comparable to the remaining gas mass shows that this galaxy is, from an evolutionary point of view, still in a very primitive state. This is confirmed by its low ([Fe/H $= -1.3$) metallicity. The small WC/WN number ratio in IC 1613 is expected from the well-established correlation between WC/WN and galaxy metallicity.

Perhaps the most surprising discovery about IC 1613 is that it contains few (if any) star clusters. The absence of globular clusters is not surprising, since an irregular galaxy with $M_V = -15.3$ is only expected to contain ∼1/2 a globular. However, the scarcity of open clusters in a galaxy that is vigorously forming young stars is unexpected. Taken at face value this result suggests (van den Bergh 1998d) that there is not a one-to-one correspondence between the rate of star formation and the rate at which open clusters are formed.

It would be of interest to obtain more detailed observations of variable star No. 39. If this object is a normal W Virginis star it would be located at a distance of 220 kpc. This would make it the most distant intergalactic star known in the Local Group.

11.2   The Wolf–Lundmark–Melotte system (= DDO 221)

11.2.1   Introduction

The Wolf–Lundmark–Melotte (WLM) system was discovered by Wolf (1923) and independently by Lundmark and by Melotte (1926). Melotte comments that the appearance of the WLM system is similar to that of NGC 6822. This conclusion is confirmed by the fact that both WLM (van den Bergh 1966a) and NGC 6822 are classified as being of DDO type Ir IV–V. A nice photograph of the WLM system is shown in de Vaucouleurs & Freeman (1972, p. 222). The most detailed photometry of this object is by Ables & Ables (1977), who also measured a globular cluster with $V = 16.56$, $B - V = 0.67$ situated about $2'$ (0.5 kpc) west of the center of this galaxy. This cluster was first noted by Humason, Mayall & Sandage (1956). Their radial velocity measurement showed that this object was a physical companion to WLM. From CCD photometry of 1,821 stars in the B and V bands Ferraro et al. (1989) conclude that star formation started in this galaxy at least 1 Gyr ago and proceeded at a more-or-less constant global rate of $\sim$3, 500 $M_\odot$ per million years until it stopped 8×10^7 yr ago. However, their conclusion that star formation in the WLM system stopped 80 Myr ago is contradicted by the presence of H II regions (Hodge & Miller 1995). Such emission regions are centered on O-type stars having ages of only a few megayears. From the total Hα emission of all H II regions in WLM, Hodge & Miller estimate the total present rate of star formation to be 1,100 $M_\odot$ per million years.

Minniti & Zijlstra (1996, 1997) have obtained a CCD color–magnitude diagram for $\sim$8,000 stars in WLM. Their data show that young blue stars are more centrally concentrated than faint (and presumably old) red stars. Furthermore, very red stars, which are probably intermediate-age AGB or carbon stars, are seen to be distributed like the young stars. These observations suggest that the WLM system contracted during its early evolution.

11.2.2   Distance and reddening

A compilation of reddening values for the WLM system is given in Minniti & Zijlstra (1997). From a review of these data they conclude that $E(V-I) = 0.03$, corresponding to $E(B - V) = 0.02$, and $A_V = 0.06$ mag. Using the Cepheid period–luminosity relation in WLM, Sandage & Carlson (1985b) found an apparent blue distance modulus of $(m - M)_B = 24.93$. In conjunction with $A_B = 0.08$ mag this yields $(m - M)_0 = 24.85$. From random-phase observations of five Cepheids in the I band, Lee, Freedman & Madore (1993c) obtain $(m - M)_0 = 24.92 \pm 0.21$, which is consistent with that derived by Sandage & Carlson. Using the position of the tip of the red giant branch derived from V and I photometry, Lee et al. find $(m - M)_0 = 24.81 \pm 0.15$. Employing the same technique, Minniti & Zijlstra (1997) obtain $(m - M)_0 = 24.75 \pm 0.1$. These distance determinations are collected in Table 11.3. On the basis of the data in this table a distance modulus of $(m - M)_0 = 24.83 \pm 0.1$ (925 kpc) will be adopted in the subsequent discussion. This distance places the Wolf–Lundmark–Melotte system firmly within the Local Group.

11.2.3   Stars and clusters

The lone globular cluster in the WLM system has $V = 16.56$, which, with the reddening and distance listed in Table 11.4, yields $M_V = -8.33$ (i.e., the sole globular

Table 11.3. *Distance modulus to the WLM system*

Method	$(m - M)_o$	References
Brightest Cepheids	24.85	Sandage & Carlson (1985b)
$\langle I \rangle$ for 5 Cepheids	$24.92^a \pm 0.21$	Lee et al. (1993a)
Tip of red giant branch	$24.81^a \pm 0.15$	Lee et al. (1993a)
Tip of red giant branch	24.75 ± 0.1	Minniti & Lee (1997)

$^a(m - M)_o = 18.5$ assumed for LMC.

Table 11.4. *Data on the WLM system*

$\alpha(2000) = 0^h\,01^m\,57\overset{s}{.}8$		$\delta(2000) = -15°27'51''$ (1)
$\ell = 75\overset{\circ}{.}85$		$b = -73\overset{\circ}{.}63$
$V_r = -120 \pm 4 \text{ km s}^{-1}$ (2)		$i = 69°$ (3)
$V = 10.42$ (3)	$B - V = 0.62$ (3)	
$(m - M)_o = 24.83 \pm 0.1$ (4)	$E(B - V) = 0.02$ (3)	$A_V = 0.06$ mag
$M_V = -14.4$ (4)	$D = 925$ kpc (4)	$D_{LG} = 0.79$ Mpc
Type = Ir IV–V	Optical size $6\overset{'}{.}5 \times 12\overset{'}{.}6$	
	$(1.7 \text{ kpc} \times 3.4 \text{ kpc})$ (5)	$r_e = 2\overset{'}{.}6$ (0.6 kpc) (3)
$M = 1.5 \times 10^8\,M_\odot$ (6)	$M_{HI} = 6.9 \times 10^7 M_\odot$ (7)	
No. globulars = 1 (3)	No. planetaries ≥ 2 (9)	$L_B/L_{H\alpha} = 160$ (10)
[Fe/H] ≈ -1.33 (8)		$12 + \log(O/H)$
		$= 7.74 \pm 0.15$ (8)

(1) Gallouet et al. (1975).
(2) de Vaucouleurs et al. (1991).
(3) Ables & Ables (1977).
(4) See Section 11.2.2.
(5) Holmberg (1958).
(6) Mateo (1998).
(7) Lisenfeld & Ferrara (1998).
(8) Skillman et al. (1989).
(9) Jacoby & Lesser (1981).
(10) Hunter et al. (1993).

cluster in this galaxy is of slightly above-average luminosity). This observation lends support to the notion that the luminosity function of globular cluster systems is universal and does not depend on the luminosities of their parent galaxies. After correcting the observed $B - V$ color of this cluster for foreground reddening one obtains $(B - V)_o = 0.65$. Such a rather blue color suggests that this cluster is probably metal poor. According to Minniti & Zijlstra (1997) the Wolf–Lundmark–Melotte system contains no other clusters with $I < 19.5(M_I = -5.4)$. This observation suggests that WLM, like IC 1613, might be cluster poor.

The colors of stars near the tip of red giant branch in the main body of WLM appear to indicate that these stars are metal poor. On the basis of their colors Lee, Freedman & Madore (1993b) estimate that [Fe/H] $= -1.6$. This value is consistent with [Fe/H] $= -1.45 \pm 0.2$ obtained by Minniti & Zijlstra (1997) from the $V - I$ color at 0.5 mag below

Table 11.5. *Number ratio of C stars to M3–M10 stars*

Galaxy	M31	M33	LMC	SMC	NGC 6822	IC 1613	WLM
C/M[a]	0.04	0.25	0.2	0.6	1.25	2.8	2.8
M_V	−21.2	−18.9	−18.5	−17.1	−16.0	−15.3	−14.4

[a]From Cook et al. (1986).

the tip of the giant branch. The observation (Skillman, Kennicutt & Hodge 1989) that $12 + \log(\text{O/H}) = 7.74 \pm 0.15$ in the H II regions in the WLM system, which typically corresponds to [Fe /H] ≈ -1.33, shows that even the most recently formed stars in this galaxy must be quite metal poor.

A color–magnitude diagram of WLM in B and V was obtained by Sandage & Carlson (1985b). They found that the brightest blue and red stars in this galaxy had $V = 17.77$ and $V = 17.23$(var), respectively, corresponding to $M_V = -7.12$ and $M_V = -7.66$(var). A more recent color–magnitude diagram in V and I (Minniti & Zijlstra 1997) shows that this galaxy also contains a number of very red (intermediate-age) carbon stars. The C star populations in M31, M33, NGC 6822, IC 1613, and WLM have been studied by Cook, Aaronson & Norris (1986), who used both $V - I$ colors and intermediate-band filters that monitor CN and TiO. These authors find that C stars outnumber M stars in the three dwarf galaxies WLM, IC 10, and NGC 6822. Furthermore, few late-M giant stars are formed in these dwarf galaxies. M33, which is of intermediate luminosity, contains more M stars and late-M giants than do the dwarfs. Finally, the supergiant galaxy M31 is found to contain many M stars and late-M giants but only a very small number of C stars. Table 11.5 and Figure 11.3 show that the carbon-to-(M3–M10) star ratio in the LMC is approximately the same as it is in M33, whereas that in the SMC is found to be intermediate between those in M33 and in the three dwarfs NGC 6822, IC 1613, and WLM. The data plotted in Figure 11.3 show a clear correlation between the C/M number ratio and galaxy luminosity (metallicity). The primary reason for this correlation is, no doubt, that less C needs to be dredged up in a metal-poor star to enable atmospheric carbon atoms to outnumber oxygen atoms. The dependence of the frequency of late-M stars on metallicity (Spinrad, Taylor and van den Bergh 1969) can be, at least in part, attributed to the giant branches' of metal-poor stellar populations being bluer (hotter) than those in metal-rich populations. It should, however, be noted that C stars do not occur in most globular clusters, even though they are metal poor. This shows that only intermediate-age giants can become C stars. The notion that the age distribution of intermediate-age stars affects the number ratio of C to O stars may also be supported by the observation that the LMC has only about half as large a C/M ratio as might have been expected from the correlation plotted in Figure 11.3. The reason for this may be that few intermediate-age potential C star ancestors were formed in the Large Cloud during the "dark ages" between 4 Gyr and 10 Gyr ago.

11.2.4   *Summary and desiderata*

The Wolf–Lundmark–Melotte system (DDO 221) is a metal-poor dwarf irregular galaxy of type Ir IV–V and luminosity $M_V = -14.4$. A summary of data on the WLM

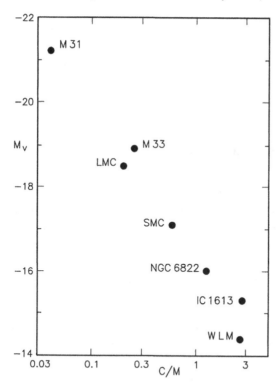

Fig. 11.3. Parent galaxy luminosity versus ratio of number of carbon stars to the number of M3–M10 stars in Local Group spiral and irregular galaxies. The figure shows that the C/M ratio decreases with decreasing parent galaxy luminosity (i.e., carbon stars are most common in dwarf galaxies).

galaxy is given in Table 11.4. The presence of a globular cluster suggests that star formation has been going on in this object for ≥ 10 Gyr. The fact that this globular is of slightly above-average luminosity supports the hypothesis that the luminosity function of globular cluster systems is independent of parent galaxy luminosity. A search for RR Lyrae fields stars in WLM could place constraints on the early history of star formation of this object. In addition, it would be of interest to see if the distance modulus, derived from RR Lyrae variables in this extreme dwarf, differs from that obtained via Cepheids, by more than the difference between the RR Lyrae and Cepheid distance moduli in the more luminous LMC.

Young blue stars, and intermediate-age carbon stars, exhibit a more centrally concentrated distribution over the face of the WLM system than do old stars. This suggests that star formation in WLM was more widely distributed originally than it is at the present time. A similar phenomenon has also been noted in the Large Magellanic Cloud (cf. Figure 6.11 and Figure 6.12).

The observation that late M-type giants are less common in the faint WLM system than they are in more luminous galaxies can mostly be explained by the red giant branches of metal-poor systems being bluer (hotter) than those of more luminous metal-rich galaxies. The high number ratio of carbon stars to M stars in WLM is, no doubt, largely due to the fact that even a modest dredge-up of carbon from the stellar interiors will make C atoms outnumber O atoms in the atmospheres of metal-poor WLM stars. The observation that the C/M number ratio in the LMC (which does not seem to have formed stars very actively between 4 Gyr and 10 Gyr ago) is smaller than might have been predicted by the data plotted in Figure 11.3 suggests that evolutionary history can also affect the number of C star progenitors and hence the C/M ratio.

11.3   The Pegasus dwarf irregular (= DDO 216)

11.3.1   Introduction

The Pegasus galaxy is a dwarf of DDO type Ir V, which was discovered on the *Palomar Sky Survey* plates by A.G. Wilson.[1] A photograph of the Pegasus dwarf galaxy is shown in Aparicio & Gallart (1995). Pegasus was regarded as a member of the Local Group by de Vaucouleurs (1975), who estimated its distance modulus to be $(m - M)_0 = 21.2$. A much larger distance modulus was later obtained by Hoessel et al. (1990), who found $(m - M)_0 = 26.22$, from what they thought to be Cepheids. However, their data showed that these "Cepheids" exhibited a large scatter about a period–luminosity relation with an unusual slope. Lee (1995a,b) subsequently showed that the variables discovered by Hoessel et al. were, in fact, located on the red giant branch of the Pegasus color–magnitude diagram. From the position of the tip of the red giant branch Lee derived a distance modulus of $(m - M)_0 = 25.13 \pm 0.11$. This is consistent with $(m - M)_0 = 24.9 \pm 0.1$ that Aparicio (1994) derived using the same method. However, more recently Gallagher et al. (1998) have derived a smaller distance modulus of $(m - M)_0 = 24.4 \pm 0.2$, corresponding to $D = 760 \pm 85$ kpc, by fitting evolutionary tracks to the observed distribution of stars in color–magnitude diagrams derived from *Hubble Space Telescope* observations in the B, V, and I bands. The main reason for the smaller distance modulus obtained by Gallagher et al. is that they find a larger reddening of $E(B - V) = 0.15 \pm 0.05$. This reddening value is (within the stated errors) consistent with the $E(B - V) = 0.2 \pm 0.1$ that Skillman, Bomans & Kobulnicky (1997) derived from spectrophotometry of the brightest H II region in the Pegasus dwarf galaxy. It should, however, be noted that these reddening values are significantly higher than the value $E(B - V) = 0.02$ that is derived from the standard Galactic reddening model of Burstein & Heiles (1984). From the H I observations of Lo et al., and the low metallicity of the Pegasus dwarf, Gallagher et al. conclude that internal reddening $E(B - V)_I < 0.03$ in the Pegasus dwarf (i.e., most of the reddening along the line of sight to this object is produced in the Galaxy). The apparent conflict between the reddening determinations cited above remains a source of concern. In the subsequent discussion a reddening of $E(B - V) = 0.15$ and a distance of $D = 760$ kpc will be tentatively adopted. It is noted in passing that this distance is similar to that of M31, from which the Pegasus dwarf is separated by $31°.0$. It follows that Pegasus might be either an outlying member of the M31 subgroup or a free-floating member of the Local Group.

---

[1] Skillman, Bomans & Kobulnicky (1997) attribute the discovery of the Pegasus galaxy to Holmberg (1958). However, Holmberg (1958, p. 100) himself credits Wilson for its discovery.

11.3.2 Stellar content

From CCD photometry in the B, V, R, and I bands Lee (1995b) finds that the Pegasus dwarf contains (1) a dominant old red giant population, (2) a small number of asymptotic giant branch stars above the red giant branch, and (3) a sparse population of massive young stars, including a few red supergiants. From the position of the tip of the red giant branch Lee estimates that [Fe/H] $= -1.5 \pm 0.2$. The conclusions on the evolutionary history of the stellar populations in the Pegasus Dwarf by Aparicio, Gallart & Bertelli (1997a) will have to be modified if the reddening toward this object is significantly larger than they assumed. In particular it is no longer clear if this object contains a significant population component with an age of ~15 Gyr. From their more recent work Gallagher et al. (1998) conclude that stars in Pegasus mainly formed 2–4 Gyr ago, although older populations with ages $\approx$8 Gyr are allowed. They find that the number of main-sequence and helium-burning blue loop stars requires the recent star formation rate to be 3–4 times lower than it was 1 Gyr ago. Gallagher et al. find that the present rate of star formation in the Perseus dwarf amounts to only ~300 $M_\odot$ per million years. At this rate the gas in the Pegasus dwarf would be exhausted in 13 Gyr. Aparicio & Gallart (1995) find that a single B0 star could excite the brightest H II region presently seen in this galaxy. Most of the star formation in Pegasus is concentrated in two centrally located clumps, while older stars form a more extended disk or halo. Even at the peak of its star forming activity the Pegasus dwarf probably remained relatively dim, with $M_V \approx -14$ (Tolstoy 1999).

From emission-line strengths in the two H II regions in the Pegasus dwarf, Skillman et al. (1997) find $12 + \log(O/H) = 7.93 \pm 0.13$. These authors conclude that N/O may be higher than that typically encountered in dwarf irregulars. They suspect that this N excess may be due to delayed nitrogen production. Such nitrogen is produced in intermediate-mass stars and ejected back into the interstellar medium by stellar winds and planetary nebulae. This excess nitrogen will build up during prolonged quiet periods between oxygen-producing bursts of star formation (Kobulnicky & Skillman 1998). As a result the N/O ratio will be highest after a long relatively quiescent period, such as seems to have occurred in the Pegasus system.

Gallagher et al. stress the fact that it would be important to know if the Pegasus system once experienced a major burst of star formation. Babul & Ferguson (1996) have argued that the faint blue galaxies observed at redshifts $0.5 < z < 1.0$, which they refer to as "boojums" (blue objects observed just undergoing moderate star formation), are dwarfs like Pegasus that underwent a major star forming events a few billion years ago. However, such starbursts might form large numbers of star clusters, many of which could have survived to the present day. No such major concentration of clusters is present in Pegasus (Hoessel & Mould 1982). In fact, Gallagher et al. argue that some of the three clusters found by Hoessel & Mould might, in fact, be galaxies in a moderately rich background cluster of galaxies. The absence of numerous star clusters in Pegasus might indicate that this galaxy did not undergo a violent burst of star formation in its recent past. However, this inference is weakened by the observation (van den Bergh 1998d) that the rates of star and cluster formation are not always tightly correlated.

Lo, Sargent & Young (1993) have mapped the 21-cm isophotes of the Pegasus dwarf. They find (for a distance of 760 kpc) a total hydrogen mass of $2.6 \times 10^6 \, M_\odot$. The optical and radio isophotes of this galaxy appear to have roughly the same orientation. Lo et al. note that there is a loose general tendency for the largest H I column densities to be associated with those regions that contain the brightest stars. The radio data give little indication of any systematic rotation and place a limit ≤ 5 km s^{-1} on the rotational velocity of this object. Application of the virial theorem to the observed 21-cm gas velocities yield a mass of $(2.7 \pm 1.1) \times 10^7 M_\odot$ for an assumed distance of 760 kpc.

11.3.3 Summary and conclusions

The Pegasus dwarf galaxy (= DDO 216) is a low-luminosity object of type Ir V that has $M_V = -12.3$. At present it is rather inactive, with most star formation concentrated in two associations located near the center of this galaxy. The older stellar population is more widely dispersed than are the young blue stars. The present rate of star formation in Pegasus is 3–4 times lower than it was ~ 1 Gyr ago. The bulk of the stars in Pegasus appear to have formed 2–4 Gyr ago, although older stars with ages ≈ 8 Gyr are not excluded by the observations. The low present level of activity of this galaxy is attested to by the fact that it has $L_B/L_{H\alpha} = 1,720$, which is one of the highest values listed by Hunter, Hawley & Gallagher (1993). Both the bulk of the stars (which formed long ago) and the young H II regions in this galaxy exhibit very low heavy element abundances. Radio observations show that the Pegasus dwarf is *not* surrounded by a huge hydrogen envelope, like those in which NGC 6822 and IC 1613 are embedded. Table 11.6 gives a summary of the observational data on the Pegasus dwarf galaxy. Detailed surface photometry and a multicolor study of the, presumably, very metal-poor Cepheid candidates in this galaxy would be of particular interest.

11.4   The Aquarius dwarf irregular (= DDO 210, = AqrDIG)

DDO 210 is a dwarf galaxy with a DDO type V, i.e., its surface brightness is so low that inspection of its blue image on the Palomar Sky Survey does not allow one to decide if it is a spiral or an irregular. Mateo (1998) classifies the Aquarius system as being of type dIr/dSph. A photograph taken with the KPNO 4-m Mayall reflector, which is reproduced in Fisher & Tully (1979), strongly suggests that it is a rather inactive dwarf irregular of low intrinsic luminosity. Because of its proximity to NGC 6822 (17°0) and to SagDIG (18°0), Fisher & Tully argued that DDO 210 might also be at a distance of only ~ 0.7 Mpc. Such a small distance, which suggests membership in the Local Group, is supported by the low radial velocity, $V_r = -131$ km s^{-1}, of DDO 210 (Lo, Sargent & Young 1993). Lee (1999) has obtained a color–magnitude diagram for this galaxy in the V and I bands. From the magnitude of the tip of the red giant branch in I he obtains a distance modulus of 25.05 ± 0.11, corresponding to a distance of (1025 ± 50) Mpc. This result differs significantly from that obtained earlier by Greggio et al. (1993). From his photometry Lee finds $E(B-V) = 0.03$ and [Fe/H] ≈ -2.0. Some stars in this galaxy appear to have formed only 30 Myr ago. From Lee's distance one finds that DDO 210 has $M_V = -11.3$. The fact that Hopp (1994) finds no H II regions in this galaxy would be consistent with a low luminosity for the brightest young blue stars in DDO 210. A search for variable stars in this galaxy by Caldwell, Schommer & Graham (1988), which employed a CCD detector at the prime focus of the CTIO 4-m Blanco telescope, did not result in the discovery of any variables.

Table 11.6. *Data on Pegasus = DDO 216*

$\alpha(2000) = 23^h\,28^m\,34^s$		$\delta(2000) = +14°44'48''$ (1)
$\ell = 94°.77$		$b = -43°.55$
$V_r = -182 \pm 3$ km s^{-1} (1)		$i = 42° \pm 10°$ (9)
$V = 12.59 \pm 0.15$ (1)	$B - V = 0.62 \pm 0.04$ (1)	$U - B = 0.12 \pm 0.05$ (1)
$(m - M)_o = 24.4 \pm 0.2$ (2)	$E(B - V) = 0.15 \pm 0.05$ (2)	$A_V = 0.47$ mag
$M_V = -12.3$	$D = 760 \pm 85$ kpc	$D_{LG} = 0.44$ Mpc
Type $=$ Ir V	Optical size $4'.6 \times 7'.7$ (1.0 kpc	
	$\times$ 1.7 kpc) (3)	
$M = (2.7 \pm 1.1)$		
$\quad \times 10^7\,M_\odot$ (4)	$M_{HI} = 2.6 \times 10^6\,M_\odot$ (4)	$L_B/L_{H\alpha} = 1720$ (5)
No. globulars $= 0$	No. planetaries $= 1$ (6)	
[Fe/H] $= -1.5 \pm 0.2$ (7)	$12 + \log(O/H)$	
	$\quad = 7.93 \pm 0.13$ (8)	

(1) de Vaucouleurs et al. (1991).
(2) Gallagher et al. (1998).
(3) Holmberg (1958).
(4) Lo et al. (1993).
(5) Hunter et al. (1993).
(6) Jacoby & Lesser (1981).
(7) Lee (1995b).
(8) Skillman et al. (1997).
(9) Mateo (1998).

Infrared photometry in the J, H, and K bands by Elias & Frogel (1985) detected only two stars (which might be foreground objects) in DDO 210. It follows that this object contains few, if any, red supergiants. This result, and the failure of Caldwell et al. to find Cepheids, would both be consistent with the hypothesis that this object is a nearby galaxy with a very small young population component. It is concluded that DDO 210 might well be a member of the Local Group, but the case is far from proven. A search for RR Lyrae stars with the *Hubble Space Telescope* might resolve the uncertainty in the distance to this galaxy. From 21-cm observations Lo et al. (1993) find that the gas in DDO 210 is somewhat more widely distributed than the stars on this galaxy. This gas shows considerable structure that exhibits no correlation with features in the optical image. The velocity field of DDO 210 appears chaotic but hints at an east–west velocity gradient. Data on the Aquarius dwarf are collected in Table 11.7.

11.5   SagDIG, the faintest irregular

The Sagittarius dwarf irregular (SagDIG), which is shown in Figure 11.4, is the faintest ($M_V = -10.7$) irregular galaxy known in the Local Group. It was discovered independently by Cesarsky et al. (1977) on plates obtained with the ESO Schmidt telescope, and by Longmore et al. (1978) on plates exposed with the UK Schmidt. On the basis of its appearance in Figure 11.4 this object can be classified as Ir V. The only two less luminous Local Group galaxies that contain young stars are the Pisces dwarf ($=$ LGS 3) at $M_V = -10.4$, and the Phoenix system with $M_V = -9.9$. In the latter two objects the young Population I is so weak that it is almost overwhelmed by the older Population II

Table 11.7. *Data on Aquarius dwarf (= DDO 210)*

$\alpha(2000) = 20^h 46^m 53^s$		$\delta(2000) = -12°\,50'\,58''$ (1,3)
$\ell = 34°05$		$b = -31°35$
$V_r = -131 \text{ km s}^{-1}$ (2)		
$V = 13.88 \pm 0.51$ (3)	$B - V = 0.12 \pm 0.16$ (3)	$U - B = -0.15 \pm 0.15$ (3)
$(m - M)_0 = 25.05 \pm 0.11$ (4)	$E(B - V) = 0.03$ (4)	$A_V = 0.10$ mag
$M_V = -11.3$	$D = 1025 \pm 50$ kpc	$D_{LG} = 1.02$ Mpc
Type = V		
$M = (1.4 \pm 0.8) \times 10^7 \, M_\odot$ (2)	$M_{HI} = 3 \times 10^6 \, M_\odot$ (1)	$M_{H_2} = M_\odot$
[Fe/H] < -1.0 (5)		

(1) Fisher & Tully (1975).
(2) Lo et al. (1993).
(3) de Vaucouleurs et al. (1991).
(4) Lee (1999).
(5) Mateo (1998).

component. The Phoenix and Pisces dwarfs are therefore classified as transitional objects of type dIr/dSph.

The distribution of stars over the face of the SagDIG is quite clumpy with evidence for at least two major associations. Presently available data do not yet allow us to establish the star formation history of SagDIG. The presence of associations of relatively bright blue stars provides evidence for at least one relatively recent burst of star formation. Hodge (1994a) estimates the present rate of star formation in SagDIG to be only 67 $M_\odot$ per million years. The carbon stars in this system (Cook 1987) appear to give evidence for two major star forming events between 1 Gyr and 10 Gyr ago. The C stars in SagDIG are more widely distributed than the bright young blue stars. In other words it appears that SagDIG contracted as it evolved. It would be particularly interesting to know if it also contains a very old population component (of which RR Lyrae stars are diagnostic). By analogy with other Local Group dwarfs one might expect the oldest stars in SagDIG to be more widely distributed over the sky than are the bright young blue stars. From the apparent brightness of the most luminous blues stars Cook (1987) estimated the distance modulus of this object to lie in the range $25.4 < (m - M)_0 < 26.0$, corresponding to $1.20 < D(\text{Mpc}) < 1.58$. Needless to say such an estimate is extremely uncertain. However, the position of SagDIG in a plot of heliocentric radial velocity versus apex angle (van den Bergh 1994a, and Chapter 18) is not inconsistent with the suggestion that this object is a bona fide member of the Local Group. Discovery and observation of RR Lyrae stars in SagDIG would provide a much securer value for its distance, than the estimate obtained from the brightest stars. Because of its low luminosity, SagDIG is expected to contain few, if any, Cepheids.

The neutral hydrogen in SagDIG has been observed by Longmore et al. (1978) and more recently by Young & Lo (1997b). The latter authors find that the H I in this galaxy is more extended than the distribution of stars. The 21-cm map of Young & Lo shows H I dimensions of $447''.\!\times 464''.$, corresponding to 2.8: $\times$ 2.9: kpc. The gas is observed to consist of a broadly distributed (hot) component with a velocity dispersion $\sigma = 10 \text{ km s}^{-1}$ and a narrow (cold) component with $\sigma = 5 \text{ km s}^{-1}$. This cold component appears to be

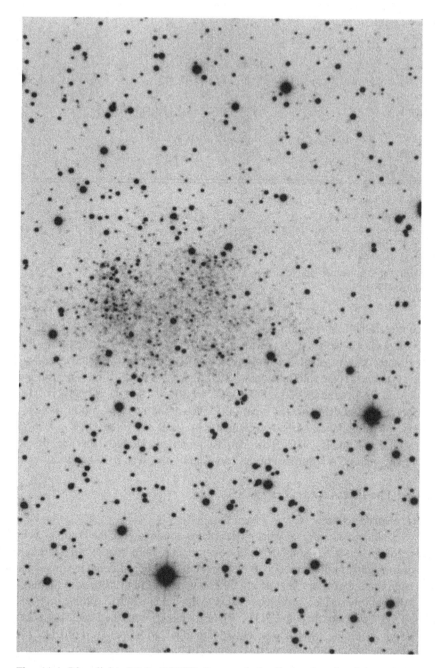

Fig. 11.4. Blue light (IIIaJ + GG385) image of the Sagittarius dwarf irregular galaxy (SagDIG) obtained with the ESO 3.6-m reflector. The distribution of the brightest blue stars is seen to be quite clumpy. Reproduced with permission from Richard West.

Table 11.8. *Data on the Sagittarius dwarf irregular (SagDIG)*

$\alpha(2000) = 19^h\,29^m\,58\overset{s}{.}9$		$\delta(2000) = -17°40'\,41''$ (1)
$\ell = 21\overset{\circ}{.}13$		$b = -16\overset{\circ}{.}23$
$V_r = -79 \pm 5\,km\,s^{-1}$ (1)		$i = 60° \pm 10°$ (5)
$V = 14.2^a$	$B - V = 0.5^a$	
$(m - M)_0 = 25.7 \pm 0.3$ (2)	$E(B - V) = 0.07$ (2)	$A_V = 0.22$ mag
$M_V = -10.7$:	$D = 1.3$: Mpc	$D_{LG} = 1.4$: Mpc
Type = Ir V	Optical size $3' \times 3'$ (1.1 kpc	
	$\times$ 1.1 kpc)	
$M = (4.5 \pm 2.4) \times 10^7 M_\odot^b$ (3)	$M_{HI} = 1.1 \times 10^7 M_\odot^b$ (3)	
No. globulars = 0		
$12 + \log(O/H) = 7.42$ (4)		

[a] $B - V = 0.5$ assumed.
[b] $D = 1.3$ Mpc assumed.
(1) Longmore et al. (1978).
(2) Cook (1987).
(3) Lo et al. (1993).
(4) Skillman, Kennicutt & Hodge (1989).
(5) Mateo (1998).

concentrated in clumps with mass $\sim 8 \times 10^5\,M_\odot$. Presumably it is this cold component that is associated with regions of recent star formation. The observations by Lo et al. (1993) show that most of the motions in the Sagittarius dwarf are chaotic. They exclude systematic rotational motions of $V(rot) \sin i > 2\,km\,s^{-1}$. For an assumed distance of 1.3 Mpc the virial mass of SagDIG is $\sim 4.5 \times 10^7\,M_\odot$ and the H I mass is $1.1 \times 10^7\,M_\odot$. The image of SagDIG shows no clear-cut evidence for dust absorption. This is, no doubt, because the abundance of heavy elements is small $12 + \log(O/H) = 7.36$ (Skillman, Terlevich & Melnick 1989, Skillman, Kennicutt & Hodge 1989) in this very low-luminosity galaxy. A summary of data on SagDIG is given in Table 11.8.

11.6   The Pisces dwarf (= LGS 3)

11.6.1   Introduction

While inspecting very deep IIIaJ plates obtained with the Palomar Schmidt telescope, Kowal, Lo & Sargent (1978) discovered three faint images, which they suspected to be intrinsically faint members of the Local Group. These Local Group Suspects (LGS) were subsequently observed at 21 cm by Thuan & Martin (1979), who found that LGS 1 was a distant low surface-brightness galaxy with a heliocentric velocity of 3,952 km s^{-1}, while LGS 2 was not detected in H I. Plates obtained with the Hale 5-m telescope (van den Bergh & Racine 1981) show a faint "smudge" at the position of LGS 2 but no excess of stars. These authors conclude that LGS 2 is probably a small Galactic reflection nebula or a dwarf galaxy that is more distant than the M31 subgroup. However, LGS 3 was found to be a bona fide dwarf with a radial velocity of -280 km s^{-1}. This object had previously been noted as a probable low-luminosity galaxy by Karachentseva (1976). A CCD image of LGS 3, obtained with the Hale 5-m telescope, is shown in Mould (1997). From its location, only $11\overset{\circ}{.}0$ from M33 in the sky, it appears probable that LGS 3 is a member of the Andromeda subgroup of the Local Group. From this assumption, and the projected mass method of Bahcall & Tremaine (1981), van den Bergh (1981a) obtained a mass of $(7.5 \pm 3.9) \times 10^{11} M_\odot$ for the Andromeda subgroup of the Local Group. More recently (see Section 18.1) Courteau & van den Bergh (1999) have found values in the range $(11.5–15) \times 10^{11} M_\odot$ for the mass of the Andromeda subgroup.

11.6.2 Stellar content

Studies of the stellar content of the Pisces dwarf have been published by Lee (1995c), by Mould (1997), and by Aparicio, Gallart & Bertelli (1997b). From its color–magnitude diagram Lee shows that LGS 3 has a well-developed giant branch and contains a small number of asymptotic giant branch stars. Furthermore, it may have several (probably young) blue stars. Mould has found two possible Cepheid variables in LGS 3. Cook & Olszewski (1989) have discovered asymptotic giant branch carbon stars in the Pisces dwarf. This shows that LGS 3 contains a significant intermediate-age population component. Aparicio et al. confirm the presence of an intermediate-age population and find that the rate of star formation in LGS 3 has either remained constant or declined slightly with time. From the location of the giant branch, in conjunction with $A_V = 0.08$ mag, these authors find a distance modulus of $(m − M)_0 = 24.54 \pm 0.21$, which corresponds to a distance of 810 kpc. From the color of the red giants 0.5 mag below the tip of the giant branch, Lee (1995c) derives a metallicity of $[Fe/H] = −2.10 \pm 0.22$. Since no H II regions have been found in LGS 3 (Hodge & Miller 1995) it is not yet possible to compare this value with the metallicity of the gas in the Pisces galaxy. Lee measured $V = 14.26$ mag within an aperture radius of $106''$ (0.4 kpc), from which the integrated magnitude of LGS 3 is derived to be $M_V = −10.36$. Lee finds that the surface brightness of this galaxy is well represented by a King profile with a core radius $r_c = 49'' \pm 3''$ (192 ± 12 pc) and a concentration parameter $c = \log(r_t/t_c) = 1.25 \pm 0.13$. The corresponding tidal radius is $\sim870''$ (3.4 kpc). Figure 11.5 shows that the star counts by Lee (1995) (except those in his outermost ring) also may be represented by an exponential disk model with a scale-length $\sim50''$ (196 pc).

11.6.3 Neutral hydrogen gas

A detailed study of the neutral hydrogen gas in the Pisces dwarf system has been undertaken by Young & Lo (1997b). They find that this galaxy is gas poor compared to SagDIG. This difference in gas content is probably the main factor that accounts for the difference in morphology between LGS 3 (dIr/dSph) and SagDIG (IrV). A plot of the neutral hydrogen density contours in the Pisces dwarf obtained by Young & Lo shows that the diameter of its hydrogen envelope is almost twice as large as the region containing most of the stars. Both the optical and radio images are seen to be slightly elongated in the north–south direction. The H I velocity profiles of LGS 3 are well described by single Gaussians with velocity dispersions in the range 5–10 km s^{-1} [i.e., there is no evidence for the existence of a two-phase (hot + cold) interstellar medium in the Pisces galaxy]. The global H I profile of LGS 3 is centered at a heliocentric velocity of $−286.5 \pm 0.25$ km s^{-1}, and has a velocity dispersion of $\sigma = 9.1 \pm 0.3$ km s^{-1}. No CO emission was detected at this velocity. The absence of clear evidence for a velocity gradient shows that $V \sin i \leq 5$ km s^{-1}. In other words the Pisces dwarf galaxy is either being viewed close to pole-on or, more probably, it is not supported by rotation at all. Assuming a distance of

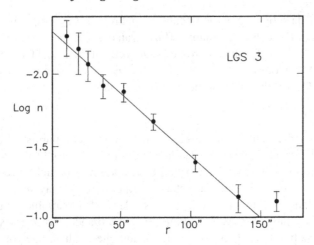

Fig. 11.5. Star counts (number per arcsec^2) by Lee (1995c) versus radius in LGS 3. The figure shows that the data (except those in the outermost ring) are adequately represented by an exponential disk model with a scale-length of 50'' (196 pc).

810 kpc, the virial mass of the innermost part of this galaxy is $2.6 \times 10^7 \, M_\odot$. This value yields a mass-to-light ratio of $\sim$25 (in solar units) (i.e., the Pisces dwarf appears to be embedded in a massive dark halo). For an assumed distance of 810 kpc, the hydrogen gas mass associated with LGS 3 is $4.2 \times 10^5 \, M_\circ$. The corresponding total (H + He) gas mass is $5.8 \times 10^5 \, M_\odot$. The low metallicity of stars in the Pisces dwarf (Lee 1995c) suggests that the gas in which it is embedded has not been greatly enriched by gas ejected from evolving stars. The dark halo of LGS 3 likely enabled this galaxy to hang onto some of this primordial material, even though much of it was probably ejected by supernovae and stellar winds. Data on the Pisces dwarf are summarized in Table 11.9.

11.7   The Phoenix dwarf galaxy

This object was discovered on an ESO blue plate by Schuster & West (1976), who opined that it might be an extremely distant globular cluster. Subsequently, Canterna & Flower (1977) obtained images of the Phoenix system with the 4-m Blanco telescope at Cerro Tololo, which demonstrated that Phoenix was, in fact, a low-luminosity galaxy. From a color–magnitude diagram in B and V, Canterna & Flower were able to show that the Phoenix dwarf contained young blue stars, in addition to a strongly populated old red giant branch. Phoenix is therefore intermediate between dwarf irregular galaxies and dwarf spheroidals. No evidence has so far been found for any H II emission associated with this galaxy. Canterna & Flower noted that three bright objects might be globular clusters. However, the fact that all three of these objects are clustered close together militates against this suggestion. Two-color CCD photometry of the Phoenix dwarf by Ortolani & Gratton (1988) showed that this galaxy contains a small young population, with an age of only $\sim$10 Myr, in addition to a dominant older population. The data by Ortolani & Gratton suggested that $(m - M)_0 = 23.5 \pm 0.5$. From BVI photometry with a CCD detector, van de Rydt, Demers & Kunkel (1991) find a distance modulus of $(m - M)_0 = 23.1 \pm 0.1$, corresponding to a distance of 417 ± 20 kpc. This places Phoenix

Table 11.9. *Data on the Pisces dwarf (= LGS 3)*

$\alpha(2000) = 01^h\,03^m\,56\overset{s}{.}5$		$\delta(2000) = +21°53'41''$ (1)
$\ell = 126°\!.77$		$b = -40°\!.88$
$V_r = -286.5 \pm 0.25$ km s^{-1}		
(2)		$i = 50°$ (5)
$V = 14.26$ (3)	$B - V = 0.80$ (1)	
$(m - M)_0 = 24.54 \pm 0.21$ (3)	$E(B - V) = 0.03$	$A_V = 0.08$ mag (4)
$M_V = -10.36 \pm 0.2$	$D = 810 \pm 80$ kpc (3)	$D_{LG} = 0.42$ Mpc
Type = dIr/dSph r_c	$r_t = 14.5' \pm 4.1'$ (3.4 kpc)	
$= 49'' \pm 3''(190 \pm 10\,\mathrm{pc})$	(3)	
$M = 2.6 \times 10^7 M_\odot$ (2)	$M_{HI} = 4.2 \times 10^5\ M_\odot$ (2)	$M_{H_2} = 0$ (2)
No. globulars $= 0$		
[Fe / H] $= -2.10 \pm 0.22$ (3)		

(1) Tikhonov & Markova (1996).
(2) Young & Lo (1997b).
(3) Lee (1995C).
(4) Burstein & Heiles (1984).
(5) Mateo (1998).

firmly within the Local Group. From their photometry of red giant stars in Phoenix, van de Rydt et al. find a metallicity of [Fe / H] $= -2.0$. More recently, Martínez-Delgado, Gallart & Aparicio (1998) have obtained CCD photometry of over 7,000 stars in this galaxy in the V and I bands. From the location of the tip of the red giant branch they conclude that $(m - M)_0 = 22.99$ (corresponding to $D = 396$ kpc) and [Fe/H] $= -1.42$. The small scatter in color along the giant branch suggests that the metallicity dispersion of stars in the Phoenix dwarf galaxy is small. Martínez-Delgado et al. also find evidence for an extended blue horizontal branch near the limit of their photometry. This result suggests that Phoenix contains an old (≥ 10 Gyr) stellar population that might include RR Lyrae variables. Martínez-Delgado et al. also believe that they have found evidence for the existence of some intermediate-age (2–10 Gyr) asymptotic branch stars. These authors also find that most recent star formation has taken place at galactocentric distances $R < 170$ pc, whereas old stars seem to dominate at $R > 170$ pc.

Early 21-cm observations of Phoenix by Carignan, Demers & Côté (1991), which were obtained with the Parks 64-m dish, show a cloud of neutral hydrogen with a velocity of $+56$ km s^{-1} that is clearly separated from a larger structure with a velocity of $\sim +140$ km s^{-1}, which appears to be associated with the Magellanic Stream. Results of higher resolution mapping with the *Very Large Array (VLA)* by Young & Lo (1997b) show a crescent-shaped H I cloud with $V = -23$ km s^{-1} along the SW rim of Phoenix at a distance of $\sim 4'.5$ from the galaxy center and a larger cloud with $V = +56$ km s^{-1} centered $\sim 9'$ south of Phoenix. Since the stellar radial velocity of the Phoenix galaxy is not known, it is not clear whether any of these clouds are physically associated with Phoenix.

A summary of presently available data on the Phoenix dwarf galaxy is given in Table 11.10. The most important presently missing piece of information is the radial velocity of the stellar component of this object. This velocity would allow one to determine

Table 11.10. *The Phoenix dwarf galaxy*

$\alpha(2000) = 01^h 51^m 03\overset{s}{.}3$		$\delta(2000) = -44°27'11''$ (1)
$\ell = 272\overset{\circ}{.}19$		$b = -68\overset{\circ}{.}95$
$V_r = ?$		
$(m - M)_0 = 22.99$ (4)	$E(B - V) = 0.02$ (2)	$A_V = 0.06$ mag
$M_V = -9.8 \pm 0.4$ (4)	$D = 396$ kpc (4)	$D_{LG} = 0.59$ Mpc
Type = dIr/dSph	Optical size $7' \times 9'$ (0.8 kpc $\times$ 1.1 kpc) (3)	
No. globulars = 3? (3)		
[Fe/H] = -1.42 (4)		

(1) Schuster & West (1976).
(2) van de Rydt et al. (1991).
(3) Canterna & Flower (1977).
(4) Martínez-Delgado et al. (1998).

if any of the hydrogen observed near Phoenix is, in fact, physically associated with it. Radial velocity information also would allow one to place this galaxy on the Local Group radial velocity versus apex distance plot. This would strengthen the determination of the velocity dispersion in the Local Group. Observations of old and intermediate-age stars would be important to place constraints on the evolutionary history of the Phoenix dwarf galaxy.

11.8   The local group member Leo A (= DDO 69)

11.8.1   *Distance to Leo A*
Leo A was discovered by Zwicky (1942) with the Palomar 0.5-m Schmidt telescope. It is classified as an Ir V galaxy in the DDO system. Leo III is another pseudonym for Leo A.[1] The distance to this galaxy has long remained controversial. From the presence of detected planetary nebulae Jacoby & Lesser (1981) placed an upper limit of 3.2 Mpc on the distance to Leo A. Sandage (1986c) derived a distance modulus of $(m - M)_0 = 26 \pm 1$, corresponding to a distance of $\sim$1.5 Mpc, from the apparent magnitudes of the three brightest stars in Leo A. However, he cautioned that this method of distance determination is almost degenerate for such low-luminosity galaxies. During CCD observations Hoessel et al. (1994) found five variable stars in Leo A. From a tentative identification of these stars as Cepheids they concluded that Leo A had $(m - M)_0 = 26.74$, corresponding to a distance of 2.2 Mpc. This would place Leo A well beyond the generally accepted limits of the Local Group. Finally, Tolstoy et al. (1998) (see Table 11.11) have used the position of the red giant clump, the helium-burning blue loops, and of the tip of the red giant branch to obtain a distance modulus of $(m - M)_0 = 24.2 \pm 0.2$, corresponding to a distance of 690 ± 60 kpc, which places Leo A firmly within the Local Group. At this distance Leo A has an absolute magnitude $M_V = -11.5$.

---

[1] Zwicky (1957, p. 225) incorrectly calls Leo II, which is a dwarf spheroidal, Leo III.

Table 11.11. *Distance to Leo A*[a]

Method	$(m - M)_0$	Remarks
Tip of red giant branch	24.5 ± 0.2	$M_I = -4$ assumed
Red clump	24.2 ± 0.2	For age of 1–2 Gyr
Red clump	23.9 ± 0.1	For age of 9–10 Gyr
He blue loops	24.1 ± 0.1	With $Z = 0.0004$

[a]From Tolstoy et al. (1998).

11.8.2 Stellar population of Leo A

A color–magnitude diagram for Leo A published by Tolstoy (1996, 1999) shows a strongly developed red giant branch of old stars that extends to $V \sim 22.3$, corresponding to $M_V \sim -2$. Furthermore, Leo A exhibits a blue main sequence population that reaches to $V \sim 20.0 (M_V \sim -4)$. From its color–magnitude diagram it is clear that Leo A, like Pisces and Phoenix, contains both old and young population components. Tolstoy (1996) concluded that either the rate of star formation in Leo A must have decreased with time or that there have been gaps in its star formation history. More recently, Tolstoy (1999) concludes that the bulk of the stellar population in Leo A is younger than 2 Gyr and that a major episode of star formation took place 900–1,500 Myr ago. This reinterpretation fits with the red clump luminosity, and with the number of anomalous Cepheids observed in this object. The presence of an older population cannot be ruled out by her data. However, recent evolutionary models for metal poor ($Z = 0.0004$) stars suggest that such an older population is limited to no more than $\sim 10\%$ of the stellar population near the center of Leo A.

From its integrated Hα emission Hodge (1998b, private communication) calculates a present rate of star formation of 32 $M_\odot$ per million years in Leo A. Strobel, Hodge & Kennicutt (1991) have detected three H II regions that are, no doubt, excited by some of these hot stars. These authors also found an unresolved emission region that they identify as a planetary nebula. For this object Skillman, Kennicutt & Hodge (1989) obtained a low heavy element abundance of $12 + \log(O/H) = 7.30 \pm 0.2$ ($\sim 2.4\%$ solar). It is not yet clear if Leo A contains any old stars. In this connection a search for RR Lyrae variables would clearly be very important.

11.8.3 Radio observations of Leo A

The first radio observations of Leo A were by Allsopp (1978) and by Lo, Sargent & Young (1993). A more detailed study, which had a resolution of 15″ on the sky, has been undertaken with the *VLA* by Young & Lo (1996a). These observations show that the optical galaxy, which has an extent of $4.5 \times 7'$ (Holmberg 1958), is embedded in a much larger hydrogen envelope with dimensions of $8.2 \times 14'$. If we adopt Tolstoy's distance of 690 kpc, the integrated 21-cm radiation of Leo A yields a total hydrogen mass of $8 \times 10^6 M_\odot$. The global H I profiles of Leo A appear to consist of a cool component, with a velocity dispersion $\sigma = 5.2$ km s^{-1} and a mass of $1.5 \times 10^6 M_\odot$, and a hot component, with $\sigma = 9.6$ km s^{-1} and a mass of $5.9 \times 10^6 M_\odot$. The hot phase of the interstellar medium in Leo A seems to be distributed

Table 11.12. *Data on Leo A (= DDO 69, Leo III)*

$\alpha(2000) = 09^h 59^m 23\overset{s}{.}0 \pm 1\overset{s}{.}0$		$\delta(2000) = +30°44'44''$
		$\pm 10''$ (1)
$\ell = 196\overset{\circ}{.}90$		$b = +52\overset{\circ}{.}41$
$V_r = +23.8 \text{ km s}^{-1}$ (2)		$i \geq 53°$ (2)
$V = 12.69$ (3)	$B - V = 0.33 \pm 0.08$ (4)	$U - B = -0.19 \pm 0.08$ (4)
$(m - M)_0 = 24.2 \pm 0.2$ (5)	$E(B - V) = 0.02$ (6)	$A_V = 0.06 \text{ mag}$ (5)
$M_V = -11.54 \pm 0.03$	$D = 690 \pm 60 \text{ kpc}$ (5)	$D_{LG} = 0.88 \text{ Mpc}$
Type = Ir V	Optical size $4\overset{'}{.}5 \times 7\overset{'}{.}0$	$(0.9 \text{ kpc} \times 1.4 \text{ kpc})$ (3)
$M < 9 \times 10^7 M_\odot$ (2)	$M_{HI} = 8 \times 10^6 M_\odot$ (2)	$M_{H_2} = 0$:
No. globulars $= 0$	No. planetaries ≥ 1 (6)	
		$12 + \log(\text{O/H})$
		$= 7.30 \pm 0.2$ (7)

(1) Allsopp (1978).
(2) Young & Lo (1996a).
(3) Holmberg (1958).
(4) de Vaucouleurs et al. (1991).
(5) Tolstoy et al. (1998).
(6) Strobel et al. (1991).
(7) Skillman et al. (1989).

globally, whereas the cool phase appears to be concentrated in the star forming regions of this galaxy. Much of the cool gas is concentrated in five major cloud complexes, which have masses of $\sim 0.6 \times 10^6 M_\odot$. These cloud complexes typically have dimensions of $\sim 1'(0.2 \text{ kpc})$ and are separated by $\sim 2\overset{'}{.}5$ (0.5 kpc). The global velocity field of Leo A appears to exhibit a weak gradient of $\sim 6 \text{ km s}^{-1}$ kpc. From velocity dispersions inside a galactocentric distance of 1.4 kpc, the virial mass of Leo A is found to be $<9 \times 10^7 M_\odot$. This upper limit is too high to place significant constraints on the amount of dark matter in Leo A. Data on Leo A are summarized in Table 11.12.

12

Spheroidal galaxies

The present section will deal with the three most luminous Local Group spheroidal galaxies: NGC 205, NGC 185, and NGC 147, all of which are companions to the Andromeda galaxy. These galaxies, which are too luminous to be called dwarf spheroidals (dSph), will be designated spheroidals (Sph). In the past such objects have been classified as being of types dE, Ep, S_1, or dE/dSph. It should, however, be emphasized that Sph and dSph galaxies are, respectively, the bright and faint representatives of the same morphological class of galaxies.

NGC 147 and NGC 185 are located at only 0.22 Mpc from our adopted center of the Local Group. This places them closer to the center of mass of the Local Group than any other presently known Group members.

12.1   The spheroidal galaxy NGC 205

12.1.1   Introduction
The luminosity of NGC 205 is similar to that of M32. However, it has a much more extended structure. Furthermore M32 rotates, but NGC 205 does not (Bender, Paquet & Nieto 1991). A good photograph of NGC 205 is shown in the *Hubble Atlas of Galaxies* (Sandage 1961). Images taken in the ultraviolet (Baade 1951) show it to contain a small number of young blue stars. Deep exposures (see Figure 12.1) reveal some extended patches of dust absorption. Moreover, inspection of the image of NGC 205 on the prints of the *Palomar Sky Survey*, or deeper plates (see, for example, Fig. 7 of Kormendy 1982), clearly show that the outer isophotes of this galaxy are distorted – presumably due to the tides exerted by M31, from which it is separated by only $37'(8.2\,\mathrm{kpc})$. According to (rather uncertain) calculations by Cepa & Beckman (1988), NGC 205 is on a retrograde orbit around M31, with a period of 2.9×10^8 yr and an inclination of $122°$. The distortion of the outer isophotes of NGC 205 is confirmed by the photometry of Hodge (1973). From CCD photometry Kim & Lee (1998) find that the ellipticity of NGC 205 rises monotonically from $\epsilon \sim 0.1$ at small radii to $\epsilon \sim 0.5$ at $r > 150''$, except for a spike with $\epsilon \sim 0.5$ for $10'' < r < 20''$. CCD photometry by Kent (1987) shows that (outside the semistellar nucleus) the radial luminosity distribution of NGC 205 is intermediate between a de Vaucouleurs $r^{1/4}$ profile and an exponential Disk. Lee (1996) has used a CCD detector to obtain detailed surface photometry of NGC 205. He finds that the semistellar nucleus of this galaxy has $B - V = 0.65$. For $r > 1\rlap{.}''5$ the integrated color decreases and reaches a minimum of $B - V = 0.60$ at $r_{\mathrm{eff}} \sim 7''$. Beyond that distance

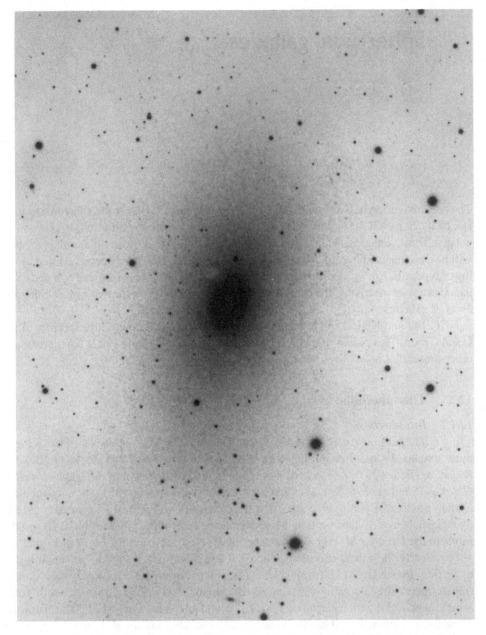

Fig. 12.1. Palomar 5-m plate in yellow light (GG14 + 103aD) of NGC 205. Note the large low-density dust clouds located NNE and S of the center of the optical image.

the integrated color increases to $B - V = 0.67$. The relatively blue color of the central region of NGC 205 is, no doubt, due to the young blue stars discovered by Baade (1951) and to the presence of numerous intermediate-age stars (Davidge 1992, Lee 1996).

According to Hodge the brightest of the young stars has a blue apparent magnitude of $B = 19.42$ ($M_B = -5.1$). The most luminous star at 1,500 Å (Bertola et al. 1995) has

Table 12.1. *Distance modulus of NGC 205*

Method	$(m - M)_0$	Reference
Planetary nebulae	24.68 ± 0.35	Ciardullo et al. (1989)
Tip red giant branch	24.54	Salaris & Cassisi (1998)
Tip red giant branch	24.3 ± 0.2	Mould et al. (1984)
RR Lyrae stars	24.65	Saha et al. (1992)

$B = 20.8$. Hodge finds that the centroid of these blue stars is situated $8''\!.5$ west and $9''\!.9$ north of the nucleus. From observations of 12 of the 24 planetary nebulae discovered by Ciardullo et al. (1989), it is found that NGC 205 has a distance modulus of $(m - M)_0 = 24.68 \pm 0.35$. This is similar to the value $(m - M)_0 = 24.54$ that Salaris & Cassisi (1998) find from the position of the tip of the red giant branch of NGC 205. Saha, Hoessel & Krist (1992) have found 30 RR Lyrae stars in NGC 205, from which they obtain a distance modulus of $(m - M)_0 = 24.65$. In Table 12.1 this value is compared with other distance determinations to NGC 205. Good agreement is found between the three distinct methods for determining the distance to this galaxy. The four distance moduli in Table 12.1 give a formal unweighted mean of $\langle (m - M)_0 \rangle = 24.54 \pm 0.09$, which differs only marginally from the value $(m - M)_0 = 24.4 \pm 0.1$ (see Section 3.2) adopted for M31. In the subsequent discussion it will be assumed that M31 and NGC 205 both have a distance of 760 kpc. In their paper Saha et al. (1992) found that the peak in the luminosity distribution for the RR Lyrae stars in NGC 205 was 0.33 mag fainter than that which Pritchet & van den Bergh (1988) obtained for the RR Lyrae variables in the halo of M31. They interpreted this difference to mean that NGC 205 is $\sim 15\%$ more distant than M31. In view of the evidence for tidal distortion of NGC 205 by M31 it seems more probable that the observed difference between the luminosity distributions of the M31 and NGC 205 RR Lyrae samples is, in fact, due to incompleteness of the Pritchet & van den Bergh sample of RR Lyrae stars at faint magnitudes. Alternatively it might be assumed that there is a real physical difference between the RR Lyrae stars in the Andromeda galaxy and in NGC 205. In addition to RR Lyrae, Saha et al. also found a number of other variables in NGC 205, which are probably Mira stars. It would be of interest to observe these objects over a few years to determine their periods. This would allow one to place them on the period–luminosity relation for Mira-type variables (van Leewen et al. 1997, Bergeat, Knapik & Rutily 1998). It would also be very worthwhile to strengthen, and confirm, the distance determination to NGC 205 by searching for Cepheids among the young population near the center of NGC 205. However, the number of young stars is so small that few, if any, Cepheids are expected.

12.1.2 Star clusters

Eight clusters appear to be associated with NGC 205. From spectrophotometry, Da Costa & Mould (1988) found that seven of these clusters are old globulars, whereas NGC 205:V is an intermediate-age cluster. From measurements in B, V, R, and I Lee (1996) finds an age of ~ 300 Myr for this object. The brightest and reddest of the seven globular clusters is NGC 205:III, for which van den Bergh (1969) measured $V = 14.93$ ($M_V = -9.6$), $B - V = 0.85$, an integrated spectral type of F8-G0, and a heliocentric radial velocity of $V_r = -377 \pm 31$ km s^{-1}. The latter value differs significantly from the value $V_r = -239 \pm 31$ km s^{-1} that he obtained for the semistellar nucleus of NGC 205. This suggests that NGC 205:III may, in fact, be an M31 globular that appears projected on NGC 205. For the remaining six NGC 205 globular clusters $\langle V \rangle = 17.30 \, (M_V = -7.2)$ and $\langle B - V \rangle = 0.64$. Da Costa & Mould find that individual NGC 205 globulars have metallicities in the range [Fe/H] $= -1.3$ to [Fe/H] $= -1.9$. The presumed M31 cluster NGC 205:III is much more metal rich and has [Fe/H] $= -1.05$. Recent *Hubble Space Telescope* observations by Cappellari et al. (1999) show that many of the UV-bright "stars" that had previously been detected by ground-based observers are, in fact, multiple star systems, open clusters, and compact associations.

12.1.3   Evolutionary history

From photometry of individual stars in the V and I bands, Davidge (1992) concludes that NGC 205 contains a large intermediate-age population that is uniformly distributed over the central part of this galaxy. His observations indicate that star formation in NGC 205 has not been continuous. In particular Davidge finds that the recent burst of star formation NW of the nucleus of NGC 205 was much less intense than the one that occurred 0.5–1.0 Gyr ago. Davidge suggests that this galaxy, during its active phase $\sim$0.8 Gyr ago, might have resembled the active Ep/amorphous galaxy NGC 3077, which is located in the M81 group. Lee (1996) supports the star formation scenario proposed by Davidge for the inner part of NGC 205, but he finds that little or no recent star formation has taken place in its outer halo. From spectrophotometry and population synthesis, Bica, Alloin & Schmidt (1990) conclude that the dominant population in the nucleus of NGC 205 has ages in the range $(1–5) \times 10^8$ yr. Jones et al. (1996) have imaged the nucleus of NGC 205 with the *Hubble Space Telescope*. They find the isophote with $I = I(0)/2$, in which $I(0)$ is the central surface brightness, has dimensions of $0\overset{''}{.}095 \times 0\overset{''}{.}107$ (0.35 pc $\times$ 0.39 pc). The luminosity within this isophote is $4.1 \times 10^4 \, L_\odot(V)$. If a black hole is present inside the semistellar core its mass must be $<1 \times 10^5 \, M_\odot$. Da Costa & Mould (1988) note that the spectrum of the nucleus of NGC 205 is similar to that of the intermediate-age cluster NGC 205:V. Jones et al. (1996) estimate [Fe/H] $= -1.44$ from the Mg$_2$ absorption line strength in the spectrum of this cluster. From photometry in the V, R, and I bands Mould, Kristian & Da Costa (1984) found that the metallicity of giant branch stars in NGC 205 is [Fe/H] $\geq -0.9 \pm 0.2$. Furthermore, Bica et al. (1990) find $-1.0 <$ [Fe/H] < -0.5 from spectrophotometry. It is of interest to note that the latter values are significantly greater than those of the NGC 205 globular clusters (which have metallicities in the range [Fe/H] $= -1.3$ to [Fe/H] $= -1.9$). Mould et al. also note that the giant branch has a significant width. This indicates that there is a dispersion of at least 0.5 dex in the [Fe/H] values of individual giant stars.

12.1.4   Interstellar material

A dozen dust clouds in NGC 205 have been mapped by Hodge (1973). For these clouds he finds that $A_V/E(B - V) = 3.4 \pm 0.6$, which is indistinguishable from the Galactic value $A_V/E(B - V) = 3.1$ (Mathis 1990). From observations at 1.1 mm Fich & Hodge (1991) find a total dust mass of $(1–3) \times 10^3 \, M_\odot$. The neutral hydrogen gas in NGC 205 has been studied by Johnson & Gottesman (1983) and by Young & Lo (1997a). The H I is found to be roughly, but not exactly, lined up with the optical major axis. It is asymmetric with respect to the nucleus of NGC 205 and extends twice as far to the south as it does to the north. The total H I mass is found to be $3.4 \times 10^5\ M_\odot$. The H I velocity centroid at $V_r = -221 \pm 1\ \mathrm{km\ s^{-1}}$ disagrees with that of the optical galaxy, for which $V_r = -244 \pm 1\ \mathrm{km\ s^{-1}}$ (Peterson & Caldwell 1993). This clear discrepancy between the velocity fields of the neutral gas and of the stars implies that the gas and stars form dynamically distinct systems with differing specific angular momenta. Both the kinematics and the distribution of gas suggest that the hydrogen is not situated in a long-lived configuration. Young & Lo find that the hydrogen has a very clumpy distribution in NGC 205. Individual clumps have dimensions of $\sim$200 pc and masses $\sim 10^4\ M_\odot$. Young & Lo (1996b) also find that the gas and dust complexes in NGC 205 have virial masses that are similar to those of giant molecular clouds in the Galaxy and in the SMC. The brightest blue stars in NGC 205 appear to be located to the north and west of the major H I concentration. Perhaps surprisingly, these blue stars are not found at the peaks of the projected hydrogen distribution. CO does, however, appear to be associated with maxima in the H I distribution. Haas (1998) appears to have found some very cold dust, with a temperature below 10 K, in NGC 205. Young (1998) draws attention to the fact that molecular clouds exist in NGC 205 in areas with remarkably low column densities. In NGC 185 and NGC 205 they are found in regions with a column density of only a few $\times 10^{20}\ \mathrm{cm^{-2}}$, whereas such molecular clouds in the Galaxy and in M31 are generally associated with H I column densities of $\sim 10^{21}\ \mathrm{cm^{-2}}$. Probably such a high column density is required in M31 and the Galaxy to shield the molecular gas from the UV radiation field produced by numerous hot stars. Such shielding is not required in NGC 185 and NGC 205, in which the UV radiation field is weak.

From observations of spectral lines in this galaxy, Carter & Sadler (1990) find that the dimensionless parameter $V(\mathrm{max})/\sigma = 0.2$, in which σ is the dispersion in the radial velocity. [For an oblate rotating spheroid a value of $V(\mathrm{max})/\sigma = 0.8$ would have been expected.] These observations show that this spheroidal galaxy is mainly flattened by its velocity anisotropy, rather than by rotation. This conclusion is confirmed by Held, Mould & de Zeeuw (1990), who place an upper limit of 25 km s^{-1} on the rotational velocity. Near the nucleus $\sigma \sim 15$ km s^{-1}, compared to $\sigma \sim 50$ km s^{-1} in its outer regions. It is remarkable that the nucleus of NGC 205 is, from a dynamical point of view, so cold. Welch, Sage & Mitchell (1998) estimate that the total mass of gas (molecular, atomic, and X-ray emitting) in NGC 205 is $\sim 10^6\ M_\odot$. They calculate that this is one order of magnitude smaller than the mass that should have been returned to the interstellar medium by evolving stars. Furthermore, NGC 205 is not rotating. Such ejected gas should therefore have zero angular momentum, whereas the gas that is presently observed in this galaxy appears to be rotationally supported. To account for these puzzling observations one might invoke a two-stage scenario. First the low angular momentum gas lost from stars was ejected by stellar winds and supernovae. Ram-pressure stripping, produced by gas in the disk of M31, could also have contributed to the ejection of gas from NGC 205 (Sofue 1994). Subsequently, fresh high angular momentum gas might have been captured from the Local Group or from the disk of M31.

12.1.5 Summary and desiderata

NGC 205 is a spheroidal companion to M31. From spectroscopic observations it is found that this galaxy is mainly flattened by its velocity anisotropy, rather than by rotation. The large observed difference between the optical systemic velocity of NGC 205 and that of the interstellar gas within it suggests that the hydrogen in this galaxy might have been captured from M31, or it might have been derived from material orbiting within the Local Group (or its Andromeda subgroup). Of the seven clusters that are physically associated with NGC 205, six are globulars and one is an object of intermediate age. The rate of star formation in the core of NGC 205 seems to have been higher 0.5–1.0 Gyr ago than it is at present. However, the dominant stellar population in the outer parts of NGC 205 appears to be quite old. In view of the discrepancy between the metallicity estimates of Jones et al. (1996), of Bica et al. (1990), and of Mould et al. (1984), it would be very desirable to obtain improved estimates for the metallicity of the dominant stellar population in NGC 205. Measurements of the metallicity of the gas in NGC 205 might also allow one to distinguish between scenarios in which the gas in this galaxy was (1) captured from M31 when NGC 205 last passed through the fundamental plane of the Andromeda galaxy $\sim 1 \times 10^8$ yr ago (Cepa & Beckman 1988) or (2) captured from intracluster space. Finally, the distance determination to NGC 205 could be strengthened by observations of Mira stars and perhaps Cepheids.

A summary of available data on NGC 205 is given in Table 12.2.

12.2   The spheroidal galaxy NGC 185

The spheroidal galaxies NGC 147 and NGC 185 play an important role in the history of modern astronomy. Baade (1963, p. 51) wrote "In August and September [of 1943] I resolved the central part of M31 and the two nearby companions in rapid succession. Mayall had told me that NGC 185 (which forms a pair with NGC 147, some 12° from M31) has a radial velocity very similar to that of M31, and that he suspected on this basis that they might be nearby and associated with the Andromeda Nebula system. So I tried NGC 185 and NGC 147, and indeed they were easy to resolve."

CCD surface photometry of NGC 185 in B, V, R, and I has been published by Kim & Lee (1998). These authors find that all colors get bluer at $r < 25''$ (80 pc). Outside this radius the ellipticity remains nearly constant with $0.20 < \epsilon < 0.25$. Inside this radius the apparent ellipticity is affected by clumps of young stars and dust clouds. These authors also show a plot of the distribution of dust clouds in the inner region of this galaxy. The distribution of H I, CO, and dust in NGC 185 are intercompared by Young (1998).

12.2.1 Distance and reddening

A photograph of NGC 185, which shows this object clearly resolved into an enormous swarm of stars of Population II, is shown in Baade (1944b). NGC 185 is located at a lower Galactic latitude ($b = -14°.5$) than M31 ($b = -21°.6$) and is therefore more highly obscured. From 21-cm observations, and an assumed gas-to-dust ratio, Burstein & Heiles (1984) find $A_V = 0.58$ mag and $E(B - V) = 0.195$. This is in good agreement with the value $E(B - V) = 0.19 \pm 0.03$ that Lee, Freedman & Madore (1993c) obtain from B, V, and I photometry of individual stars in this galaxy. With this reddening Lee et al. find that the position of the red giant branch in NGC 185 yields a distance modulus of $(m - M)_0 = 23.96 \pm 0.21$. Using the same technique, Salaris & Cassini (1998) and Martínez-Delgado & Aparicio (1998) obtain similar values of $(m - M)_0$. From observations of 151 RR Lyrae stars Saha & Hoessel (1990) obtain $(m - M)_0 = 23.79 \pm 0.25$. A rediscussion of this result by Lee et al., which takes into

Table 12.2. *Data on NGC 205*

$\alpha(2000) = 00^h 40^m 22\overset{s}{.}5$		$\alpha(2000) = +41° 41' 11''$ (1)
$\ell = 120\overset{\circ}{.}72$		$b = -21\overset{\circ}{.}14$
$V_r = -244 \pm 1 \text{ km s}^{-1}$ (12)		$\epsilon = 0.35 - 0.50$ (4)
$V = 8.06$ (2)	$B - V = 0.67$ (3)	$U - B = 0.18$ (4)
$(m - M)_0 = 24.4$ (5)	$E(B - V) = 0.035$ (6)	$A_V = 0.11$ mag
$M_V = -16.45$	$D = 760$ kpc	$D_{LG} = 0.30$ Mpc
Type = Sph	Size $16' \times 26' (3.5$ kpc	
	$\times 5.7$ kpc) (7)	
$M = 7.5 \times 10^8 M_\odot$ (10)	$M_{HI} = 3.4 \times 10^5 M_\odot$ (11)	$M_{H_2} = 2.4 \times 10^5 M_\odot$ (17)
No. globulars = 6 (3, 15)	No. planetaries = 24 (16)	Nova rate <1 yr^{-1} (13)
[Fe/H] $\lesssim 0.9$ (8)	$12 + \log(\text{O/H}) = 8.60$ (14)	

(1) de Vaucouleurs et al. (1991).
(2) Kent (1987).
(3) Lee (1996).
(4) Hodge (1973).
(5) See Section 12.1.1.
(6) Burstein & Heiles (1984).
(7) Holmberg (1958).
(8) Mould et al. (1984).
(9) Johnson & Gottesman (1983).
(10) Held et al. (1990).
(11) Young & Lo (1997a).
(12) Peterson & Caldwell (1993).
(13) Qiao et al. (1997).
(14) Richer & McCall (1995).
(15) van den Bergh (1969).
(16) Ciardullo et al. (1989).
(17) Welch et al. (1998).

account the dependence of M_V (RR) on [Fe/H], yields $(m - M)_0 = 24.02 \pm 0.25$. The data on the distance to NGC 185 are summarized in Table 12.3.

12.2.2 *Stellar populations*

From the color of the red giant branch, Lee et al. estimate that [Fe/H] $= -1.23 \pm 0.16$. Using the same technique, Martínez-Delgado & Aparicio (1998) obtain an average metallicity $\langle$[Fe/H]$\rangle = -1.43 \pm 0.15$. Lee et al. find that the dispersion of the red giant branch suggests that the metallicities of individual red giant stars scatter over the range $-1.6 <$ [Fe/H] < -0.9. Martínez-Delgado & Aparicio believe that they may have detected a radial decrease in both the mean metallicity and in the metallicity dispersion in NGC 185. Not unexpectedly Da Costa & Mould (1988) find a lower mean metallicity $\langle$[Fe/H]$\rangle = -1.65 \pm 0.25$ for the globular clusters in NGC 185. Metallicities of individual globulars are found to range from [Fe/H] $= -1.2$ to [Fe/H] $= -2.5$. From observations of planetary nebulae, Richer & McCall (1995) and Kaler (1995) find $12 + \log(\text{O/H})$ values of 8.2 and 7.9, respectively. Detailed studies of the stellar populations in NGC 185 have been published by Lee, Freedman & Madore (1993c) and by Martínez-Delgado

Table 12.3. *Distance determinations to NGC 185*

Method	$(m - M)_o$	References
RR Lyrae	23.79 ± 0.25	Saha & Hoessel (1990)
RR Lyrae	24.02 ± 0.25	Lee et al. (1993c)
Tip red giant branch	23.96 ± 0.21	Lee et al. (1993c)
Tip red giant branch	23.95 ± 0.10	Martínez-Delgado & Aparicio (1998)
Tip red giant branch	24.12	Salaris & Cassini (1998)

and Aparicio (1998). From the presence of a number of stars above the red giant branch of Population II, Lee et al. conclude that NGC 185 contains a significant intermediate-age stellar population. Of seven cluster suspects Geisler et al. (1999) find six to be probable globular clusters, whereas one appears to be a distant background elliptical. For clusters III, IV, and V these authors find metallicities [Fe/H] of approximately -1.6, -1.9, and -1.5, respectively.

12.2.3   Stellar kinematics

The kinematics of NGC 185 have been studied by Bender, Paquet & Nieto (1991), who find an internal velocity dispersion of 24 ± 2 km s^{-1} and a rotational velocity of 1.2 ± 1.1 km s^{-1} between $r = 0''$ and $r = 100''$. For a rotationally flattened galaxy one would expect

$$v/\sigma = [\epsilon/(1 - \epsilon)]^{1/2}. \tag{12.1}$$

Since $\epsilon = 0.20$ in the inner part of NGC 185, one would have expected to observe a rotational velocity of 12 km s^{-1}, rather than the observed value of 1.2 km s^{-1}. It follows that NGC 185 is not rotationally supported (i.e., its flattening is mainly due to velocity anisotropy). From photometry and velocity dispersion measurements of the mass-to-light ratio (in solar units) over the inner $100''$ (320 pc), Bender et al. find $M/L = 5 \pm 2$. A rather similar result, $M/L \approx 3$, is obtained by Lee et al. (1993c). The strength of Hβ absorption, and of [O III] emission, shows that star formation is still taking place near the center of NGC 185 at the present time. From CCD surface photometry Kent (1987) finds that the innermost part of the radial luminosity profile of NGC 185 is affected by dust absorption in this star forming region. From its photometric profile the core radius of NGC 185 is $r_c = 60''$ (190 pc).

12.2.4   Interstellar material

The hydrogen gas in NGC 185 has been studied by Johnson & Gottesman (1983), who find an H I mass of 1.3×10^5 $M_\odot$. Wiklind & Rydbeck (1986) have also detected $\sim 1 \times 10^5$ $M_\odot$ of CO in this galaxy. More recently higher resolution observations of NGC 185 have been published by Young & Lo (1997a). Their observations suggest that the stars and hydrogen in this galaxy have similar kinematics, indicating that the gas may have been derived from stellar winds and planetary nebulae. The gas in NGC 185 is extremely clumpy on scales of ~ 200 pc, with individual clumps having masses $\sim 10^4$ $M_\odot$. An extended region of Hα + [N II] emission with a diameter of ~ 50 pc is observed at the H I column density peak, which occurs close to the center of this galaxy. Brandt et al. (1997) were not able to detect any X-ray sources in NGC 185 with the *ROSAT* high-resolution imager.

12.2.5 Summary and conclusions

NGC 185 and NGC 147 are physical companions (van den Bergh 1998c) that are located in front of M31, but within the Andromeda subgroup of the Local Group. The metallicity of the dominant stellar population in NGC 185 is presently not yet well determined. However, it does appear to be higher than that of the NGC 185 globular clusters. Additional studies of the metallicity of the stars in NGC 185 would clearly be of great value. The presence of RR Lyrae variables demonstrates that star formation began $\gtrsim$10 Gyr ago. The existence of significant numbers of stars above the red giant branch of Population II attests to the presence of a strong intermediate-age population. It would be of interest to also search for, and study, the long-period variables that are probably associated with this intermediate-age stellar component. The presence of Hβ absorption, and [O III] emission lines, in the central region of NGC 185 shows that star formation is presently still proceeding at a low level in this galaxy.

The rotational velocity of NGC 185 is found to be an order of magnitude too low to account for the observed flattening of this galaxy. This demonstrates that the flattening of NGC 185 is mainly due to velocity anisotropy. The kinematics of the interstellar gas in NGC 185 is similar to that of the stellar system within which it is embedded. This suggests that the observed gas might have been injected into interstellar space by stellar winds and planetary nebulae. High-precision radial velocities of the six globular clusters in NGC 185 might provide some insight into the kinematics of the oldest stellar population component in this galaxy. A summary of the data on NGC 185 is given in Table 12.4.

12.3   The spheroidal galaxy NGC 147

12.3.1 Introduction

In his pioneering paper on stellar populations, Baade (1944b) published a beautiful print of NGC 147, which clearly shows the resolution of this object into a swarm of red giant stars of Population I. (To the best of my knowledge this is the only original of a photographic print that was ever bound into a volume of *The Astrophysical Journal.*) Intercomparison of Baade's images of NGC 147 and NGC 185 shows a striking small dust patch in NGC 185, whereas NGC 147 is apparently dust free. In this connection it is of interest to note that Young & Lo (1997a) were only able to set a firm upper limit of 3,000 $M_\odot$ on the total amount of H I present in this galaxy. The lack of interstellar material in NGC 147 is confirmed by Sage, Welch & Mitchell (1998), who place an upper limit of 4,100 $M_\odot$ on the amount of H$_2$ in NGC 147. Both Young & Lo and Sage et al. point out that it is puzzling that NGC 147 is dust and gas free, whereas its companion NGC 185 contains significant amounts of both gas and dust. A similar dichotomy exists between the two close companions of the Andromeda galaxy. M32 is gas free, whereas NGC 205 contains significant amounts of gas and dust.

Bender, Paquet & Nieto (1991) have found an internal velocity dispersion of $\sigma = 23 \pm 5$ km s^{-1} in NGC 147. Substituting this value, together with $\epsilon = 0.35$ (which is

Table 12.4. *Data on NGC 185*

$\alpha(2000) = 00^h\,38^m\,58\overset{s}{.}0$		$\alpha(2000) = +48°\,20'\,18''$ (1)
$\ell = 120\overset{\circ}{.}79$		$b = -14\overset{\circ}{.}48$
$V_r = -202 \pm 3\,\text{km s}^{-1}$ (5)		$\epsilon = 0.22$ (12)
$V = 9.13$ (7)	$B - V = 0.92 \pm 0.01$ (1)	$U - B = 0.39 \pm 0.03$ (1)
$(m - M)_0 = 24.1 \pm 0.1$ (13)	$E(B - V) = 0.19 \pm 0.03$ (4)	$A_V = 0.58$ (2)
$M_V = -15.55$	$D = 660 \pm 30\,\text{kpc}$	$D_{LG} = 0.22\,\text{Mpc}$
Type = Spheroidal	Size = $12' \times 13'(2.3\,\text{kpc}$	
	$\times 2.5\,\text{kpc})$ (3)	$r_c = 60''$ (190 pc) (5)
$M = 6.6 \times 10^8\,M_\odot$ (5)	$M_{HI} = 1.3 \times 10^5\,M_\odot$ (10)	$M_{H_2} \sim 1.2 \times 10^5\,M_\odot$ (11)
No. globulars = 6 (9)	No. planetaries =4 (8)	
[Fe/H] = -1.23 ± 0.16 (4)	$12 + \log\,(O/H) = 7.9$ (6)	

(1) de Vaucouleurs et al. (1991).
(2) Burstein & Heiles (1984).
(3) Holmberg (1958).
(4) Lee et al. (1993c).
(5) Bender et al. (1991).
(6) Kaler (1995).
(7) Kent (1987).
(8) Ciardullo et al. (1989).
(9) Ford et al. (1977).
(10) Johnson & Gottesman (1993).
(11) Wiklind & Rydbeck (1986).
(12) Hodge (1963).
(13) See Section 12.3.2.

representative of the region with $r < 100''$), into Eq. 12.1 yields a rotational velocity of 17 km s^{-1}. This value is significantly greater than the rotational velocity of 6.5 ± 1.1 km s^{-1} that Bender et al. actually measured between galactocentric distances $r = 20''$ and $r = 100''$. This difference shows that the flattening of NGC 147 is mainly due to its anisotropic velocity dispersion.

12.3.2   *Distance and reddening*

The reddening of NGC 147 is discussed by Mould, Kristian & Da Costa (1983). They conclude that $E(B - V) = 0.17$. This value is similar to $E(B - V) = 0.19 \pm 0.03$ that Lee, Freedman & Madore (1993c) obtained for its close companion NGC 185. Recent distance determinations of NGC 147 are collected in Table 12.5. Individual distance estimates for NGC 147 and NGC 185 are consistent with the assumption that these objects are equidistant. From the positions of the tips of the red giant branches in their color–magnitude diagrams, Salaris & Cassisi (1998) find true distance moduli of $(m - M)_0 = 24.27$ and $(m - M)_0 = 24.12$ for NGC 147 and NGC 185, respectively. Saha, Hoessel & Mossman (1990) obtained a mean apparent magnitude, on the Gunn system, of $\langle g \rangle = 25.25$ for the RR Lyrae variables in NGC 147, compared to $\langle g \rangle = 25.20$ (Saha & Hoessel 1990) for the RR Lyrae in NGC 185. The corresponding distance moduli (cf. Lee et al. 1993c) are $(m - M)_0 = 24.07$ for NGC 147 and $(m - M)_0 = 24.02$ for NGC 185. On the basis of these data it will assumed that both NGC 147 and NGC 185 have $(m - M)_0 = 24.1 \pm 0.1$, corresponding to a distance of 660 ± 30 kpc. The adopted

Table 12.5. *Distance to NGC 147*

Method	$(m - M)_0$	References
RR Lyrae	23.92 ± 0.25	Saha et al. (1990)
Red giants	24.2 ± 0.3	Mould et al. (1984)
Red giants	24.3	Davidge (1994)
Horizontal branch	24.44 ± 0.04	Han et al. (1997)
Tip red giant branch	24.37 ± 0.06^a	Han et al. (1997)
Tip red giant branch	24.36 ± 0.06^b	Han et al. (1997)
Tip red giant branch	24.27	Salaris & Cassisi (1998)

[a] Outer region.
[b] Inner region.

true distance modulus of NGC 147 and NGC 185 is a few tenths of a magnitude smaller than the individual distance moduli of M31 that are listed in Table 3.1. Taken at face value this suggests that the NGC 147/NGC 185 pair is located in front of the center of the Andromeda subgroup of the Local Group.

NGC 147 and NGC 185 are separated by only 0.972 on the sky (see Figure 12.2). In addition, Bender, Paquet and Nieto (1991) find that their radial velocities differ by only 9 ± 5 km s^{-1}. Davis et al. (1995) show that for a gravitationally bound pair of galaxies

$$\left(\Delta V_r^2 R_p / 2GM \right) < \sin^2 \alpha \, \cos \alpha < 0.385 \qquad (12.2)$$

in which ΔV_r is the velocity difference between the galaxies, R_p is their projected separation, and α is the angle between the true line joining these two masses and the plane of the sky. If we assume a distance of 660 kpc, Eq. 12.2 shows that NGC 147 and NGC 185 must have a combined mass $> 2.7 \times 10^8 \, M_\odot$ if they are to form a gravitationally bound pair. From the luminosities quoted in Tables 12.4 and 12.6 it is seen that this corresponds to $M/L_B > 1.3$ (in solar units) for stability. This limit is easily exceeded by the mass-mass-to light ratios $M/L_B = 7 \pm 3$ for NGC 147 and $M/L_B = 5 \pm 2$ for NGC 185 that Bender, Paquet and Nieto (1991) found by combining surface photometry and spectroscopic internal velocity measurements for these galaxies. It is therefore concluded (van den Bergh 1998c) that NGC 147 and NGC 185 form a gravitationally bound pair. This conclusion conflicts with one deduced earlier by Ford, Jacoby & Jenner (1977) from the radial velocities of a small number of planetary nebulae in NGC 147 and NGC 185.

It is noted in passing that NGC 147/NGC 185 and the LMC/SMC are the only physical binary systems with comparable masses in the Local Group. It is, perhaps, no coincidence that the components of both of these pairs have similar morphological types. NGC 147 and NGC 185 are both spheroidals, and the Large Cloud and Small Cloud are each Magellanic irregulars.

12.3.3 *Evolution and stellar populations*

From photometry with a CCD detector Kent (1987) finds that the surface brightness of NGC 147 is well represented by an exponential disk with a scale-length

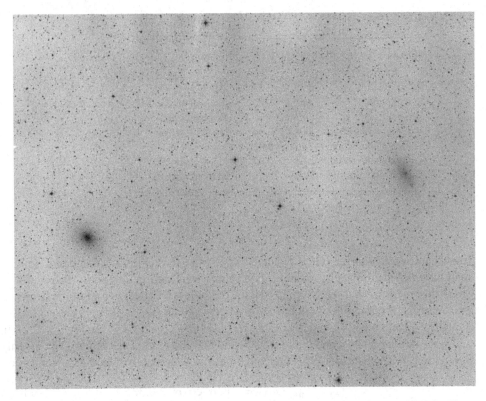

Fig. 12.2. Palomar 1.2-m Schmidt photograph of NGC 147 (right) and NGC 185 (left). These two Local Group spheroidal galaxies are separated on the sky by only 0.972 degrees. Note that the surface brightness of NGC 147 is much lower than that of NGC 185.

of 143" (0.46 kpc). However, he notes that the center of this galaxy cannot be measured because of the presence of a star and a possible globular cluster. Hodge (1976) writes that "at a distance of only 15 arcsec from the adopted center is a globular cluster ... which is barely within the uncertainty of the position of the center. Possibly this object is a nucleus of the system rather than a globular cluster, as apparently Baade (1951) believed possible." It would be of interest to obtain a high-resolution spectrum of this object to see if it has the low-velocity dispersion expected for a genuine star cluster, or if it exhibits the larger velocity dispersion that is diagnostic of a true nucleus that contains a central black hole. Hodge finds that NGC 147 contains three other globular cluster suspects. These four clusters have $\langle V \rangle = 18.53$, which with $A_V = 0.53$ and $(m - M)_0 = 24.1$ yields $\langle M_V \rangle = -6.1$. According to Da Costa & Mould (1988) clusters NGC 147:I and NGC 147:III have metallicities of [Fe/H] $= -1.9$ and [Fe/H] $= -2.5$, respectively.

From V and I photometry of individual stars Han et al. (1997) find that the mean metallicity of stars ranges from $\langle$[Fe/H]$\rangle = -0.91$ near the center of NGC 147 to $\langle$[Fe/H]$\rangle = -1.0$ in its outer regions. The metallicities of individual stars show a clear radial variation in the sense that the largest dispersion is seen at small radii. The discovery of RR Lyrae variables in NGC 147 by Saha, Hoessel and Mossman (1990) provides prima facie evidence for the existence of stars with ages $\gtrsim 10$ Gyr. From the period spread

Table 12.6. *Summary of data on NGC 147*

$\alpha(2000) = 00^h\,33^m\,11\overset{s}{.}6$		$\alpha(2000) = +48° 30' 28''$ (1)
$\ell = 119\overset{\circ}{.}82$		$b = -14\overset{\circ}{.}25$
$V_r = -193 \pm 3$ km s^{-1} (2)		$\epsilon = 0.44$ (3)
$V = 9.52$ (4)	$B - V = 0.91$ (1)	$U - B = 0.32$ (5)
$(m - M)_o = 24.1 \pm 0.1$ (6)	$E(B - V) = 0.17$ (7)	$A_V = 0.53$
$M_V = -15.1 \pm 0.1$	$D = 660 \pm 30$ kpc	$D_{LG} = 0.22$ Mpc
Type = Spheroidal	Size 12' × 18'(2.3 kpc	
	× 3.4 kpc) (8)	
	Disk scale-length	
	= 143'' (0.46 kpc) (4)	
$M = (5.5 \pm 2.3) \times 10^7\,M_\odot$ (2)	$M_{HI} < 3 \times 10^3\,M_\odot$ (9)	$M_{H_2} < 4.1 \times 10^3\,M_\odot$ (10)
No. globulars = 4 (3)	No. planetaries = 5 (11)	
[Fe/H] = −0.91 (core) (12)		
[Fe/H] = −1.0 (envelope) (12)		

(1) de Vaucouleurs et al. (1991).
(2) Bender et al. (1991).
(3) Hodge (1976).
(4) Kent (1987).
(5) Sandage (1972).
(6) See Table 12.5.
(7) Mould et al. (1983).
(8) Holmberg (1958).
(9) Young & Lo (1997a).
(10) Sage et al. (1998).
(11) Ford et al. (1977).
(12) Han et al. (1997).

of these variables Hoessel et al. estimate a metallicity range from [Fe/H] ≈ -1 to [Fe/H] ≈ -2. The color–magnitude diagram of NGC 147 shows the following features: (1) a well-defined red giant branch, from which Han et al. derive a true distance modulus of $(m - M)_o = 24.36$, and (2) a strong red clump in the NGC 147 color–magnitude diagram, from which a similar distance modulus is derived (but see Cole 1998 for caveats that apply to such distance estimates). Furthermore, (3) the existence of a small number of extended asymptotic branch objects shows that stars with ages of several Gyr are present. Davidge (1994) estimates that such intermediate-age stars contribute 2–3% of the total V light of NGC 147. Finally, (4) the absence of main-sequence stars brighter than $M_V = -1$ shows that no significant amount of star formation has occurred during the past 1 Gyr. The distribution of extended asymptotic branch stars indicates that the younger stars are more centrally concentrated than the majority of older stars.

Brandt et al. (1997) used *ROSAT* to search for X-ray sources in NGC 147, but none were found.

12.3.4 *Summary and conclusions*

NGC 147 and NGC 185 exhibit a radial velocity difference of only 9 ± 4 km s^{-1}. In conjunction with mass estimates of these galaxies derived from surface photometry and internal velocity dispersions, it is shown that these two galaxies form a bound physical pair. The conclusion that these objects form a physical pair is supported by their small (projected) separation of 11 kpc and by the fact that the luminosities of their red giant branches (corrected for reddening) differ by only ~0.15 mag (Salaris & Cassini 1998). It is of interest to note that NGC 147 and NGC 185 are both spheroidal galaxies. Perhaps it is no coincidence that the LMC and SMC, which are the only other Local Group members with comparable luminosities (masses), also have similar morphological types.

If the observed flattening of NGC 147 were entirely due to rotation, then Eq. 12.1 predicts that its rotational velocity between $r = 20''$ and $r = 100''$ should be $V(\text{rot}) = 17$ km s^{-1}. In fact the observed rotational velocity over this range in galactocentric distance is only $V(\text{rot}) = (6.5 \pm 1.1)$ km s^{-1}. This shows that the observed flattening of NGC 147 is mainly due to anisotropic stellar velocities.

As expected from its relatively low luminosity ($M_V = -15.1$), the bulk of the stars in NGC 147 have a low metallicity of [Fe/H] ≈ -1.0. However, the globular clusters in this object are even more metal poor. The existence of RR Lyrae variables in NGC 147 shows that stars started to form in this galaxy $\gtrsim 10$ Gyr ago. The bulk of the stars in this object are probably several billion years old. However, the absence of blue main-sequence stars brighter than $M_V = -1$ shows that no significant amount of star formation has taken place in this galaxy during the past 1 Gyr. Baade (1953) first drew attention to the fact that NGC 147 "shows no absorption patches and is apparently transparent up to its very center, because extragalactic nebulae shine through it." In fact the contrast between NGC 147 (which is dust free and contains no OB stars) and NGC 185 (which contains both young blue stars and dust patches) led Baade to his famous dictum: No dust, no Population I. It remains a profound mystery why NGC 147 has no gas or dust, while its sister galaxy NGC 185 contains significant amounts of both. It would be important to investigate the long-period red variables in NGC 147 that were discovered half a century ago (Baade 1951). It would also be of interest to study the metal abundances in some of the planetary nebulae that Ford, Jacoby & Jenner (1977) discovered in this galaxy. Finally, observations of the color–magnitude diagrams of the globular clusters associated with NGC 147 could be determined with the *Hubble Space Telescope*.

A summary of the data on NGC 147 is given in Table 12.6.

13

The most luminous dwarf spheroidal galaxies

13.1 Introduction

Inspection of Table 2.1, which lists the known members of the Local Group, shows that dwarf spheroidals are the most common type of galaxy in the Local Group. Observations of rich clusters (Trentham 1998a,b) appear to show that the faint ends of their luminosity functions are steep. Moreover, the colors of these faint galaxies suggest that they are mostly dwarf spheroidals. Since dwarf spheroidals are ubiquitous in poor clusters, and even more frequent in rich clusters, it appears safe to conclude that such dwarfs are the most common type of galaxy in the Universe. The fact that they were not discovered until 1938 is entirely due to their low luminosity. Their discoverer (Shapley 1943, p. 142) writes

> Perhaps not a great deal can be or needs to be known about them. They are relatively simple. And already they have made their greatest contribution by revealing themselves as members of our family of galaxies, and by possessing such low luminosities that they increase to six (out of eleven) the number of dwarfs among us. This result is upsetting, because it implies that our former knowledge and assumptions concerning the average galaxy may need serious modification. Moreover, the estimates of the total number of external organizations, and the total mass of the Metagalaxy will be involved in the reconsideration. Two hazy patches [the Sculptor and Fornax dwarfs] on a photograph have put us in a fog.

The suggestion that dwarf spheroidals might be distributed uniformly throughout the Local Group (Ambartsumian 1962) appears to conflict with the observation that strong clusterings of such objects are now known around the Milky Way system and the Andromeda galaxy. It should, however, be noted that our knowledge of the distribution of faintest dwarf spheroidals may be biased by observational selection. The searches that have resulted in the discovery of six dwarf spheroidals near M31 (van den Bergh 1972a,b; Armandroff, Davies & Jacoby 1998; Karachentsev & Karachentseva 1999) were confined to a relatively small area near the Andromeda galaxy. Furthermore, Figure 13.1, which is based on data compiled by Mateo (1998), shows a deficiency of low surface-brightness dwarf spheroidal (dSph) companions for $R_{GC} > 100$ kpc. Distant low-surface brightness Galactic satellites are probably absent because such objects cannot be discovered as concentrations of very faint individual stars on Schmidt sky survey plates.

That dwarf spheroidals are galaxies, rather than oversized globular clusters, is attested to by the fact that many of them are dark matter dominated (Mateo 1997), whereas

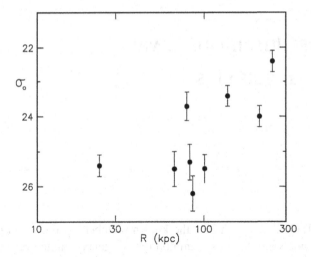

Fig. 13.1. Central surface brightness of dwarf spheroidal companions to the Milky Way system as a function of Galactocentric distance. The apparent correlation of σ_0 and R_{GC} is probably due to the fact that it is difficult to discover unresolved very low surface-brightness dSph galaxies beyond $R_{GC} > 100$ kpc.

globulars appear to contain little (or no) dark matter (Heggie, Griest & Hut 1993). Furthermore, the Fornax and Sagittarius dwarfs, which are the two most luminous dwarf spheroidals, are themselves embedded in small swarms of globular clusters.

For excellent reviews on dSph galaxies the reader is referred to Zinn (1993), Gallagher & Wyse (1994), Mateo (1998), and Da Costa (1998).

13.2   The Fornax dwarf

13.2.1   Discovery and large-scale structure
The discovery of the first two dwarf spheroidal galaxies was described as follows by Shapley (1943, p. 141):

> We were taken by surprise in 1938 when Harvard plates unexpectedly, and almost accidentally, yielded two sidereal specimens of an entirely new type. Presumably the gamut of galaxies had already been run. All forms had long been fully described. There were spirals, spheroidals, irregulars, with many variations on the spiral theme. The newly found organizations in Sculptor and Fornax did not seem essential in order to fill in a natural sequence; they were not logically necessary. On the contrary, they introduced some doubt into the picture we had sketched – they suggested that we may be farther than we think from understanding the world of galaxies.

The large-scale structure of the Fornax dwarf has been discussed by Hodge (1961a), Eskridge (1988), Demers, Irwin & Kunkel (1994), and by Stetson, Hesser & Smecker-Hane (1998). These authors find that this object is asymmetrical, with the densest region offset by 1.57 north and 2.70 east from the centroid determined at large radii. Stetson et al. find that the region of highest density, which is dominated by stars of below-average age, is boomerang shaped. The observed asymmetry of Fornax is most likely due to the patchy distribution of star formation during the most recent phase of its

Table 13.1. *Data on the Fornax dwarf galaxy*

$\alpha(2000)^a = 02^h\,39^m\,53\overset{s}{.}1$		$\delta(2000)^a = -34°\,30'\,16''$ (1)
$\ell = 237\overset{\circ}{.}24$		$b = -65\overset{\circ}{.}66$
$V_r = 53.0 \pm 1.8\,\mathrm{km\,s^{-1}}$ (2)		$\epsilon = 0.30 \pm 0.01$ (3)
$V = 7.3$ (4)	$B - V = 0.63 \pm 0.05$ (5)	$U - B = 0.08$ (5)
$(m - M)_o = 20.70 \pm 0.12$ (1)	$E(B - V) = 0.03 \pm 0.01$ (5)	$A_V = 0.09$ mag
$M_V = -13.1$ (6)	$D = 138 \pm 8$ kpc	$D_{LG} = 0.45$ Mpc
Type = dSph	scale-length	
	$= 10'.35\ (0.42\ \mathrm{kpc})$ (5,7)	
$R_c = 13'.8 \pm 1'.2\ (0.55\ \mathrm{kpc})$ (3,6)	$R_t = 71'.1 \pm 4'.0\ (2.85\ \mathrm{kpc})$ (3)	
$M = 6.8 \times 10^7 M_\odot$ (5)	$M_{HI} \lesssim 10^2 M_\odot$ (10)	
No. globulars = 5 (1,8)	No. planetaries = 1 (9)	
[Fe/H] = -1.3 ± 0.2 (1)	$12 + \log(\mathrm{O/H}) \geq 7.98$ (7)	

[a] Centroid for area with radius 14'.5. The region with highest stellar density, with radius 3'.6, is centered at $\alpha = 02^h40^m07\overset{s}{.}7$, $\delta = -34°\,29'\,10''$.

(1) Stetson et al. (1998).
(2) Mateo et al. (1991).
(3) Irwin & Hatzidimitriou (1993).
(4) See Section 13.2.1.
(5) Mateo (1998).
(6) Demers et al. (1994).
(7) Caldwell et al. (1992).
(8) Hodge (1961a).
(9) Danziger et al. (1978).
(10) Young (1999).

evolution. Detailed information on the structural parameters of the Fornax dSph galaxy are collected in Table 13.1.

Because of its low surface brightness and large size the integrated luminosity of the Fornax galaxy is both difficult to determine and slightly controversial (Demers et al. 1994). Caldwell et al. (1992) obtained $V(\mathrm{total}) = 6.9 \pm 0.5$, whereas Mateo (1998) in his review adopts $V(\mathrm{total}) = 7.6 \pm 0.3$. The corresponding absolute magnitudes are $M_V = -13.9$ and $M_V = -13.2$, respectively. This uncertainty in the luminosity of the Fornax dwarf results in a factor of two uncertainty in the specific globular cluster frequency for this galaxy. From precise radial velocities of 44 individual stars Mateo et al. (1991) find $M/L = 12 \pm 4$ in solar units.[1] From these velocities they also obtain a central density of $0.07 \pm 0.03\ M_\odot\ \mathrm{pc^{-3}}$. This value is comparable to that found in gas-rich dwarf irregulars but lower than that in most other dwarf spheroidals. Mateo et al. (1991) find no evidence for significant rotation about the minor axis of this system.

13.2.2 *Evolutionary history*

The most detailed color–magnitude diagram of Fornax has been published by Stetson et al. (1998). These authors find a sequence of several dozen very red luminous stars extending down toward the red, from the tip of the red giant branch. These objects

---

[1] Since the $B - V$ color of Fornax is similar to that of the Sun $M/L_B \approx M/L_V$.

are suspected of being double-shell intermediate-age carbon stars. A well-developed red giant clump in the Fornax color–magnitude diagram shows that this object contains a large population of stars with ages between 1 Gyr and 10 Gyr. A curious, and as yet unexplained, clustering of a few dozen stars is found just above the well-defined tip of the red giant branch. A range of sparsely populated giant branch tips fans upwards to the blue of the principal red giant branch. These stars probably represent the oldest, and most metal poor, population component(s) in the Fornax dwarf spheroidal. The extended horizontal branch of the Fornax system is found to lie 0.1 mag to 0.2 mag below the centroid of the red giant clump. Finally, a sequence of young ($T \sim 10^8$ yr) blue stars extends up to $V \sim 19 (M_V \sim -2)$. Stetson et al. find that these blue main sequence stars exhibit a clearly flattened distribution on the sky. Young blue stars in Fornax are more centrally concentrated than are the intermediate-age red clump stars. In summary it is concluded that Fornax contains three population components: (1) old widely dispersed stars that include RR Lyrae variables and horizontal-branch stars, (2) a slightly more centrally concentrated intermediate-age population that includes red clump stars and luminous objects that appear to be oxygen-rich M stars, and (3) a small young stellar population that has a flattened spatial distribution. These young stars are more centrally concentrated than either the old or the intermediate-age population. The very reddest carbon stars appear to be as centrally concentrated as the youngest blue stars.

13.2.3   Globular clusters

The first two globular clusters in Fornax (including NGC 1049) were discovered by Baade & Hubble (1939). A listing of all five presently known Fornax globulars is given in Hodge (1961b). A sixth cluster listed by him seems to be a clustering of distant background galaxies (Stetson et al. 1998). Integrated spectra and UBV photometry of the Fornax globulars by van den Bergh (1969) showed that these objects were all quite metal poor. The mean radial velocity of the Fornax globulars is $\langle V_r \rangle = +54 \pm 10$ km s^{-1}, which is in good agreement with the stellar systemic velocity of $+53 \pm 2$ km s^{-1} (Mateo et al. 1991). The structure of the Fornax globulars has been measured by Rodgers & Roberts (1994). These authors find that Clusters No. 1 and No. 2, which fit truncated King (1966) models, have similar core radii, which are twice as large as those of clusters Nos. 3, 4, and 5 (which do not appear to fit truncated profiles). Rodgers & Roberts conjecture that clusters Nos. 1 and 2 have interacted with each other dynamically, while the other clusters have had no such interactions during their lifetimes. In fact, with slightly different inter-action parameters, these two Fornax globular clusters might have merged (van den Bergh 1996b). This brings home the point that the globular cluster systems of dwarf spheroidal galaxies (which have low velocity dispersions) are likely to have non-negligible merger rates. It is noted in passing that the core radii of clusters No. 1 and No. 5 given by Smith et al. (1996) are inconsistent with those published by Rodgers & Roberts (1994).

The color–magnitude diagrams of globular clusters associated with the Fornax dwarf galaxy have been studied by Beauchamp et al. (1995) (Nos. 2 and 4), Buonanno et al. (1996) (Nos. 1, 3, 4, and 5), Marconi et al. (1999) (No. 4), Smith et al. (1996) (Nos. 1 and 5), Jørgensen & Jimenez (1997) (Nos. 1 and 3), and by Smith, Rich & Neil (1997) (No. 3). Smith et al. (1996) find that clusters No. 1 and No. 5 (which are both very metal poor) form a second-parameter pair, with No. 1 having a red horizontal branch, and No. 5 a hor-izontal branch of intermediate color. These authors also note that cluster No. 1 appears to have an anomalously low ratio of horizontal branch to red giant branch stars. Within the accuracy of their data the level of the horizontal branches in both clusters is equal to that of the Fornax field stars. According to Smith et al. (1997), cluster No. 3 also has a horizontal branch that is of intermediate color. Buonanno et al. (1996) show that this cluster has both a red giant clump and a blue horizontal branch tail. Van den Bergh (1996b) has speculated that this peculiar horizontal branch morphology might have been produced by the merger of two globulars of differing horizontal branch type. Rich (1998) and Smith, Rich & Neil (1998) point out that the Fornax globular clusters differ systematically from those of typical Galactic globulars. At a given horizontal branch type, which is characterized by $(B - R)/(B + V + R)^2$, Fornax clusters are more metal-poor than their Galactic counterparts. Alternatively, one might say that the Fornax clusters have unusually red horizontal branches for their metallicity. They might therefore be regarded as extreme examples of the NGC 7006 type of "second-parameter" clusters. Buonanno et al. (1998) have used the *Hubble Space Telescope* to show that Fornax clusters Nos. 1, 2, 3, and 5 have ages that are identical to within 1 Gyr. Furthermore, this age is indistinguishable from that of the very old metal poor Galactic globular cluster M92. In other words these four Fornax globulars are coeval with both NGC 2419 and the metal poor clusters in the main body of the Galactic halo. However, Marconi et al. (1999) find that Fornax cluster No. 4, which has [Fe/H] ≈ -2.0, is significantly younger than the other Fornax globulars.

Dubath, Meylan, and Mayor (1992) have measured the velocity dispersion in clusters Nos. 3, 4, and 5, from which they find mass-to-light ratios of $M/L_V = 3.2 \pm 0.8$, 1.7 ± 0.8, and 5.2 ± 1.9 (in solar units), respectively. Within their errors these values are indistinguishable from those of Galactic globular clusters. Buonanno et al. (1996) note that the globular clusters (except, perhaps, No. 4) appear to be significantly older than that of the bulk of the field stars in the Fornax system.

Published integrated absolute magnitudes for Fornax range from $M_V = -13.2$ to $M_V = -13.9$. Since Fornax contains five bona fide globular clusters the corresponding specific cluster frequencies (Harris & van den Bergh 1981) are $S = 26$ and $S = 13$, respectively. The cause of this high cluster frequency is not known. One might speculate that the reason for this high specific frequency is that Fornax has a low density, so that bulge shocks are weak. This might contribute to the survival of globular clusters that would have been destroyed in normal elliptical galaxies. An argument against this hypothesis (van den Bergh 1998b) is that the faintest Local Group dwarf spheroidals appear to have a low specific globular cluster frequency. Alternatively, it might be assumed that the physical environment in (some) dwarf spheroidal galaxies is particularly conducive to the formation of globular clusters. Miller et al. (1998) note that (1) the specific globular cluster frequency in early-type dwarf galaxies increases with decreasing luminosity and (2) that $\langle S \rangle = 6.4 \pm 1.2$ in nucleated systems, versus a lower value $\langle S \rangle = 3.3 \pm 0.5$ in early-type galaxies without nuclei. Possibly this difference is related to the fact that nucleated dwarfs are most common near the centers of clusters of galaxies.

The properties of individual globulars in Fornax are summarized in Table 13.2. The mean magnitude of these clusters is $\langle V \rangle = 13.7 \pm 0.5$, corresponding to $M_V = -7.1 \pm 0.5$. Note that this value is, within its error bars, indistinguishable from $\langle M_V \rangle \sim -7.4$ in giant galaxies (Harris 1991).

---

[2] B, V, and R are the number of blue, variable, and red horizontal branch stars, respectively.

Table 13.2. *Fornax globular clusters^{a,f}*

No.	α(2000)	δ(2000)	Conc. Cl.	R_c	V^c	$B - V^c$	$U - B^c$	[Fe/H]^d	σ(km s^{-1})^e	$\frac{B-R^d}{B+V+R}$
1	02^{h}37^{m}05^s	−34°10′10″	XI	5″.9	15.6	0.51	0.03	−2.20	⋯	−0.2
2	02 38 40	−34 48 05	VIII	5.9	13.6	0.68	0.03	−1.78	⋯	0.38
3^b	02 39 53	−34 16 09	V	3.4	12.5	0.64	0.03	−1.96	8.8	0.50
4	02 40 10	−34 32 10	IV	3.9	13.4	0.75	0.01	−1.35^e	5.1	⋯
5	02 42 21	−34 06 07	III	3.7	13.4	0.61	0.04	−2.20	7.0	0.44

^{a}Positions and core radii from Rogers & Roberts (1994).
^{b}= NGC 1049.
^{c}From van den Bergh (1969), except No. 1 from Hodge (1969).
^{d}Buonanno et al. (1998).
^{e}Dubath et al. (1992).
^{f}From *Hubble Space Telescope* images, Stetson, Hesser & Smecker-Hane (1998) find that Shapley's (1939) cluster Fornax 6 is, in fact, an asterism in which about half the members appear to belong to a distant compact cluster of galaxies.

13.2.4 Summary and conclusions

The discovery of the Fornax and Sculptor galaxies was important because it introduced astronomers to a populous new class of galaxies that falls outside the popular Hubble (1936) classification scheme. The luminosity of the Fornax dwarf still remains uncertain at the 0.7-mag level. This introduces an uncertainty of a factor of two in the unusually high specific globular cluster frequency for the Fornax dwarf. The environment in dwarf spheroidals like Fornax is particularly favorable for the merger of globular clusters. It is also noted that two of the metal-poor Fornax globulars appear to have differing horizontal branch morphologies (i.e., they present evidence for a "second-parameter" effect). Table 13.2 shows that cluster No. 2, which has the lowest central concentration of light, also has the reddest horizontal branch. From measurements of the stellar velocity dispersion for field stars in Fornax the mass-to-light ratio (in solar units) is found to be $M/L = 12 \pm 4$; that is, the Fornax dwarf appears to contain a significant amount of dark matter. Because Fornax is quite distant, measurement of the core radii of its globular clusters is difficult, which probably accounts for the discrepancies between different published values of r_c. It would be of interest to repeat such measurements with the *Hubble Space Telescope*.

An X-ray image of Fornax obtained with the *ROSAT* satellite by Gizis, Mould & Djorgovski (1993) reveals no sources that can be associated with this galaxy. They conclude that the low-density environment of this dwarf does not produce a population of accreting neutron stars through star–star collisions. Furthermore, their observed upper limit on the diffuse X-ray emission from Fornax places an upper limit of $10^5 \, M_\odot$ on the amount of 10^7 K gas that might be present in this dwarf galaxy.

13.3   The Sagittarius dwarf spheroidal

13.3.1 Introduction

The Sagittarius dwarf galaxy was discovered serendipitously by Ibata, Gilmore & Irwin (1994, 1995). These authors had used blue and red plates, obtained with the UK Schmidt telescope, to isolate a sample of candidate K and M giants in the general direction of the Galactic center. Follow-up spectroscopy of these objects showed that most of them exhibited the broad radial velocity dispersion expected for stars located in the bulge of the Galaxy. However, fields near $\ell = 5°, b = -15°$ exhibited a puzzling excess of giant stars with heliocentric velocities $V_r \approx +150 \, \mathrm{km \, s^{-1}}$, which could not be accounted for by reasonable models of the Galactic bulge or thick Disk. Follow-up observations showed that these excess giants belonged to a new dwarf spheroidal galaxy that had previously remained hidden among the rich population of foreground stars belonging to the Galactic disk and bulge. The Sagittarius dwarf is located at a distance of $\sim$24 kpc. Alcock et al. (1997a) have used MACHO observations of RR Lyrae stars in Sagittarius to establish the fact that the main body of this object is both elongated and inclined to the line of sight. From deep photometry in the V and I bands Mateo, Olszewski, & Morrison (1998) appear to have detected stars associated with the Sagittarius system out to an angular distance of 34°. This shows that the outer portions of this object resemble a stream, rather than an extension of the inner ellipsoidal regions of this galaxy. At even larger angular separations (Irwin 1999), the clusters Pal 2 and Pal 12 have both velocities and positions on the sky that would be consistent with their having been tidally stripped from Sagittarius.

These observations suggest that Sagittarius is currently being torn apart by Galactic tidal forces. The suspicion that the excess of red stars with $V_r \approx +150 \, km \, s^{-1}$ near $\ell = 5°$, $b = -15°$ is due to a dwarf spheroidal was strengthened by (1) the discovery of five carbon stars (which are relatively rare in the nuclear bulge of the Galaxy) and (2) the observation that the metallicity distribution of the K-type giants with $V_r \approx +150 \, km \, s^{-1}$ differs from that of Galactic bulge stars in the sense that it contains a larger fraction of metal-poor objects. Finally, Ibata et al. noted that the globular clusters NGC 6715 (= M54), Arp 2, Terzan 7, and Terzan 8 are grouped around the Sagittarius dwarf. The radial velocities and distances of these objects are consistent with the hypothesis that they are satellites of Sagittarius (Da Costa & Armandroff 1995). In this respect Sagittarius resembles the Fornax dwarf spheroidal (see Section 13.2), which has five satellite globular clusters. If one assumes that Fornax and Sagittarius have the same specific globular cluster frequency of $S = 29$, then with M_V (For) $= -13.1$ it follows that M_V (Sgr) $= -12.9$. However, the specific globular cluster frequency of Fornax is unusually high.[3] For a five times lower value of S the luminosity of the Sagittarius dwarf would be M_V (Sgr) $= -14.6$. In the subsequent discussion it will be assumed that M_V (Sgr) $= -13.8 \pm 0.9$. This luminosity is twice that of Fornax (see Table 13.1).

Ibata et al. (1997) find a stellar velocity dispersion of $(11.4 \pm 0.7) \, km \, s^{-1}$ for the stars in Sgr. This value is consistent with $\sigma = 12 \, km \, s^{-1}$ that is estimated by Da Costa & Armandroff (1995) from the radial velocity dispersion of the four globulars associated with the Sagittarius dwarf. By application of the virial theorem to the observed stellar velocity dispersion in Sgr, Ibata et al. (1997) find a mass of $\sim 1.5 \times 10^8 \, M_\odot$ for the Sagittarius dwarf. This value is twice as large as the $6.8 \times 10^7 \, M_\odot$ mass that Stetson et al. (1998) obtain for Fornax. Taken at face value these results suggest that Fornax has 5.8 globulars per $10^8 \, M_\odot$, compared to 3.3 globulars per $10^8 \, M_\odot$ for Sagittarius. In other words Fornax appears to contain almost twice as many globular clusters per unit mass as does Sagittarius. However, a mass value obtained by application of the virial theorem to a tidally disrupting galaxy should clearly be taken with a grain of salt! An additional complication is that Pal 2 and Pal 12 may originally have been Sagittarius clusters that were later stripped off by tidal interactions with the Galaxy (Irwin 1999). A summary of the data on the Sagittarius dwarf is given in Table 13.3.

13.3.2 The Sagittarius globular clusters

The fact that four globular clusters appear to be associated with the Sagittarius dwarf spheroidal was already noted by Ibata et al. (1994). The data on these globulars, which are collected in Table 13.4, show two remarkable features: (1) The clusters Ter 7, Ter 8, and Arp 2 are significantly fainter than the average Galactic globular, for which $\langle M_V \rangle = -7.1$ (Harris 1991), and (2) the cluster NGC 6715 (= M54), which

---

[3] The combined luminosity of the 14 known Local Group dSph galaxies that are less luminous than Fornax and Sagittarius is $M_V = -13.5$ (see Table 19.1). The total number of globular clusters that one expects to be associated with these galaxies is $0.26 \, S$, where S is the mean specific globular cluster frequency of faint globular clusters. In fact not a single globular cluster appears to be associated with these faint Local Group dwarf spheroidals. From Poisson statistics the probability of observing no globular clusters in our sample is 0.77 for $S = 1$, 0.27 for $S = 5$, and 0.006 for $S = 20$. These data strongly suggest that typical dSph galaxies have $S < 5$.

Table 13.3. *Data on the Sagittarius dwarf*

$\alpha(2000) = 18^h\ 55^m\ 04\overset{s}{.}3$		$\delta(2000) = -30°\ 28'\ 42''$ (1)
$\ell = 05\overset{\circ}{.}61$		$b = -14\overset{\circ}{.}09$
$V_r = +142.1 \pm 0.5$ km s^{-1} (1)		$\epsilon = ?$ (2)
$V = ?$ (3)		$B - V \sim 0.7$ (4)
$(m - M)_0 = 17.02$ (5)	$E(B - V) = 0.15$ (1)	$A_V = 0.46$ mag
$M_V = -13.8 \pm 0.9$ (5)	$D = 24.4$ kpc	$D_{LG} = 0.46$ Mpc
Type = dSph	$R_{GC} = 18.6$ kpc (1)	$R_h = 1\overset{\circ}{.}25$ (0.5 kpc) (6)
$M \sim 1.5 \times 10^8\ M_\odot$ (7)	$M_{HI} < 1 \times 10^4\ M_\odot$ (8)	
No. globulars =4	No. planetaries = 2 (9)	
[Fe/H] $\sim -1.1 \pm 0.3$ (10)	$12 + \log(O/H) = 8.35$ (11)	

(1) Adopted radial velocity, position, reddening, and distance are those of NGC 6715.
(2) Tidally distorted.
(3) See Section 13.3.1.
(4) Assumed similar to other dSph galaxies.
(5) Harris (1997).
(6) Ibata & Lewis (1998).
(7) Ibata et al. (1997).
(8) Koribalski et al. (1994).
(9) Walsh et al. (1997).
(10) Mateo et al. (1995).
(11) Zijlstra et al. (1997).

Table 13.4. *Globular clusters associated with Sagittarius*

Cluster	ℓ	b	$(m - M)_0{}^b$	$M_V{}^b$	$V_r{}^c$(km s^{-1})	[Fe/H]^d	$\frac{B-R^e}{B+V+R}$
M 54^a	5°.6	−14°.1	17.49	−9.96	+142 ± 1	−1.59	+0.20
Ter 7	3.4	−20.1	16.88	−5.00	+166 ± 4	−0.36	−0.86
Arp 2	8.5	−20.8	17.65	−5.24	+115 ± 3	−1.70	+0.53
Ter 8	5.8	−24.6	17.40	−5.06	+132 ± 8	−1.99	+0.82

^{a}= NGC 6715.
^{b}Harris (1997).
^{c}Da Costa & Armandroff (1995).
^{d}Layden & Sarajedini (1997).
^{e}Smith et al. (1998).

has $M_V = -9.96$, is much brighter than the average Galactic globular. In fact NGC 6715 was, before it was found to be associated with Sgr, thought to be the second brightest[4] Galactic globular. Because of its great luminosity some authors have suggested that M54 might be the nucleus of the Sagittarius dwarf galaxy. However, Sarajedini (1998a) has pointed out that this cluster is bluer than the bulk of the stellar population in Sgr, which

---

[4] The brightest Galactic globular is ω Centauri (= NGC 5139), with $M_V = -10.24$.

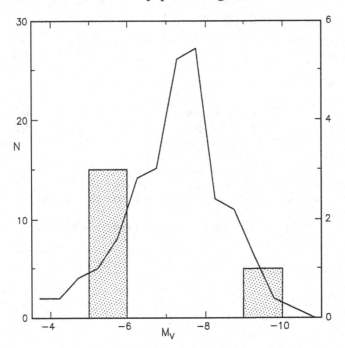

Fig. 13.2. Comparison between the luminosity distribution of all Galactic globular clusters (curve, scale at left) and of the globular clusters associated with the Sagittarius dwarf (histogram, scale at right). A Kolmogorov–Smirnov test shows that there is only a 10% probability that these two distributions were drawn from the same parent population. Note the similarity of the Sagittarius cluster luminosity function to that of Galactic halo globular clusters with $R_{GC} > 80$ kpc (see Fig. 13.3).

possibly militates against this suggestion.[5] The other three globulars associated with Sgr are all of below-average luminosity with $M_V = -5.24$ (Arp 2), $M_V = -5.00$ (Terzan 7), and $M_V = -5.06$ (Terzan 8). The luminosity function of the clusters associated with Sgr (see Figure 13.2) is reminiscent of that in the outermost Galactic halo (see Figure 13.3), in which NGC 2419 has $M_V = -9.53$, while the other clusters have $-6.0 < M_V < -4.0$. A Kolmogorov–Smirnov test shows that there is only a 4% probability that the Galactic halo clusters with $R_{GC} > 80$ kpc were drawn from a parent population with the same luminosity distribution as the clusters with $R_{GC} < 80$ kpc. By the same token it is found that there is only a 10% probability that the luminosity function of the globular clusters associated with the Sagittarius system was drawn from the same parent population as all Galactic globulars. A second similarity between the outer Galactic halo globulars and the Sagittarius cluster system is that both exhibit a much larger age range than is encountered in the inner Galactic halo.[6] According to Montegriffo et al. (1998), Arp 2 is

---

[5] In a study of the spheroidal galaxies in the Virgo cluster (van den Bergh 1986), it was found that the fraction of such objects that are nucleated drops from ~100% at $M_B = -17$ to only ~10% near $M_B = -12$. In other words the a priori probability that a dwarf as faint as Sgr would be nucleated is rather small, unless Sagittarius was originally much more luminous than it is now.

[6] The cluster Palomar 12, which according to Irwin (1999) may have been tidally stripped from the Sagittarius system, appears to be 3–4 Gyr younger than the majority of Galactic globulars (Rosenberg et al. 1998a). In this respect it resembles the young Sagittarius globular clusters Arp 2 and Ter 7.

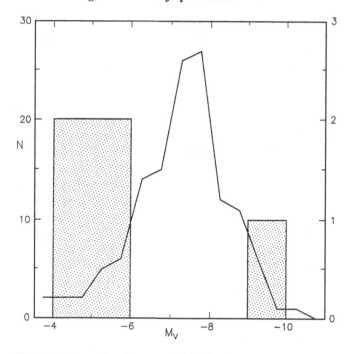

Fig. 13.3. Comparison between the luminosity distribution of all Galactic globulars with $R_{GC} < 80$ kpc (curve, scale at left) and that of the globular clusters with $R_{GC} > 80$ kpc (histogram, scale at right). A Kolmogorov–Smirnov test shows that there is only a 4% probability that these two distributions were drawn from the same parent population.

~4 Gyr younger than M54 and Ter 8, while Ter 7 is 6 to 7 Gyr younger. Determinations of [α/Fe] for stars in the Sgr clusters would allow one to strengthen the age determinations for these objects.

The cluster M54 has an age that is similar to that of the Fornax globulars (see Section 13.2.3), the outer halo cluster NGC 2419 (Harris et al. 1997), and metal-poor clusters in the inner halo of the Galaxy. In contrast, the outer halo clusters Palomar 3, Palomar 4, and Eridanus are younger by up to 2 Gyr (Stetson et al. 1998). The fact that the faint cluster Ter 8 is metal poor ([Fe/H = -2.0) demonstrates that there is not a one-to-one relationship between metallicity and luminosity in Sgr. Table 13.4 shows that the metallicity of Terzan 7 is an order of magnitude greater than that of the other globular clusters associated with the Sagittarius dwarf. Moreover, the age of Ter 7 appears to be similar to that of the bulk of the field stars in Sgr. This suggests that the Sgr dwarf spheroidal galaxy was able to retain and enrich gas for a period that may have been as long as 6 Gyr. However, the gas in Sgr subsequently seems to have been lost, and no stars appear to have ages ≲4 Gyr (Montegriffo et al. 1998). Smith, Rich & Neil (1998) note that the three metal-poor Sagittarius globulars have bluer horizontal branches than do those in the (metal poor) Fornax globulars.

The fact that (1) the globular clusters associated with both the Sagittarius dwarf and the outer Galactic halo at $R_{GC} > 80$ kpc appear to have similar bimodal luminosity functions and (2) that the globulars in Sgr and in the outer halo have a larger age range than do those at $R_{GC} < 80$ kpc might be taken to indicate that Sgr is a Searle–Zinn (1978) fragment that fell inward toward the center of the Galaxy. An argument in favor of the idea that the Sagittarius dwarf originated in the outer halo is provided by the [Fe/H] versus horizontal branch type diagram of Smith, Rich & Neil (1998). These authors show that in such a plot the Sgr globular clusters Arp 2, Terzan 8, and M54 (but not Terzan 7!) fall in the same region as do Galactic halo globular clusters. An argument against this hypothesis is, however, that the Sagittarius dwarf presently appears to be on a short-period ($\sim$0.7 Gyr) orbit (Velázquez & White 1995; Ibata et al. 1997; Ibata & Lewis 1998). Possibly Sgr was scattered into a short-period orbit after a close gravitational interaction with the Magellanic Clouds (Zhao 1998). However, such a close encounter has a low a priori probability. Finally, it is noted that only one (Arp 2) of the globular clusters associated with the Sagittarius system has the large diameter that is diagnostic of outer halo clusters (van den Bergh & Morbey 1984b). In summary it appears that one must bring in a Scottish "not proven" verdict on the hypothesis that Sgr started its existence as a Searle–Zinn fragment.

13.3.3 Evolutionary history

Studies of the stellar population of the Sagittarius dwarf galaxy are rendered difficult because this object is located behind the rich stellar population of the Galactic bulge. Nevertheless, Marconi et al. (1998) have been able to draw some tentative conclusions about the nature of the stellar population in the Sagittarius dwarf. These authors find that individual stars in Sgr have metallicities in the range $-1.58 \lesssim$ [Fe/H] $\lesssim -0.71$ and lie on isochrones indicating that the dominant population has an age of $\simeq$10 Gyr. This dominant population appears to be quite similar to that in the globular Terzan 7, which suggests that this cluster may have formed near the end of the major burst of star formation in this galaxy. Fahlman et al. (1996) have obtained a color–magnitude diagram for main-sequence stars in both the globular cluster NGC 6715 and the field population in the Sgr dwarf. They find that the Sgr field star main sequence is almost as blue as that of the metal-poor cluster NGC 6715. If the field population is relatively metal rich, then it follows that it must be younger than that in the globular cluster. From V and I photometry in three large fields Bellazini, Ferraro & Buonanno (1999) conclude that star formation in Sgr started early and ended abruptly $\sim$8 Gyr ago, probably at the time when the gas in this object was suddenly depleted. They do, however, claim to see some evidence for the continuation of star formation at a low rate until $\sim$1 Gyr ago.

Mateo et al. (1995) have discovered seven RR Lyrae variables in a small field in Sgr, which are probably physically associated with this galaxy because they all have mean apparent magnitudes similar to that of the Sagittarius horizontal branch. Large numbers of additional RR Lyrae stars in Sgr have been discovered by Alard (1996). The existence of these variables shows that there is a component of the field population of Sgr that has an age $\gtrsim$ 10 Gyr. From their observations Mateo et al. estimate that Sgr contains $1,930 \pm 730$ RR Lyrae stars if $M_V(\mathrm{Sgr}) = -13$ and 310 ± 120 RR Lyrae variables if $M_V(\mathrm{Sgr}) = -11$. Bellazini, Ferraro & Buananno (1999) have obtained color–magnitude diagrams that seem to confirm the presence of some very metal poor stars in the Sagittarius dwarf.

Whitelock, Irwin & Catchpole (1996) have found four variable stars that are probably semiregular and/or Mira type variables, with periods of 150–300 days, that are members of Sgr. These authors have also found 71 late-type giants that are probably associated with the Sgr dwarf galaxy. These objects are either carbon stars, or probable carbon stars, with a wide range of properties. The reddest appear to be similar to the C stars in Fornax, while the bluest may resemble the carbon stars in the Sculptor, Draco, and Ursa Minor systems. The cool Sgr giants appear to lie on a giant branch with a metallicity that is slightly lower than that of 47 Tucanae, which has [Fe/H] = −0.76. Ng (1997) has suggested that the carbon stars discovered by Azzopardi et al. (1991) in the Galactic bulge might, in fact, have been formed in the (partly disintegrated) Sagittarius dwarf galaxy. However, Whitelock (1999) has pointed out that the distribution of the C stars in the Galactic bulge, in a $J - H$ versus $H - K$ color–color diagram, is not like that of the carbon stars in the Sagittarius dwarf. This observation shows that the C stars in Sagittarius are both fainter and bluer than those in the nuclear bulge of the Galaxy. This observation militates against Ng's suggestion that the C stars in the Galactic bulge are, in some way, physically associated with those in the Sagittarius dwarf system. Alternatively, it might be argued that the C stars in the Galactic bulge were formed during the last disk crossing when the metal-poor gas associated with Sagittarius slammed into the metal-rich disk gas. Whitelock has found 30 C stars in Sagittarius, of which eight are probably Mira variables, and seven appear to be small-amplitude semiregular variables.

13.3.4 Interstellar matter

Bland-Hawthorn et al. (1998) have discussed the possible association of an H I cloud in Aquila with the Sagittarius dwarf. Neither its radial velocity nor its ~35° distance from Sgr argue for a close physical linkage between these objects. The planetary nebulae He 2-436 and Wray 16-423, which appear to be associated with Sgr, have been observed by Zijlstra, Dudziak & Walsh (1997), who find that they have 12 + log(O/H) values of 8.3 and 8.4, respectively. This is lower than the solar value of 8.87 by ~0.5 dex. Walsh et al. (1997) find that the O, Ne, and Ar abundances in the Sgr planetaries are, within their errors, identical to those of the planetary nebula in Fornax (Maran et al. 1984).

13.3.5 Tidal interactions

The observations of the Sagittarius dwarf by Ibata et al. (1994, 1995) clearly show that it is greatly elongated as a result of one, or more, tidal interactions with the Milky Way system. More recently Ibata et al. (1997) were able to detect this galaxy over an area of 8° × 22°, corresponding to linear dimensions of 2.2 kpc × 8.0 kpc. Mateo, Olszewski & Morrison (1998) seem to have traced it to an angular distance of 34°. From numerical simulations Velázquez & White (1995) concluded that Sgr must be on a relatively short-period orbit with perigalacticon at 10 kpc, apogalacticon at ~50 kpc, and a period of ~760 Myr. Their simulations suggest that the presently observed distortion of Sgr might have occurred predominantly during the last pericentric passage, rather than on the present one. A possible problem with their model is that it is not clear whether an object as fragile as Sgr would have been able to survive 10–20 tidal encounters with the Galaxy. More recently Ibata et al. (1997) have measured a mean proper motion of 2.1 ± 0.7 mas yr^{-1}, corresponding to 250 ± 90 km s^{-1} toward the Galactic plane, for stars that appear to be physical members of the Sgr system. The corresponding space velocity components are $U = 232 \pm 60$ km s^{-1} toward the Galactic center and $W = 194 \pm 60$ km s^{-1} toward the Galactic pole. Unfortunately the V velocity component, which points in the direction of Galactic rotation, is not significantly constrained by the proper motion observations. By assuming $V = 0$ km s^{-1}, Ibata et al. find an orbital period <1 Gyr. Zhao (1998) has suggested that Sagittarius might have been scattered into its present orbit by an encounter with the LMC + SMC. Additional orbit modeling, which uses the observed half-light radius $r_h = 1°25$ of Sgr as an additional constraint, has been performed by Ibata & Lewis (1998), who (neglecting a possible gravitational encounter with the LMC + SMC) again find a relatively small orbit with an initial period of ~1 Gyr and a present period of ~0.7 Gyr. Ibata et al. (1997) use the virial theorem, in conjunction with a stellar velocity dispersion $\sigma = 11.4 \pm 0.7$ km s^{-1} and a characteristic major axis diameter of 9.6 kpc, to obtain a mass of ~$1.5 \times 10^8 \, M_\odot$ for the Sagittarius dwarf. Ibata & Razoumov (1998) point out that a higher mass of ~$10^9 \, M_\odot$ for Sgr would have produced significant tidal distortions of the outer H I Disk of the Galaxy. Irwin (1999) estimates that as many as half of the stars in the Sagittarius dwarf may have been lost as a result of tidal interactions with the Galaxy. Furthermore, the positions of the globular clusters Palomar 2 and Palomar 12 would be consistent with their having been torn from Sagittarius by Galactic tides. In this connection it is of interest to note that Arp 2, Pal 12, and Terzan 7 are all "young" globular clusters. The fact that one half to one third of all young globular clusters presently known in the Galaxy appear to be associated with Sagittarius suggests that *late infall and capture of dwarf spheroidals probably accounted for no more than two or three times the mass of Sagittarius*. In particular such late infall may account for carbon stars observed in the Galactic halo. With a three-dimensional Local Group velocity dispersion of 108 ± 15 km s^{-1} (see Section 19.2), the crossing time for the Galaxy (assuming a radius of 100 kpc) is <2×10^9 yr. This suggests that accretion during the first 2 Gyr of the Galactic lifetime should be regarded as part of the Galaxy formation process, whereas later accretion is more properly described as infall. It should, perhaps, be emphasized that early infall might have included mergers with fragments of significant mass.

13.3.6 Summary

The Sagittarius dwarf spheroidal is the nearest external galaxy. Its elongated structure shows that it has suffered severe tidal distortions, resulting from one or more gravitational interactions with the Milky Way system. The presence of RR Lyrae stars and blue horizontal branch stars show that star formation in Sagittarius started >10 Gyr ago. The main population component in Sgr appears to be similar to that in the globular cluster Terzan 7, for which Chaboyer, Demarque & Sarajedini (1996) find an age of 7.2 ± 0.2 Gyr. After this time, most of the gas in the Sagittarius dwarf appears to have been ejected by supernova-driven winds, or via ram pressure exerted by the Galactic disk. Later, some star formation may, however, have continued at a low level until ~1 Gyr ago.

The other three globular clusters associated with Sgr appear to have ages comparable to those of typical inner-halo Galactic globulars. The Sagittarius dwarf resembles the outer halo of the Galaxy, in which the globular clusters also exhibit a discernible age spread. Another similarity between the halo at $R_{GC} > 80$ kpc and Sgr is that they both hint at a bimodal luminosity function for globular clusters. These similarities make it tempting to suggest that the Sagittarius dwarf galaxy represents the stellar remnant of a Searle–Zinn fragment that originated in the outermost part of the Galactic halo. However, this conjecture appears to conflict with orbital modeling, which suggests that Sgr is on an orbit with R (apogalacticon) $\sim$60 kpc and a period of only $\sim$0.7 Gyr. A possible escape from this dilemma is provided by the hypothesis that Sgr was scattered into a shorter-period orbit by an encounter with the LMC + SMC. The observation that stars were forming in Sgr as recently as 7 Gyr ago shows that its orbit prior to that time did not take it through the Galactic plane, where gas would have been swept out by ram pressure. If Sgr has been on a short-period orbit since then it would have passed through the Galaxy $\sim$10 times. It is not clear if a fragile dSph galaxy could have survived so many tidal encounters. The fact that one half to one third of all known young Galactic globular clusters are associated with Sagittarius suggests that late infall does not account for a major fraction of the stars in the Galactic halo.

14

Dwarf spheroidals in the Andromeda subgroup

14.1 Introduction

While inspecting a mosaic of nine deep IIIaJ plates of M31 and its environment taken with the Palomar 1.2-m Schmidt telescope, van den Bergh (1972a) discovered three faint nebulous patches that were immediately suspected of being dwarf spheroidal companions to the Andromeda galaxy. These objects were dubbed Andromeda I, Andromeda II, and Andromeda III. Subsequently van den Bergh (1972b) used the Palomar 5-m Hale reflector to resolve And III into stars that had approximately the same V magnitude as the brightest stars in NGC 185 and NGC 205. Later van den Bergh (1974b) showed that And I and And II resolve at about the same magnitude as And III. These observations indicated that And I, And II, and And III are all located at about the same distance as M31. The suspicion that they are physically associated with M31 was strengthened by a Palomar Schmidt survey of an area of ~700 square degrees[1] surrounding the Andromeda galaxy, which showed that And I, II, and III are concentrated toward M31. Of a fourth candidate van den Bergh (1972a) wrote: "The object And IV is probably not a dwarf spheroidal galaxy. It is smaller and bluer than the other objects in Table 1 [i.e., And I, II & III]. Furthermore, And IV has a much higher surface brightness than do the other galaxies in this table. This suggests that And IV, which is located very close to M31, might be a relatively old star cloud in the outer Disk of M31." This suspicion was subsequently confirmed by Jones (1993), who obtained a color–magnitude diagram of And IV that showed that its stars belonged to Population I. An image of And IV is shown in Figure 4 of Caldwell et al. (1992).

Recently the digitized images of the second *Palomar Sky Survey* (Reid et al. 1991) have been used by Armandroff, Davies and Jacoby (1998), and by Karachentsev & Karachentseva (1999), to search for additional faint objects that are probably situated within the Andromeda subgroup of the Local Group. At the time of writing (1999) this digitized survey was still incomplete, so that additional very faint members of the Andromeda subgroup of the Local Group probably remain to be discovered.

---

[1] The field searched had a radius of ~15°, corresponding to $R(M31) \approx 200\ (D/760)$ kpc. A sphere with a radius of 200 kpc around the Galactic center would include the LMC, the SMC, and six dwarf spheroidals. However, Leo I and Leo II would be excluded.

234

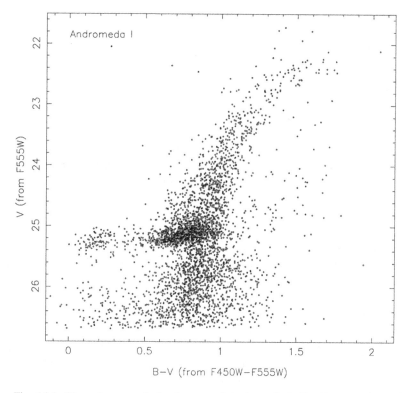

Fig. 14.1. The color–magnitude diagram of Andromeda I (Da Costa et al. 1996) shows that this dwarf spheroidal has a strong population on the red side of the horizontal branch. Reproduced with the kind permission of Taft Armandroff.

14.2 Andromeda I

And I appears almost spherical in projection. Caldwell et al. (1992) show that, for $r > 40''$, the brightness profile of And I resembles an exponential disk. The disk scale-length is $83''.8 \pm 0''.9$, which corresponds to 330 pc. Da Costa et al. (1996) have used the *Hubble Space Telescope* to obtain a beautiful color–magnitude diagram of And I, which is shown in Figure 14.1. This color–magnitude diagram shows a horizontal branch with a strong red clump and a weak population of blue horizontal-branch stars. Its red horizontal branch, in conjunction with a metallicity [Fe/H] $= -1.45 \pm 0.2$ (which is estimated by comparing the And I red giant branch with those of Galactic globulars of known metallicity), suggests that Andromeda I might be related to Galactic "second-parameter" clusters such as NGC 7006 and the distant Galactic satellite Leo II (Demers & Irwin 1993). Da Costa et al. have found four cluster-type variables that are, no doubt, associated with the population component observed to the blue of the RR Lyrae gap. Adopting $V(\text{HB}) = 25.25$, $A_V = 0.13$ mag, and $M_V(\text{RR}) = +0.7$ for [Fe/H] $= -1.45$, Da Costa et al. (1996) obtain a true distance modulus $(m - M)_0 = 24.55 \pm 0.08$, corresponding to a distance of 810 ± 30 kpc. This agrees well with the distance $D = 790 \pm 60$ kpc that Mould & Kristian (1990) had previously derived from the I magnitude of the tip of the

And I red giant branch. And I is located at an angular distance of $3°31$ from the nucleus of M31, which corresponds to a projected linear separation of $46.2(D/800)$ kpc. The observational data on And I are summarized in Table 14.1. According to Armandroff (1998), the bulk of the stellar population in And I has an age of $\sim$10 Gyr.

14.3   Andromeda II

This dwarf spheroidal galaxy is located at a distance of $10°30$ from the nucleus of the Andromeda galaxy and $4°61$ from the nucleus of the Triangulum galaxy. Even though And II is closest to M33 it might, in fact, be a satellite of M31. This is so because M31 has a deeper potential well than does M33. Alternatively, And II could be a free-floating member of the Andromeda subgroup of the Local Group. An accurate radial velocity for the stars in And II might throw some light on this problem. Observations of And II have been reported by Aaronson et al. (1985), who found that this object has an asymptotic branch extending to $M_{bol} \approx -4.5$. They concluded that And II contains at least one, and perhaps two, carbon stars. This strongly suggests that And II resembles many of the dwarf spheroidals associated with the Galaxy, which also contain intermediate-age stars. A color–magnitude diagram for And II on the Gunn g,r system was obtained by König et al. (1993) with the Palomar 5-m reflector. From a comparison with the color–magnitude diagrams of Galactic globular clusters of known metallicity, König et al. derive a rather uncertain distance modulus of $(m - M)_0 = 23.8 \pm 0.4$ and a metallicity range of $-2.03 < $ [Fe/H] $ < -1.47$. Taken at face value these results would seem to indicate that And II is located in front of M31, and hence it is not physically associated with M33, which is more distant than M31. More recent *Hubble Space Telescope* observations by Armandroff (1998) yield a distance $D \approx 700$ kpc for And II. This suggests that this object lies in front of the Andromeda galaxy. It also indicates that And II is probably associated with M31, rather than with M33, which is more distant than the Andromeda galaxy. Armandroff also finds that $\langle$[Fe/H]$\rangle \approx -1.3$ for And II and that the metallicity dispersion in And II is $\sigma = 0.35$ dex. This is larger than the value $\sigma = 0.20$ dex found for And I. The fraction of blue horizontal branch stars is 18% in And II, compared to 13% in And I. Presently available data on And II are summarized in Table 14.2.

14.4   Andromeda III

The dwarf spheroidal galaxy And III was discovered by van den Bergh (1972a) on IIIaJ plates obtained with the Palomar 1.2-m Schmidt telescope. Subsequently he (van den Bergh 1972b) was able to resolve this object with the Hale 5-m telescope into a flattened swarm of individual stars, with magnitudes comparable to those of the brightest stars in the M31 companions NGC 185 and NGC 205. These observations established that And III was a dwarf spheroidal situated at roughly the same distance as the Andromeda galaxy. A photograph of And III is shown in Figure 14.2. And III is located at an angular distance of $4°98$ from the nucleus of M31, which corresponds to a projected distance of 66 kpc. The most detailed study of And III has been undertaken by Armandroff et al. (1993) with the Kitt Peak 4-m telescope. By comparing the color–magnitude diagram of this object, with those of Galactic globular clusters of known metallicity, these authors find that And III has a metallicity [Fe/ H] $= -2.0 \pm 0.15$. They also find some evidence for a small (σ[Fe/ H] ~ 0.2 dex) metallicity dispersion among the stars in And III. Unfortunately, these ground-based observations do not go deep enough to establish the

Table 14.1. *Andromeda I*[a]

$\alpha(2000) = 00^h\, 45^m\, 43^s$		$\delta(2000) = +38°\, 00'\, 24''$ (1)
$\ell = 121°69$		$b = -24°85$
		$\epsilon = 0.0$ (2)
$V = 12.75 \pm 0.07$ (2)		$B - V = 0.80 \pm 0.02$ (2)
$(m - M)_o = 24.55 \pm 0.08$ (3)	$E(B - V) = 0.04$ (3)	$A_V = 0.13$ (3)
$M_V = -11.8 \pm 0.1$ (2,3)	$D = 810 \pm 30$ kpc (3)	$D_{LG} = 0.36$ Mpc
Type = dSph	scale-length	
No. globulars = 0	$= 83''8 \pm 0''9$ (330 pc) (2)	
[Fe/H] = -1.45 ± 0.2 (3)		

[a]Located at a distance of 3°31 from M31.
(1) Mateo (1998).
(2) Caldwell et al. (1992).
(3) Da Costa et al. (1996).

Table 14.2. *Andromeda II*[a]

$\alpha(2000) = 01^h\, 16^m\, 27^s$		$\delta(2000) = +33°\, 25'\, 42''$ (1)
$\ell = 128°91$		$b = -29°15$
		$\epsilon = 0.3$ (2)
$V = 12.71 \pm 0.16$ (2)		
$(m - M)_o = 24.22$ (3)	$E(B - V) = 0.08$ (3)	$A_V = 0.25$ (3)
$M_V = -11.8$ (2)	$D \approx 700$ kpc (3)	$D_{LG} = 0.26$ Mpc
Type = dSph	scale-length	
No. globulars = 0	$= 94''1 \pm 2''0$ (320 pc)	
[Fe/H] ≈ -1.3 (3)		

[a]Located at a distance of 10°30 from M31, and 4°61 from M33.
(1) Mateo (1998).
(2) Caldwell et al. (1992).
(3) Armandroff (1998).

morphology of the horizontal branch of this dwarf spheroidal galaxy. From the I magnitude of the tip of the red giant branch of And III, Armandroff et al. derive a true distance modulus of $(m - M)_o = 24.4 \pm 0.2$, corresponding to a distance of 760 kpc. Comparison with Table 3.1 shows that this distance is, within its errors, identical to that of M31. From counts of asymptotic branch stars Armandroff et al. conclude that $(10 \pm 10)\%$ of the luminosity of And III is produced by intermediate-age stars with ages ranging from 3 Gyr to 10 Gyr. Available data on And III are collected in Table 14.3.

14.5   The recently discovered dwarf Andromeda V

Armandroff, Davies & Jacoby (1998) have used a digitized version of the second *Palomar Sky Survey* (Reid et al. 1991) to search an area of 1,550 square degrees around M31 for low surface-brightness companions to the Andromeda galaxy. This survey easily

Table 14.3. *Andromeda III*[a]

$\alpha(2000) = 00^h\,35^m\,17^s$		$\delta(2000) = +36°\,30'\,30''$ (1)
$\ell = 119°31$		$b = -26°25$
		$\epsilon = 0.6$ (2)
$V = 14.21 \pm 0.08$ (2)		
$(m-M)_0 = 24.4 \pm 0.2$ (3)	$E(B-V) = 0.05$ (3)	$A_V = 0.16$ mag
$M_V = -10.2$ (2)	$D = 760 \pm 70$ kpc	$D_{LG} = 0.31$ Mpc
Type $=$ dSph	scale-length	
No. globulars $= 0$	$= 45''0 \pm 0''8$ (165 pc)	
[Fe/H] $= -2.0 \pm 0.15$ (3)		

[a]Located at a distance of 4°98 from M31.
(1) Mateo (1998).
(2) Caldwell et al. (1992).
(3) Armandroff et al. (1993).

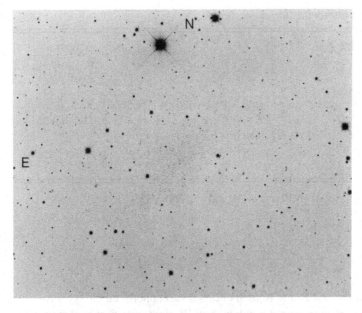

Fig. 14.2. KPNO 4-m telescope photograph of the dwarf spheroidal galaxy Andromeda III. Reproduced with the kind permission of Taft Armandroff.

recovered the already known dSph galaxies And II and And III and also showed a new object that was dubbed And V (See Figure 14.3). [As was already noted in Section 14.1, And IV (van den Bergh 1972a, Jones 1993) is probably a relatively young star cloud associated with the outer disk of M31.] Because the entire region of the second *Palomar Sky Survey* around M31 has yet to be digitized, And I was not recovered by Armandroff et al. Other faint companions to M31 may also still be lurking in fields that have not yet been digitized. The main problems encountered in searches for objects with very low surface brightness, on such very deep surveys, is that some faint patches might be due to

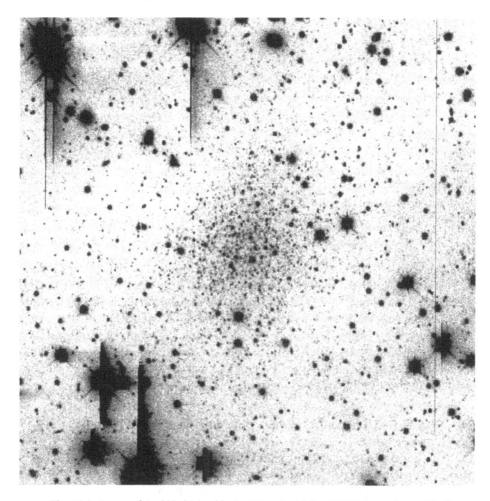

Fig. 14.3. Image of And V obtained in the V band with the KPNO 4-m reflector. North is at the top, and east to the right. Reproduced with the kind permission of Taft Armandroff.

either (1) ghosts produced by internal reflections in the optics of the Schmidt telescope or (2) to high-latitude emission or reflection nebulae. Ghosts may be eliminated by taking a second exposure centered on a point that is offset from the first position, while high-latitude nebulosity can only be eliminated by obtaining large reflector images, which are capable of resolving dSph galaxies into stars. Such images show that And V is indeed a dwarf spheroidal and that the tip of its giant branch is located at $I = 20.85 \pm 0.10$. This yields a true distance modulus $(m - M)_0 = 24.55 \pm 0.12$, corresponding to a distance of 810 ± 45 kpc. Comparison with the data in Table 3.1 suggests that And V has a distance that is similar to, or slightly greater than, that of its parent galaxy M31. From a comparison of the I versus $V - I$ color–magnitude diagram of And V with that of Galactic globular clusters of known metallicity, Armandroff et al. conclude that And V has a metallicity of [Fe/H] ≈ -1.5. Their color–magnitude diagram also shows that And V does not contain a prominent intermediate-age population. In this respect, it is similar to

Table 14.4. *Data on Andromeda V[a]*

$\alpha(2000) = 01^h\ 10^m\ 17\overset{s}{.}1$		$\delta(2000) = +47°\ 37'\ 41''$
$\ell = 126\overset{\circ}{.}22$		$b = -15\overset{\circ}{.}12$
		$\epsilon = 0.15$
$V \approx 15$		
$(m - M)_0 = 24.55 \pm 0.12$	$E(B - V) = 0.16 \pm 0.03$	$A_V = 0.50$ mag
$M_V \approx -10.2$	$D = 810 \pm 45$ kpc	$D_{LG} = 0.36$ Mpc
Type = dSph	size = $135'' \times 155''$ (0.5 kpc	
No. globulars = 0	$\times$ 0.6 kpc)	
[Fe/H] ≈ -1.5	No. planetaries = 0	

[a] All data from Armandroff et al. (1998).
Located at a distance of $8\overset{\circ}{.}03$ from M31.

And I and And III. Not unexpectedly, And V (which exhibits no evidence for recent star formation) has not been detected as a source of either IR or radiocontinuum radiation. No search for neutral hydrogen in And V has yet been undertaken. The central surface brightness of And V is lower than that of And I, And II, and And III, which explains why it was not detected during the survey of van den Bergh (1972a). Armandroff et al. (1998) estimate that $M_V = -10.2$ for And V. According to Armandroff (1998), And V contains no planetary nebulae and no upper asymptotic branch stars. A summary of presently available data on Andromeda V is given in Table 14.4.

14.6   Additional probable and possible members of the Andromeda subgroup

Searches of the (presently still incomplete) digitized version of the *Palomar Sky Survey* (Armandroff, Davies & Jacoby 1998; Karachentsev & Karachentseva 1999) have already turned up a number of additional probable and possible new members of the Andromeda subgroup of the Local Group.

14.6.1   *Pegasus dwarf spheroidal (= And VI)*

Grebel & Guhathakurta (1999) have used the Keck II 10-m telescope to obtain CCD images in the V and I passbands of the Pegasus dwarf spheroidal galaxy. They find that the tip of the red giant branch occurs at $I = 20.45 \pm 0.05$ and that the reddening in this direction is $E(B - V) = 0.04 \pm 0.01$. The corresponding true distance modulus is $(m - M)_0 = 24.60 \pm 0.2$. This value is, within observational errors, indistinguishable from that of M31. A similar distance, $(m - M)_0 = 24.5 \pm 0.2$, has been found by Hopp et al. (1999) by using the same technique. In conjunction with the observation that And VI is located only 20° from the nucleus of M31, this strongly suggests that And VI is a physical member of the Andromeda subgroup of the Local Group. From the $(V - I)_0$ color of its giant branch Grebel & Guhathakurta estimate a metallicity of $\langle[Fe/H]\rangle = -1.3 \pm 0.3$. No obvious radial gradient in metallicity is observed in the Pegasus dwarf. The central surface brightness of And VI in the V band is found to be $\sigma_0 \simeq 24.5$ mag arcsec^{-2}, which is about a magnitude fainter than that of And I, And II, and And III. The observational data on And VI are collected in Table 14.5.

Table 14.5. *Data on Pegasus dwarf spheroidal (= And VI)*[a]

$\alpha(2000) = 23^h\,51^m\,39\overset{s}{.}0$		$\delta(2000) = +24°\,35'\,42''$ (1)
$\ell = 106\overset{\circ}{.}01$		$b = -36\overset{\circ}{.}30$
		$\epsilon = 0.3 \pm 0.1$
$V = 14.1 \pm 0.2$ (2)	$B - V = 0.5$ (3)	
$(m - M)_0 = 24.6 \pm 0.2$ (2)	$E(B - V) = 0.04 \pm 0.01$	$A_V = 0.12$
$M_V = -10.6 \pm 0.3$	$D = 830 \pm 80$ kpc	$D_{LG} = 0.43$ Mpc
Type = dSph	size = $2' \times 4'$ (0.4 kpc	
No. globulars $= 0$ (2)	$\times$ 0.8 kpc) (1)	
[Fe/H] $= -1.3 \pm 0.3$ (2)	No. planetaries $= 0$ (2)	

[a]Located at a distance of $19\overset{\circ}{.}76$ from M31.
(1) Karachentsev & Karachentseva (1999).
(2) Grebel & Guhathakurta (1999).

14.6.2 The Cassiopeia dwarf spheroidal (= And VII)

From V and I photometry of And VII, Grebel & Guhathakurta (1999) find that the tip of the red giant branch occurs at $I = 20.60 \pm 0.05$. Adopting $E(B - V) = 0.17 \pm 0.03$, we obtain a true distance modulus of $(m - M)_0 = 24.4 \pm 0.2$, which corresponds to a distance of 760 ± 70 kpc. This places And VII at the same distance as M31 (which is located 16° away). No globular clusters or H II regions are found to be associated with this dwarf galaxy. From the $(V - I)_0$ color of the red giant branch of And VII, Grebel & Guhathakurta estimate the mean metallicity of this system to be $\langle[\text{Fe/H}]\rangle = -1.4 \pm 0.3$. The observational data on this object are summarized in Table 14.6.

14.6.3 Dwarf in Camelopardalis

Another candidate dwarf member of the Andromeda subgroup of the Local Group (see Table 14.7) has been reported by Karachentsev (1994). Follow-up observations by Grebel (1999) show that this object has a dwarf spheroidal-like red giant branch that seems to exhibit a large intrinsic scatter. Karachentsev, Markova & Andersen (1999) estimate $I = 22.4 \pm 0.2$ for the tip of the red giant branch in this galaxy. From this value these authors obtain a distance of 1.64 Mpc for Cam A. They conclude that "Cam A may be considered as a new peripheral member of the Local Group." Karachentsev et al. find that the youngest stellar population of this galaxy seems to have an age of $\sim$100 Myr. Because of its large distance Cam A has not been included in the list of probable Local Group members given in Table 2.1.

14.6.4 Two Andromeda subgroup suspects

Two additional dwarf galaxy suspects have been found by Karachentsev & Karachentseva (1999) during the course of their search on the prints of the *Palomar Sky Survey* of an area with a radius of 22° around M31. The first of these is located at $\alpha = 23^h\,08^m\,22\overset{s}{.}8$, $\delta = +56°\,30'\,1''$ (J 2000) and has a very low surface brightness. The possibility that this small and very faint object at $b = -3\overset{\circ}{.}56$ is a Galactic reflection nebula cannot yet be excluded. The other object found by Karachentsev & Karachentseva is faint granulated, irregular, and blue. It is located at $\alpha = 23^h\,45^m\,33\overset{s}{.}0$, $\delta = +38°42'\,40''$ (J 2000) at $b = -22\overset{\circ}{.}38$.

Table 14.6. *Data on Cassiopeia (= AND VII)*[a]

$\alpha(2000) = 23^h 26^m 31^s$		$\delta(2000) = +50° 41' 31''$ (1)
$\ell = 109°46$		$b = -9°95$
		$\epsilon = 0.3 \pm 0.1$
$V = 15.2 \pm 0.5$ (2,3)		
$(m - M)_o = 24.2 \pm 0.2$ (2)	$E(B - V) = 0.17 \pm 0.03$ (2)	$A_V = 0.53$
$M_V = -9.5 \pm 0.5$	$D = 760 \pm 70$ kpc	$D_{LG} = 0.28$ Mpc
Type $= $ dSph	size $= 2'0 \times 2'5$ (0.41 kpc	
No. globulars $= 0$ (2)	$\times$ 0.52 kpc) (1)	
[Fe/ H] $= -1.4 \pm 0.3$ (2)		

[a]Located at a distance of 16°18 from M31.
(1) Karachentsev & Karachentseva (1999).
(2) Grebel & Guhathakurta (1999).
(3) $B - V = 0.8$ assumed.

Table 14.7. *Data on Camelopardalis A*

$\alpha(2000) = 04^h 25^m 16°3$	$\delta(2000) = +72° 48' 21''$ (1)
$\ell = 137°25$	$b = +16°20$
$(m - M)_o = 26.1$ (2)	$D = 1.64$ Mpc (2)
Type $= $ dSph (?) (2,3)	
No. globulars $= 0$ (2)	

(1) Karachentsev (1994).
(2) Karachentsev et al. (1999).
(3) Guthakurta & Grebel (1999).

14.7   Summary

The discovery of three dwarf spheroidal companions to M31 in 1972 served to increase the similarity between the Milky Way system (which is known to have eight or nine dwarf spheroidal satellites) and the Andromeda galaxy. However, it remains a bit puzzling that M31, which is more luminous than the galaxy, appears to have fewer dSph companions. The recent discovery of And V, And VI, and And VII now lessens this anomaly. Furthermore, additional faint galaxies, which may belong to the Andromeda subgroup of the Local Group, have recently been found. Perhaps the most striking feature of And I, And II, And III, And V, And VI, and And VII is that they have absolute magnitudes, metallicities, structures, and stellar age distributions that closely resemble those of the dSph companions to the Milky Way system. The discovery of the new dSph companions to M31 strengthens and confirms the suspicion that dwarf spheroidals are the most common type of galaxy in the Universe. Perhaps the most intriguing remaining question is how such low-mass objects were able to retain gas, and continue to form stars, over periods comparable to the Hubble time.

15

Faint dwarf spheroidals

Faint dwarf spheroidals are probably the most common type of galaxy in the Universe. However, they contribute only a small fraction of the total mass and luminosity of all galaxies. The number of dwarf spheroidals has been increasing slowly with time, as dwarf irregulars exhaust their supply of interstellar gas. An additional source of new Ir galaxies, which evolve into dSph galaxies, is provided by objects formed in the tidal tails produced during the interactions of giant spirals (Dottori, Mirabel & Duc 1994). However such "young" dSph galaxies will have low M/L ratios because dark matter that is pulled out of giant galaxies is stretched (Kormendy 1998). As a result fragments pulled out of more luminous galaxies will have a lower density than their progenitors. Furthermore, the initial dark matter density in giants was already lower than it is in dwarf galaxies, even before it was tidally stretched. Finally, dSph galaxies formed from gas that was pulled out of giant galaxies are expected to be more metal rich than most dwarf spheroidals.

The data in Table 2.1 show that of the 18 known Local Group galaxies fainter than $M_V = -14.0$, 13 (72%) are dSph, two (11%) are dIr, two (11%) are of intermediate (dSph/dIr) type, and one (6%) is of unknown type. Excellent reviews on these faint galaxies have been given by Da Costa (1998) and by Mateo (1998).

15.1   The dwarf spheroidal Leo I

Leo I was discovered by Harrington (see Wilson 1955) on plates of the *Palomar Sky Survey*. The Leo I system is located close to the first magnitude star α Leonis, so that it is sometimes referred to as the Regulus system. Owing to the presence of this bright nearby star, scattered light presents a problem for some types of observations of faint objects in Leo I. Spectra of 33 red giants in this galaxy have been obtained by Mateo et al. (1998), who find a mean radial velocity $V_r = +287.0 \pm 1.9$ km s^{-1} and a velocity dispersion of $\sigma = 8.8 \pm 1.3$ km s^{-1}. From this velocity dispersion the minimum mass of Leo I is found to be $(2.0 \pm 1.0) \times 10^7 M_\odot$. In conjunction with $M_V = -11.9$ (see Table 15.1) this yields $M/L_V > 4$ (in solar units); that is, it is marginally larger than the value $M/L_V = 1.2$–1.7 that Pryor and Meylan (1993) obtain for globular clusters with a low central concentration of light. These results are, however, somewhat misleading because Leo I contains a dominant intermediate-age population that will fade as it ages. It would therefore be more meaningful to look at the asymptotic mass to light ratio, for which Mateo et al. find M/L_V in the range 6–13. These results suggest, but do not yet prove, that dark matter may provide a significant contribution to the total mass

Table 15.1. *The Leo I (= Regulus) dwarf spheroidal*

$\alpha(2000) = 10^h\,08^m\,26\fs7$		$\delta(2000) = +12°\,18'\,29''$ (1)
$\ell = 225\fdg98$		$b = +49\fdg11$
$V_r = +287.0 \pm 1.9$ km s^{-1} (2)		$\epsilon = 0.21 \pm 0.03$ (3)
$V = 10.2 \pm 0.1$ (4)		
$(m - M)_0 = 22.0 \pm 0.2$	$E(B - V) = 0.02$ (5)	$A_V = 0.06$ mag
$M_V = -11.9 \pm 0.3$	$D = 250 \pm 25$ kpc (1,6)	$D_{LG} = 0.61$ Mpc
Type = dSph	scale-length	
	$= 2\farcm38$ (172 pc) (7)	
$M \gtrsim (2.0 \pm 1.0) \times 10^7\,M_\odot$ (2)		$M_{HI} < 10^4\,M_\odot$ (8)
No. globulars = 0		
[Fe/H] = -2.0 ± 0.1 (1)		

(1) Lee et al. (1993d).
(2) Mateo et al. (1998).
(3) Irwin & Hatzidimitriou (1995).
(4) Caldwell et al. (1992).
(5) Burstein & Heiles (1984).
(6) Demers et al. (1994).
(7) From data tabulated by Irwin & Hatzidimitriou (1993).
(8) Carignan et al. (1998).

of Leo I. The radial velocities of red giants in the Regulus system give no evidence for significant rotation of this object. Mateo et al. point out that presently available observations do not entirely rule out rotation. This is so because (1) Leo I exhibits a relatively large central velocity dispersion, and (2) the Mateo et al. velocity sample subtends only a limited angular extent. The data of Mateo et al. (1998) would only have been able to detect rotation of Leo I if this object had a rapidly rising rotation curve of moderate amplitude.

Because of its large radial velocity, and great distance from the Sun, Leo I contributes significantly to mass estimates for the Galaxy (Zaritsky et al. 1989). If Leo I is bound, $M(\text{Galaxy}) = 12 \times 10^{11}\,M_\odot$, whereas $M(\text{Galaxy}) = 4 \times 10^{11}\,M_\odot$ if it is unbound.[1] For the latter case Byrd et al. (1994) find that Leo I left the vicinity of M31 ∼9 Gyr ago. Using a tidal radius of 12$\farcm$6 (920 pc), Irwin & Hatzidimitriou (1995) find a minimum Galactocentric distance of ∼70 kpc. However, it should be emphasized that the star count data are very noisy, so that the numerical value of the tidal cutoff radius remains quite uncertain. Presently available data do not strongly favor the hypothesis (Lynden-Bell & Lynden-Bell 1995) that Leo I belongs to one of the suspected tidal debris streams in the outer Galactic halo. Over the range $1' < r < 10'$ the radial surface density profile of Irwin & Hatzidimitriou may be represented by an exponential disk, with a rather uncertain scale-length of 2$\farcm$38 (172 pc). However, for $r < 1'$ the counts fall below such a disk model. Demers et al. (1995) find no evidence for nonrandom clumpings of stars in Leo I.

---

[1] The Galactic mass of $1.2 \times 10^{12}\,M_\odot$, which is obtained if Leo I is bound, is approximately half of the total Local Group mass of $(2.3 \pm 0.6) \times 10^{12}\,M_\odot$ obtained by Courteau & van den Bergh (1999) from the velocity dispersion of Local Group members.

The metallicity of Leo I remains somewhat controversial. From the strength of calcium and magnesium lines in two stars, Suntzeff et al. (1986) found [Fe/H] = -1.5 ± 0.25. From the color of its giant branch, Demers, Irwin & Gambu (1994) obtained [Fe/H] = -1.6. A much lower value, [Fe/H] = -2.0 ± 0.1, was, however, found by Lee et al. (1993d), from a comparison of the $V - I$ color–magnitude diagram of Leo I with those of Galactic globular clusters of known metallicity. The discovery of carbon stars in Leo I (Aaronson, Olszewski & Hodge 1983), and of $\sim$50 asymptotic giant branch stars above the tip of the red giant branch, shows that Leo I contains an intermediate-age stellar population. From the I magnitude of the tip of the red giant branch Lee et al. derived a distance of 270 ± 10 kpc. However, Demers et al. (1994) find $D = 205 \pm 25$ kpc from the magnitudes of the carbon stars and red clump giants. Santolamazza (1998) has used the characteristics of the anomalous Cepheids in Leo I to derive a somewhat speculative distance of 218 ± 30 kpc. In the subsequent discussion a distance of 250 ± 25 kpc will be adopted.

Gallart et al. (1998, 1999) have used a color–magnitude diagram of the Regulus system obtained with the *Hubble Space Telescope* to study the evolutionary history of this object. They conclude that Leo I experienced a major increase in its rate of star formation between 6 Gyr and 2 Gyr ago. Some prior star formation possibly occurred during another episode of enhanced activity that lasted 2–3 Gyr. Finally, a low level of star formation continued until 200–500 Myr ago. Caputo et al. (1999) find no evidence for RR Lyrae variables in Leo I. These results support the suspicion by Lee et al. (1993d) that Leo I might be the youngest dwarf spheroidal galaxy associated with the Milky Way system. It is therefore slightly unexpected that Carignan et al. (1998) have placed a lower limit of $<10^4 \, M_\odot$ on the amount of neutral hydrogen in Leo I. In addition, Bowen et al. (1997) measured a low column density of $N(\text{H\,I}) < 10^{17}$ at cm^{-2} toward three background galaxies/quasars. However, these observations do not rule out the possibility that Leo I might be embedded in a sphere (or shell) of ionized gas with a mass up to that of the entire galaxy.

In view of the fact that star formation in Leo I appears to have continued up until quite recently, Gallart et al. (1999) make the interesting suggestion that some of the variables found in Leo I by Hodge & Wright (1978), which have periods in the range 0.8–2.4 days, might be very short-period classical Cepheids, rather than anomalous Cepheids. More observations of these interesting objects would clearly be very desirable.

15.2 The Carina dwarf spheroidal

The Carina dwarf was discovered on plates obtained with the ESO/SRC Schmidt by Cannon, Hawarden & Tritton (1977). From observations of 23 photometrically selected red giant members of the Carina dwarf galaxy, Mateo et al. (1993) found a mean heliocentric radial velocity $V_r = +223.1 \pm 1.8$ km s^{-1} and a central velocity dispersion $\sigma = 6.8 \pm 1.6$ km s^{-1}. From this dispersion they calculate that $M = 1.1 \times 10^7 \, M_\odot$. In conjunction with an absolute magnitude $M_V = -9.4$ (see Table 15.2), this yields $M/L_V \approx 23$ in solar units. Taken at face value this result suggests that the mass of Carina is dominated by dark matter.

From the star counts of Irwin & Hatzidimitriou (1995) it is found that the surface density, over the range $2' < r < 16'$, can be approximately represented by an exponential with a scale-length of 6$'.$5 (0.2 kpc). This result is, however, somewhat uncertain because of the relatively high surface density of Galactic foreground stars in this field at $b = -22°$. Irwin & Hatzidimitriou find a tidal radius of 28$'.$8 $\pm$ 3$'.$6 (840 pc). Outside this radius Kuhn,

Table 15.2. *The Carina dwarf spheroidal*

$\alpha(2000) = 06^h\,41^m\,36\overset{s}{.}7$		$\delta(2000) = -50°\,57'\,58''$ (1)
$\ell = 260\overset{\circ}{.}11$		$b = -22\overset{\circ}{.}22$
$V_r = +223.1 \pm 1.8$ km s^{-1} (2)		$\epsilon = 0.33 \pm 0.05$ (3)
$V = 10.6$ (4,5)		
$(m - M)_0 = 20.0 \pm 0.1$ (4,5)	$E(B - V) = 0.05$ (4,5)	$A_V = 0.14$ mag
$M_V = -9.4$ (7)	$D = 100$ kpc	$D_{LG} = 0.51$ Mpc
Type = dSph	scale-length = $6\overset{'}{.}5$ (190 pc) (9)	
$M = 1.1 \times 10^7\,M_\odot$ (2)		$M_{HI} \lesssim 10^3\,M_\odot$ (6)
No. globulars = 0		
[Fe/H] = -1.9 (8)		

(1) Cannon et al. (1977).
(2) Mateo et al. (1993).
(3) Irwin & Hatzidimitriou (1995).
(4) Smecker-Hane et al. (1994).
(5) Mighell (1997).
(6) Mould et al. (1990b).
(7) Demers et al. (1983).
(8) Da Costa (1994).
(9) From data tabulated by Irwin & Hatzidimitriou (1995).

Smith & Hawley (1996) find some evidence for RR Lyrae stars, as well as stars on a color–magnitude diagram similar to that of Carina, in a tidal tail that extends as far as 2° (3.5 kpc) from the center of this galaxy. The surface density of stars in this putative tail is about 1% of that in the center of Carina.

From the location of the horizontal branch, and of the tip of the red giant branch, Smecker-Hane et al. (1994) obtain a distance modulus of $(m - M)_0 = 20.09 \pm 0.06$. This is consistent with the value $(m - M)_0 = 20.06 \pm 0.12$ that Mateo, Hurley-Keller & Nemec (1998) derived from the period–luminosity relation for dwarf Cepheids in Carina. A slightly larger distance modulus, $(m - M)_0 = 20.18 \pm 0.08$, was derived from RR Lyrae stars by Kuhn et al. (1996). A smaller distance modulus of $(m - M)_0 = 19.87 \pm 0.11$ was, however, derived from V and I photometry by Mighell (1997). The difference between these distance moduli is, at least in part, due to the fact that Mighell adopts $A_V = 0.18$ mag, while Smecker-Hane et al. found $A_V = 0.08$ mag. In the following discussion values of $(m - M)_0 = 20.0 \pm 0.1$ and $A_V = 0.14$ will be adopted.

Mould & Aaronson (1983) showed that the bulk of the stars in the Carina dwarf have ages of 7.5 ± 1.5 Gyr. These stars are responsible for the strong red giant clump seen in the color–magnitude diagram of Carina. A second, much older, population is found on the horizontal branch and in the RR Lyrae instability strip. The discovery of numerous RR Lyrae stars in the Carina dwarf (Saha, Monet & Seitzer 1986) shows that at least a few percent of the stellar population in this object must be old. It is of interest to note that the mean period of the variables of Bailey's types a and b is $\langle P_{ab} \rangle = 0.62$, which is intermediate between that of Galactic globular clusters belonging to Oosterhoff's Types I and II. This shows that Galactic globulars and dwarf spheroidals exhibit different systematics and therefore had different evolutionary histories.

Poretti (1999) has reanalyzed the light curves of 20 large-amplitude, short-period, variable stars in the Carina galaxy that had originally been found by Mateo, Hurley-Keller & Nemec (1998). Poretti finds that six of these dwarf Cepheids are first-overtone pulsators, while 13 appear to be pulsating in their fundamental mode. (At a given frequency the overtone pulsators are ~0.28 mag brighter than the fundamental mode pulsators.)

The existence of an intermediate-age population in Carina was also supported by the discovery of two carbon stars (Cannon, Niss & Nørgaard-Nielsen 1981). A beautiful illustration of the complex evolutionary history of the Carina dwarf is provided by the *B* and *I* photometry of Smecker-Hane et al. (1994). More recently similar results have been obtained by Hurley-Keller, Mateo & Nemec (1998), who derive a rather complex evolutionary history for Carina. They find that this object underwent significant episodes of star formation at ~15 Gyr, ~7 Gyr, and ~3 Gyr ago. These authors conclude that at least half of all of this star formation occurred during the burst that took place ~7 Gyr ago. Only 10–15% of the star formation in Carina happened during the initial burst that occurred ~15 Gyr ago. The remaining ~30% of the star formation in Carina took place during the most recent burst ~3 Gyr ago. From spectra of 15 giants, Da Costa (1994) found a mean metallicity of $\langle[Fe/H]\rangle = -1.88 \pm 0.08$. However, one of these stars was observed to have a significantly lower metallicity of $[Fe/H] = -2.3$. Presumably this object belonged to an older stellar population.

Hurley-Keller et al. (1998) speculate that Carina might have been as bright as $M_V = -16$ during the star burst that took place ~7 Gyr (corresponding to $z \approx 0.5$) ago. These authors find no evidence for a significant radial variation in the evolutionary history of the Carina dwarf. This confirms a similar conclusion by Smecker-Hane et al. (1996). It is of interest to note that there appears to have been a "dark age" with very little star formation, between the initial (~15 Gyr) and the main (~7 Gyr) starbursts. This situation is reminiscent of the even longer period of low star forming activity in the Large Magellanic Cloud (see Section 6.7). Mould et al. (1990b) find that Carina presently contains $<10^3 \, M_\odot$ of neutral hydrogen gas, so that little or no future star formation is expected. Alternatively, some neutral gas might remain outside the 15' half-power beamwidth used by Mould et al.

15.3 The Sculptor system

The dwarf spheroidal galaxy in Sculptor was discovered by Shapley (1938) on a plate taken during a routine survey with the Bruce telescope in 1935. However, two years passed (Shapley 1943, p. 144) until it was examined under the microscope. Then it took another year before confirmation, with a different telescope, showed that the Bruce image was not a faint spurious photographic artifact. In his discovery paper Shapley describes it as follows: "A large rich cluster with remarkable characteristics appears on photographs received from the Boyden Station ... nothing quite like it is known." More detailed studies by Baade and Hubble (1939) resulted in the discovery of some 40 variables, the majority of which were RR Lyraes. The globular cluster–like color–magnitude diagram of the members of the Sculptor system was first investigated by Hodge (1965). More recently it was reinvestigated by Kunkel & Demers (1977), who estimated the metallicity of this dwarf spheroidal to be $[Fe/H] = -1.9$. Detailed studies of the variables in Sculptor have been published by van Agt (1978) and by the OGLE team (Kaluzny et al. 1995). Kaluzny et al. derived light curves for 226 RR Lyrae variables and three anomalous Cepheids. From $\langle V(RR) \rangle$ they obtained a distance modulus of $(m - M)_0 = 19.71$ for the Sculptor system. From the sharp cutoff of the periods of Bailey's type ab RR Lyrae stars at $P = 0.475$ days, Kaluzny et al. conclude that the Sculptor RR Lyrae stars have [Fe/H] ≤ -1.7. For the Sculptor variables it is found that $\langle P_{ab} \rangle = 0.586$ days; that is, Sculptor is intermediate between Galactic globular clusters of Oosterhoff's Type I (with $\langle P_{ab} \rangle \approx 0.55$ days) and Oosterhoff's Type II (with $\langle P_{ab} \rangle \approx 0.65$ days). The Sculptor horizontal branch is found to be well populated on both sides of the RR Lyrae gap. From the width of the red giant branch Kaluzny et al. estimate the metallicities of Sculptor members to lie in the range $-2.2 \leq$ [Fe/H] ≤ -1.6.

Shetrone, Briley & Brewer (1998) have found that the star Sculptor No. 314 is an incipient CH star. It exhibits both strong CN bands and strong CH bands. This rules out any similarity to the strong CN stars found in some Galactic globular clusters. In spectra of two Sculptor giants, Norris & Bessell (1978) found [Fe/H] values of -2.2 and -1.5, respectively. Using photometry on the Washington system, van de Rydt et al. (1994) derived [Fe/H] $= -1.6 \pm 0.3$. That the dispersion found by Kaluzny et al. is intrinsic, and not due to inclusion of field stars in the color–magnitude diagram, was convincingly demonstrated by Schweitzer et al. (1995). She and her colleagues showed that the red giant branch for stars with membership probability $p > 90\%$ was still significantly broadened. From these data they obtained marginally significant proper motions of $\mu_\alpha = +0\rlap{.}''036 \pm 0\rlap{.}''022$ yr^{-1} and $\mu_\delta = +0\rlap{.}''043 \pm 0\rlap{.}''025$ yr^{-1}. In conjunction with Sculptor's radial velocity this yields a space velocity of 220 ± 125 km s^{-1}, which rules out a motion along the Magellanic Stream. Queloz, Dubath & Pasquini (1995) have obtained radial velocities for 23 K giants in the Sculptor system, which yield $\langle V_r \rangle = +109.9 \pm 1.4$ km s^{-1} and $\sigma = 6.2 \pm 1.1$ km s^{-1}. From this radial velocity dispersion, and $M_V = -10.9 \pm 0.5$ (Caldwell et al. 1992), they obtain $M/L_V = 13 \pm 6$ in solar units. This result does *not* provide unambiguous evidence for the presence of dark matter in Sculptor. Queloz et al. also find that two of the 23 K giants had variable radial velocities, from which they conclude that the fraction of binaries in Sculptor may be as high as $\sim 20\%$.

Hurley-Keller, Mateo & Grebel (1999) have mapped most of Sculptor in B and R. They find significant radial population effects in this object. Blue horizontal branch stars, which are presumably old, occur at all radii, whereas (younger) red horizontal branch stars only occur near the center. A not too numerous intermediate-age population is also seen near the center of the Sculptor system.

The structure of the Sculptor dwarf spheroidal galaxy has been studied in great detail by Eskridge (1988a), who fits his star counts to a King profile with a tidal radius $r_t = 95' \pm 10'$ (2.4 kpc) and a core radius $r_c = 8\rlap{.}''9$ (225 pc). This tidal radius is much larger than the value $r_t = 46' \pm 3'$ (along the major axis) obtained by Hodge (1961c). An intermediate value $r_t = 76' \pm 5'$ was derived by Irwin & Hatzidimitriou (1995). Part of the reason for this difference in tidal radii is that Eskridge let the ellipticity and position angle vary in the inner regions but kept it fixed in the outer regions (in which the signal-to-noise ratio is small) whereas Irwin & Hatzidimitriou kept them fixed in both regions. With Hodge's value of r_t Innanen & Papp (1979) found (see Figure 15.1) that $\sim 10\%$ of van Agt's RR Lyrae variables in Sculptor lie outside its tidal radius. For $r_t = 95'$ only 2–3% of the RR Lyrae stars lie beyond the Sculptor tidal limit.

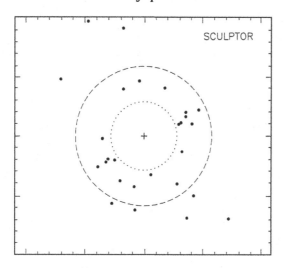

Fig. 15.1. Location of the Sculptor RR Lyrae variables relative to circles with radii of 46′ and 95′. The figure shows that the number of extratidal variables is decreased by a factor of three or four by increasing the adopted tidal cutoff radius. Figure adapted from Innanen & Papp (1979).

Demers & Battinelli (1998) have found nearly 200 blue stars brighter than the main-sequence turnoff in the Sculptor color–magnitude diagram. Isochrone fitting implies that some of these stars (if they are not blue stragglers) might be as young as 1 Gyr. In this connection it is of interest to note that Carignan et al. (1998) have discovered $\geq 3 \times 10^4 \, M_\odot$ of hydrogen associated with Sculptor. This gas is concentrated in two clouds (see Figure 15.2) that have a mean radial velocity of $+110 \pm 1.5$ km s^{-1}, which is identical to the mean velocity of the K giants in Sculptor (Queloz et al. 1995). The velocity of the NE cloud is 119 ± 2 km s^{-1}, and that of the SW cloud is 104 ± 11 km s^{-1}. The 21-cm isophotes of these clouds resemble those of the radio isophotes of extragalactic radio sources. If they were ejected from the center of Sculptor at 10 km s^{-1}, then they would have been able to reach their present positions in only $\sim$40 Myr. High-dispersion absorption-line studies of two quasars located behind Sculptor (Zinnecker 1994) might provide additional information on the kinematics and composition of the gas associated with the Sculptor dwarf spheroidal. A summary of data on the Sculptor system is given in Table 15.3.

15.4　The Draco dwarf spheroidal

The Draco system was discovered by Wilson (1955) on the plates of the *Palomar Sky Survey*. It was subsequently studied in great detail by Baade & Swope (1961), who discovered over 260 variables in this system. Most of these turned out to be RR Lyrae stars, which were found to have a mean period $\langle P_{ab} \rangle = 0.611$ days (van Agt 1978), which makes them intermediate between those in Galactic globular clusters of Oosterhoff (1939) Type I ($\langle P_{ab} \rangle \approx 0.55$ days) and Oosterhoff Type II ($\langle P_{ab} \rangle \approx 0.65$ days). Baade & Swope obtained a color–magnitude diagram for Draco that showed a red giant

Table 15.3. *The sculptor dwarf spheroidal*

$\alpha(2000) = 01^h\ 00^m\ 04^s\!3$		$\delta(2000) = -33°\ 42'\ 51''$ (1)
$\ell = 287°\!69$		$b = -83°\!16$
$V_r = +109.9 \pm 1.4\ \text{km s}^{-1}$ (2)		$\epsilon = 0.32 \pm 0.03$ (1)
$V = 8.8 \pm 0.5$ (3)		
$(m - M)_o = 19.71$ (4)	$E(B - V) = 0.00$ (5)	$A_V = 0.0$ mag
$M_V = -9.8$	$D = 87$ kpc	$D_{LG} = 0.44$ Mpc
Type = dSph	scale-length = 8.′5 (215 pc) (7)	
$M = (1.6 \pm 1.2) \times 10^7\ M_\odot$		$M_{HI} \geq 3.0 \times 10^4\ M_\odot$ (6)
No. globulars = 0		
[Fe/H] = -2.2 to -1.6 (4)		

(1) Irwin & Hatzidimitriou (1995).
(2) Queloz et al. (1995).
(3) Caldwell et al. (1992).
(4) Kaluzny et al. (1995).
(5) Burstein & Heiles (1984).
(6) Carignan et al. (1998).
(7) From data in Irwin & Hatzidimitriou (1995).

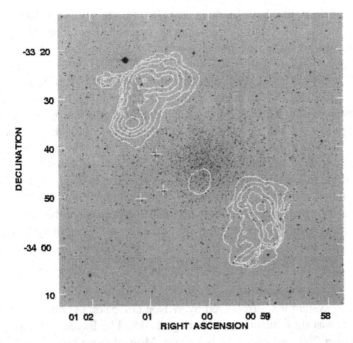

Fig. 15.2. Observed H I isophotes by Carignan et al. (1998) superposed on an optical image of the Sculptor system. The morphological similarity of the hydrogen isophotes to those of extragalactic radio sources might encourage one to speculate that both H I clouds were ejected from this galaxy. Reproduced with the kind permission of Claude Carignan.

branch extending to $V = 17.0$, a well-developed asymptotic giant branch, and a horizontal branch at $V = 20.0$. The horizontal branch was strongly populated to the red of the RR Lyrae gap, but it also exhibited a dozen stars to the blue of the gap. The features of the Draco color–magnitude diagram, first outlined by Baade & Swope, have been confirmed by Stetson (1980), who used proper motion observations to eliminate foreground stars from the Draco color–magnitude diagram. Recently a deep color–magnitude diagram of the Draco system has been obtained with the *Hubble Space Telescope* by Grillmair et al. (1998). Their data yield no evidence for multiple main-sequence turnoffs, suggesting that most of the star formation in Draco took place during a single short burst. However, the discovery of three carbon stars by Aaronson (1983) shows that Draco must contain a few objects of intermediate age. A comparison of the color magnitude diagram of Draco with that of old metal-poor Galactic globulars shows that Draco is 1.6 ± 2.5 Gyr older than M92. In other words, Draco was among the first stellar systems to have been formed in the outer halo of the proto-Galaxy [at about the same time as NGC 2419 (Harris et al. 1997)]. Grillmair et al. find that the best-fitting isochrone for their Draco observations has [Fe/H] $= -2.2$ and an age of 16 Gyr. The observation that Draco has a red horizontal branch, a low metallicity, and a great age shows that the second-parameter effect cannot be (entirely) an age phenomenon.

Both Zinn (1978) and Lehnert et al. (1992) find that spectra of individual stars in Draco exhibit a significant metallicity range. Lehnert et al. suggest that the abundances may fall into two groups with [Fe/H] $= -2.3 \pm 0.2$ and [Fe/H] $= -1.6 \pm 0.2$, respectively. It would be important to obtain spectra of a large number of individual stars to check if the metallicity distribution in Draco is, indeed, bimodal. From studies of the light curves of RR Lyrae variable stars in Draco Silbermann, Harris & Smith (1996) confirm the existence of a metallicity range from relatively metal rich, like M3, to quite metal poor, like M92. Shetrone, Bolte & Stetson (1998) have obtained high-resolution observations of four giants in Draco, for which they find metal abundances in the range $-3.0 <$ [Fe/H] < 1.5. The abundance ratios that are found for these objects are typical of those found in stars of Population II, although the [Ca/Fe] ratio may be more nearly solar. Furthermore, [Ba/Fe] exhibits a steeper slope with [Fe/H] than has been found for stars in Galactic globular clusters and in the Galactic halo. Shetrone et al. find a hint of a deep-mixing pattern in one of their stars, which suggests that Draco giants may be more similar to globular cluster giants than they are to field halo stars. If confirmed, this result would show that the Galactic halo was not built up by disintegration of Draco-like dSph galaxies.

Armandroff, Olszewski & Pryor (1995) have obtained a heliocentric radial velocity of -293.3 ± 1.0 km s^{-1} for Draco. A similar result has been obtained by Hargreaves et al. (1996). The radial velocity observations give no evidence for significant rotation of the Draco system.[2] Olszewski, Pryor & Armandroff (1996) conclude that the fraction of binaries among stars near the main-sequence turnoff is 3 to 5 times greater in Draco than it is in the solar vicinity. After taking into account the effect of binaries, the internal velocity dispersion of Draco reduces from 9.9 ± 1.4 km s^{-1} to 8.2 ± 1.3 km s^{-1}. From the

---

[2] Since Draco is among the closest dSph companions of the Galaxy, it is possible that tidal effects could cause streaming that might mimic rotation.

Table 15.4. *The Draco dwarf spheroidal*

$\alpha = (2000) = 17^h\,20^m\,18\overset{s}{.}6$		$\delta(2000) = +57°\,55'\,06''$ (1)
$\ell = 86\overset{\circ}{.}37$		$b = +34\overset{\circ}{.}71$
$V_r = -293.3 \pm 1.0$ km s^{-1} (2)		$\epsilon = 0.29 \pm 0.01$ (1)
$V = 11.0 \pm 0.5$ (1)		
$(m - M)_0 = 19.48$ (3)	$E(B - V) = 0.03$ (1)	$A_V = 0.09$ mag
$M_V = -8.6 \pm 0.5$	$D = 79$ kpc	$D_{LG} = 0.43$ Mpc
Type = dSph	scale-length = $6\overset{.}{.}1$ (140 pc) (6)	
$M = 2.2 \times 10^7\,M_\odot$ (5)		$M_{HI} \leq 50\,M_\odot$ (4)
No. globulars = 0		
[Fe/H] = -2.2 (3)		

(1) Irwin & Hatzidimitriou (1995).
(2) Armandroff et al. (1995).
(3) Grillmair et al. (1998).
(4) Young (1999).
(5) From $L = 2.4 \times 10^5\,L_\odot$ and $M/L_V = 90$.
(6) From data in Irwin & Hatzidimitriou (1995).

latter value, and a King model with $r_c = 9\overset{.}{.}0$ (205 pc) and $r_t = 28\overset{.}{.}3$ (650 pc) (parameters from Irwin & Hatzidimitriou 1995), one obtains a total mass of $2.2 \times 10^7\,M_\odot$ for Draco, and hence $M/L = 90$. This high value clearly shows that the dynamics of the Draco dwarf spheroidal galaxy are dominated by dark matter. Smith, Kuhn & Hawley (1997) have measured the color–magnitude diagram for field stars outside the $28'.3$ tidal radius of Draco. The aim of this program was to search for stars that might have been pulled from this galaxy by interaction(s) with the Milky Way. Smith et al. claim to have found evidence for a faint stream of tidal debris, with a color–magnitude diagram resembling that of Draco, out to a distance of $\sim 3°$ from the center of the Draco system. Scholz & Irwin (1994) have measured the proper motions of stars in Draco and conclude that its motion is (within large error bars) consistent with a polar orbit along the direction of the Magellanic Stream.

A very low upper limit, of only a few tens of solar masses of neutral hydrogen gas, has been set by Young (1999). A summary of data on the Draco system is given in Table 15.4.

15.5   The recently discovered dwarf in Tucana

The Tucana system was first cataloged by Corwin, de Vaucouleurs & de Vaucouleurs (1985). However, Lavery (1990) later noticed that this object "is resolved into stars on CCD images and is a likely new Local Group member." The suspicion that this object was probably a relatively nearby dwarf spheroidal was soon confirmed by the color–magnitude diagram of Lavery & Mighell (1992). Lavery & Mighell emphasized the fact that this object, which is located ~ 0.9 Mpc from the Galaxy and $108\overset{\circ}{.}3$ from M31, is not a member of either the Andromeda or Milky Way subgroups of the Local Group. It therefore probably evolved in isolation during much of its lifetime. A photograph of this object has been published by Saviane, Held & Piotto (1996). It shows Tucana to be a flattened dwarf spheroidal with an axial ratio of about 2:1. From CCD photometry of

Table 15.5. *The Tucana dwarf spheroidal*

$\alpha(2000) = 22^h 41^m 48\overset{s}{.}9$		$\delta(2000) = -64° 25' 21''$ (1)
$\ell = 322\overset{\circ}{.}91$		$b = -47\overset{\circ}{.}37$
		$\epsilon = 0.56$ (3)
$V = 15.15 \pm 0.18$ (3)	$B - V = 0.70 \pm 0.05$ (2)	
$(m - M)_0 = 24.70 \pm 0.15$ (5)	$E(B - V) = 0.00 \pm 0.01$ (4)	$A_V = 0.00$ mag
$M_V = -9.55 \pm 0.23$	$D = 870 \pm 60$ kpc	$D_{LG} = 1.10$ Mpc
Type = dSph	scale-length	
	$= 31'' \pm 3''$ (130 pc) (3)	
No. globulars = 0	$M_{HI} = 1.5 \times 10^6 M_\odot$ (6)	
[Fe/H] = -1.8 ± 0.2 (3)		

(1) Lavery (1990).
(2) Da Costa (1994).
(3) Saviane et al. (1996).
(4) Burstein & Heiles (1982).
(5) See Section 15.5.
(6) Carignan et al. (1998).

$\sim$360 stars in V and I these authors conclude that $(m - M)_0 = 24.69 \pm 0.16$. This agrees well with the values $(m - M)_0 = 24.72 \pm 0.2$ and $(m - M)_0 = 24.8 \pm 0.2$ derived by Castellani, Marconi & Buonanno (1996ab) and by Da Costa (1994), respectively. In the subsequent discussion it will be assumed that $(m - M)_0 = 24.70 \pm 0.15$, corresponding to $D = 870 \pm 60$ kpc.

From a comparison between the V versus $B - V$ color–magnitude diagram of the Tucana dwarf with those of Galactic globular clusters of known metallicity, Da Costa (1994) concludes that Tucana has [Fe/H] $= -1.8 \pm 0.3$. Saviane et al. find [Fe/H] $= -1.8 \pm 0.2$, in good agreement with Da Costa (1994), but slightly higher than the value [Fe/H] $= -1.56 \pm 0.2$ adopted by Castellani et al. (1996a,b). Saviane et al. also find that the observed width of the red giant branch of Tucana is no greater than that expected from observational errors. They therefore conclude that the majority of stars in this object exhibit little dispersion in metallicity. According to Da Costa (1998), *Hubble Space Telescope* observations show that the Tucana dwarf contains RR Lyrae variables and both red and blue horizontal branch stars. The red horizontal branch stars might belong to a relatively young population component with an age of 3–4 Gyr. Available data on Tucana are summarized in Table 15.5.

15.6 The dwarf spheroidal Leo II

Leo II[3] was discovered by Harrington & Wilson (1950) on some of the first plates obtained with the 1.2-m Palomar Schmidt telescope. Subsequent studies of this object have been published by Demers & Harris (1983), Demers & Irwin (1993), Lee (1995c), and Mighell & Rich (1996a,b). From the I magnitude of the tip of the red giant branch Lee obtains a distance modulus of $(m - M)_0 = 21.66 \pm 0.21$. Mighell &

---

[3] This object is incorrectly called Leo III in Zwicky (1957, p. 225).

Table 15.6. *The Leo II dwarf spheroidal*

$\alpha(2000) = 11^{\mathrm{h}}\,13^{\mathrm{m}}\,27\overset{\mathrm{s}}{.}4$		$\delta(2000) = +22° 09' 40''$ (1)
$\ell = 22\overset{\circ}{.}14$		$b = +67\overset{\circ}{.}23$
$V_{\mathrm{r}} = +76.0 \pm 1.3\,\mathrm{km\,s^{-1}}$ (2)		$\epsilon = 0.13 \pm 0.05$ (3)
$V = 11.62 \pm 0.25$ (4)	$B - V = 0.65 \pm 0.15$ (4)	
$(m - M)_{\mathrm{o}} = 21.60 \pm 0.15$ (5,6)	$E(B - V) = 0.027 \pm 0.015$ (5)	$A_V = 0.08$ mag
$M_V = -10.07 \pm 0.29$	$D = 210 \pm 15\,\mathrm{kpc}$	$D_{\mathrm{LG}} = 0.57\,\mathrm{Mpc}$
Type = dSph	scale-length = $18\overset{''}{.}(110\,\mathrm{pc})$ (7)	
$M = (1.1 \pm 0.3) \times 10^7\,M_\odot$ (2)	$M_{\mathrm{HI}} \lesssim 200\,M_\odot$ (8)	
No. globulars = 0		
[Fe/H] = -1.60 ± 0.25 (5)		

(1) Harnington & Wilson (1950).
(2) Vogt et al. (1995).
(3) Irwin & Hatzidimitriou (1995).
(4) Hodge (1982).
(5) Mighell & Rich (1996b).
(6) Lee (1995d).
(7) From data in Irwin & Hatzidimitriou (1995).
(8) Young (1999).

Rich (1996a) find $(m - M)_{\mathrm{o}} = 21.55 \pm 0.18$ from the magnitude level of the horizontal branch. In the subsequent discussion it will be assumed that $(m - M)_{\mathrm{o}} = 21.60 \pm 0.15$, corresponding to a distance of 210 ± 15 kpc.

Demers & Irwin showed that Leo II has a horizontal branch that is heavily populated to the red of the RR Lyrae gap, but with some (presumably old and metal-poor) stars to the blue of it. From a comparison of the red giant branch of Leo II with those Galactic globulars of known metallicity, Mighell & Rich find [Fe/H] = -1.60 ± 0.25. Their deep color–magnitude diagram, obtained with the *Hubble Space Telescope*, shows that the main-sequence turnoff region of the Leo II color–magnitude diagram may be represented by an isochrone with [Fe/H] = -1.6 and an age of 9.6 Gyr. For [Fe/H] = -1.9 the corresponding age would be 10.5 Gyr. Mighell & Rich conclude that star formation in Leo II started 14 ± 1 Gyr ago, reached its peak 9 ± 1 Gyr ago, and that its rate has been negligibly small for the past ~7 Gyr. The existence of an intermediate-age population in Leo II is attested to by the presence of a few carbon stars (Lee 1995c, and references therein).

Vogt et al. (1995) have used the Keck telescope to obtain radial velocities of 31 red giants in Leo II. From their observations[4] these authors obtain $\langle V_{\mathrm{r}} \rangle = +76.0 \pm 1.3$ km s^{-1} and a velocity dispersion $\sigma = 6.7 \pm 1.1$ km s^{-1}, from which they find a total mass of $(1.1 \pm 0.3) \times 10^7\,M_\odot$. In conjunction with the luminosity of Leo II (see Table 15.6) this yields $M/L_V \sim 12$ (in solar units). In their paper Vogt et al. derive a global value $M/L_V = 11.2 \pm 3.8$, from which they conclude that Leo II probably contains a significant

---

[4] I am indebted to Jim Hesser for pointing out to me that this velocity of +76 km s^{-1} lies outside the -90 km s^{-1} to +25 km s^{-1} range that Knapp, Kerr & Bowers (1978) searched for hydrogen line emission. A recent search by Young (1999), at the correct radial velocity, yields an upper limit of $\leq 200\,M_\odot$ to the H I mass in Leo II.

dark matter component. Because Leo II is so remote from the Galaxy (and from other members of the Local Group), it seems unlikely that tidal effects can have made a significant contribution to the observed velocity dispersion in Leo II. From star counts Irwin & Hatzidimitriou (1995) find that Leo II has a tidal radius of $r_t = 8'.7 \pm 0'.9$ (530 pc).

15.7 The Sextans dwarf spheroidal

The Sextans system was discovered by Irwin et al. (1990) on the plates of the UK Schmidt survey. The discoverers found that Sextans has a globular cluster–like color–magnitude diagram. From spectroscopy of 43 radial velocity members of the Sextans system, Suntzeff et al. (1993) find [Fe/H] $= -2.05 \pm 0.04$ and $\langle V_r \rangle = +227.9 \pm 1.8$ km s^{-1}. A similar value, $\langle V_r \rangle = +224.4 \pm 1.6$ km s^{-1}, has been reported by Hargreaves et al. (1994c). The metallicity value of Suntzeff et al. has been independently confirmed by Geisler & Sarajedini (1996), who found [Fe/H] $= -2.0 \pm 0.1$ from photometry on the Washington system. However, a lower value, [Fe/H] $= -1.6 \pm 0.2$, is found by Mateo, Fischer & Krzeminski (1995). Suntzeff et al. find that the intrinsic metallicity dispersion for the stars that they observed in Sextans is 0.19 ± 0.02 dex. In other words these data provide unambiguous evidence for a significant range in metallicity among the physical members of the Sculptor system. Suntzeff et al. (1993) find a true velocity dispersion of $\sigma = 6.2 \pm 0.9$ km s^{-1} for the stars in Sextans, from which they derive a total mass of $(2.6 \pm 0.7) \times 10^7 \, M_\odot$. In conjunction with the (uncertain!) total luminosity of Sextans listed in Table 15.7, this yields $M/L_V \sim 50$, indicating that the mass of the Sextans system is dominated by dark matter. From radial velocities of 26 stars Hargreaves et al. (1994c) find no evidence for systemic rotation of the Sextans system. At a galactocentric distance of 300 pc, their observations place an upper limit of 0.4 km s^{-1} on the rotational velocity of Sextans around its minor axis. The cluster color–magnitude diagram of

Table 15.7. *The Sextans dwarf spheroidal*

$\alpha(2000) = 10^h \, 13^m \, 02\overset{s}{.}9$		$\delta(2000) = -01° \, 36' \, 52''$ (1)
$\ell = 243°\!.50$		$b = +42°\!.27$
$V_r = +225.9 \pm 1.2$ km s^{-1} (2,8)		$\epsilon = 0.35$ (3)
$V = 10.3 \pm 0.7$ (4)		
$(m - M)_0 = 19.7 \pm 0.3$ (4)	$E(B - V) = 0.037 \pm 0.027$ (5)	$A_V = 0.11$ mag
$M_V = -9.5 \pm 0.8$	$D = 86 \pm 6$ kpc	$D_{LG} = 0.51$ Mpc
Type $=$ dSph	scale-length $= 14'\!.0(350$ pc) (6)	
$M = (2.6 \pm 0.7) \times 10^7 \, M_\odot$ (2)		$M_{HI} < 130 \, M_\odot$ (7)
No. globulars $= 0$		
[Fe/H] $= -2.0 \pm 0.1$ (5)		

(1) Irwin et al. (1990).
(2) Suntzeff et al. (1993).
(3) Irwin & Hatzidimitriou (1993).
(4) Mateo et al. (1991).
(5) Geisler & Sarajedini (1996).
(6) From plot in Irwin et al. (1990).
(7) Carignan et al. (1998).
(8) Hargreaves et al. (1994c).

Table 15.8. *Anomalous Cepheids in dwarf spheroidals*

Galaxy	M_V	$N_{AC}{}^a$	S^b	$\langle P_{ab}\rangle^c$
Ursa Minor	−8.5	7	2.8	0.638
Draco	−8.6	5	1.8	0.611
Carina	−9.4	7	1.2	0.620
Sextans	−9.5	6	1.0	0.606
Sculptor	−9.8	4	0.5	0.586
Leo II	−10.1	4	0.4	...
Leo I	−11.9	12	0.02	...
Sagittarius	?	1	...	...
Fornax	−13.1	1	0.01	...

[a] Number of anomalous Cepheids from Mateo et al. (1995).
[b] Number per $M_V = -7.5$ of parent galaxy luminosity.
[c] Mean period of type ab RR Lyrae variables in days.

the Sextans system exhibits a predominantly red horizontal branch at $V = 20.35 \pm 0.2$, which extends into the RR Lyrae instability strip. From its luminosity Mateo et al. (1991) derived a true distance modulus of $(m - M)_0 = 19.7 \pm 0.2$. This is consistent with the value $(m - M)_0 = 19.67 \pm 0.15$, corresponding to $D = 86 \pm 6$ kpc, that Mateo, Fischer & Krzeminski (1995) find from the pulsating variables in Sextans. From a study of an $18' \times 18'$ field in Sextans these authors were able to identify 36 RR Lyrae variables, six anomalous Cepheids, and one long-period variable. Mateo et al. find $\langle P_{ab}\rangle = 0.606$ days for the RR Lyrae stars of Bailey types a and b, which makes this system intermediate between the Galactic globular clusters of Oosterhoff Types I and II. In this respect Sextans resembles Sculptor, Draco, and Carina, which have $\langle P_{ab}\rangle = 0.586, 0.611$, and 0.620 days, respectively. Mateo et al. note that the specific frequency of anomalous Cepheids (i.e., evolved intermediate-age pulsators) appears to anticorrelate with parent galaxy luminosity (see Table 15.8). The reason for this correlation is presently not understood.

A deep color–magnitude diagram published by Mateo, Fischer & Krzeminski (1995) shows a number of stars above the main-sequence turnoff. Relative to the number of main-sequence stars these "blue stragglers" are almost an order of magnitude more frequent than they are in low-density Galactic globular clusters. This result suggests that these objects are probably of intermediate age. Mateo et al. estimate that as much as ∼25% of the population of Sextans may belong to this intermediate-age population. These objects might be only 2 Gyr to 4 Gyr old. No carbon stars associated with this intermediate-age population component are known. Presently the amount of neutral hydrogen gas in Sextans is $<130\,M_\odot$ (Carignan et al. 1998), so that little or no future star formation is expected.

Using multicolor photometry, Gould et al. (1992) have detected five metal-poor dwarf stars beyond the tidal radius of Sextans. These dwarfs might have been stripped from this object. However, spectroscopic confirmation of this association will be difficult, because they are presently still too faint for accurate radial velocity measurements.

15.8 The Ursa Minor dwarf spheroidal

The faint dwarf spheroidal galaxy in Ursa Minor was discovered on the plates of the *Palomar Sky Survey* by Wilson (1955). From radial velocities of 167 members of this system, Armandroff, Olszewski & Pryor (1995) obtain a mean heliocentric radial velocity of $\langle V_r \rangle = -247.4 \pm 1.0$ km s^{-1}. Furthermore, their data appear to show that the Ursa Minor system is rotating with an amplitude of $\sim$3 km s^{-1}. However, the observation by Kleyna et. al. (1998) that UMi exhibits a statistically significant asymmetry suggests that the apparent rotation of this object might be due to nonequilibrium (or tidal) effects. From their velocity observations Armandroff et al. find an internal velocity dispersion of $\sigma = 10.4 \pm 0.9$ km s^{-1}. By combining this dispersion with the structural parameters of the UMi system, they obtain a mass-to-light ratio $M/L_V = 77 \pm 13$ in solar units. A slightly smaller M/L_V value has been derived independently, from observations of 35 giant stars in UMi, by Hargreaves et al. (1994b). These observations show that the dynamics of the UMi system are dominated by dark matter. Moreover, Olszewski (1997) has demonstrated that the velocity dispersion in UMi does not drop with increasing radius, which shows that the dark matter distribution is more extended than the distribution of stars. Olszewski, Pryor & Armandroff (1996) find that the fraction of binary stars near the main-sequence turnoff in UMi is three to five times larger than it is in the solar neighborhood.

Kleyna et al. (1998) have covered UMi with a mosaic of CCD images to obtain a complete color–magnitude diagram of this system in the V and I bands. From a comparison of this color–magnitude diagram with that of the well-studied metal-poor globular cluster M92, these authors find that the total integrated magnitude of UMi is $M_V = -8.87 \pm 0.25$. For this luminosity Kleyna et al. find that the Ursa Minor system has a mass-to-light ratio $M/L = 70^{+30}_{-20}$ in solar units. A search for neutral hydrogen in UMi by Knapp, Kerr & Bowers (1978) over the velocity range -110 km s^{-1} to -10 km s^{-1} found less than 280 $M_\odot$ of gas. This is not surprising since UMi is now known (Armandroff et al. 1995) to have $V_r = -247$ km s^{-1}.

Color–magnitude diagrams for the Ursa Minor system have been published by Olszewski & Aaronson (1985), by Geisler & Sarajedini (1996), by Martinez-Delgado & Aparicio (1997), by Kleyna et al. (1998), and by Burke & Mighell (1998). These diagrams show that UMi is a very metal-poor system, with a sparsely populated giant branch and a rather blue horizontal branch. There appears to be no evidence for stars younger than $\sim$15 Gyr near the main-sequence turnoff. However, Armandroff et al. have found four carbon stars in the Ursa Minor system, which shows that this galaxy must contain a small intermediate-age population component. Olszewski & Aaronson find that a good fit to the color–magnitude diagram is provided by isochrones with heavy element abundance $Z = 0.0001$, corresponding to [Fe/H] $= -2.2$. From a study of the variable star population of UMi, Gay (1998) finds that the RR Lyrae variables in Ursa Minor system have an Oosterhoff Type II period distribution. The number of anomalous Cepheids is small enough to be accounted for by mass transfer. Assuming that the variables are tracers of the main stellar population in UMi leads to the conclusion that most of the members of this dwarf galaxy were made during a single star forming event.

The proper motion of UMi has been studied by Scholz & Irwin (1994) on Schmidt plates with a 35-year baseline and by Schweitzer, Cudworth & Majewski (1997) on large

Table 15.9. *The Ursa Minor dwarf spheroidal*

$\alpha(2000) = 15^{\mathrm{h}}\,08^{\mathrm{m}}\,49\overset{s}{.}2$		$\delta(2000) = +67°\,06'\,38''$ (1)
$\ell = 104\overset{\circ}{.}88$		$b = +44\overset{\circ}{.}90$
$V_r = -247.4 \pm 1.0\,\mathrm{km\,s^{-1}}$ (2)		$\epsilon = 0.56 \pm 0.05$ (3)
$V = 10.6 \pm 0.5$ (3)		
$(m - M)_o = 19.0 \pm 0.1$ (4)	$E(B - V) = 0.03$ (4)	$A_V = 0.09$ mag
$M_V = -8.87 \pm 0.25$ (8)	$D = 63$ kpc	$D_{LG} = 0.43$ Mpc
Type = dSph	scale-length $= 11\overset{'}{.}2$ (205 pc) (5)	
$M = 1.7 \times 10^7\,M_\odot$ (6)		$M_{HI} < 280\,M_\odot$? (2,7)
No. globulars = 0		
[Fe/H] $= -2.2 \pm 0.2$ (3)		

(1) Wilson (1955).
(2) Armandroff et al. (1995).
(3) Irwin & Hatzidimitriou (1995).
(4) Olszewski & Aaronson (1985).
(5) From data in Irwin & Hatzidimitriou (1995).
(6) From $L = 2.2 \times 10^5\,L_\odot$ and $M/L_V = 77$.
(7) Knapp et al. (1978).
(8) Kleyna et al. (1998).

reflector plates with a 42-year baseline. The latter (more accurate) data show that UMi has a space velocity, in the Galactic rest frame, of 209 ± 20 km s^{-1} in a direction that is almost along the direction of the major axis of this very flattened object. The Ursa Minor system appears to be orbiting the Galaxy in nearly the same plane as the Magellanic Stream, and in the same sense as the Large Magellanic Cloud. Grillmair & Irwin (1998) have selected the field stars with colors and magnitudes similar to those in UMi itself. They find some evidence for surface density enhancements on large scales, which might be associated with tidal tails. A summary of data on the Ursa Minor system is given in Table 15.9.

15.9   Summary and conclusions

Figure 15.3 shows that the dwarf spheroidals with the lowest luminosities tend to have the highest mass-to-light ratios. Furthermore, the masses of known dSph systems are all $>1 \times 10^7\,M_\odot$. It is presently not clear if this is an observational selection effect, or if there is a real lower limit to the masses of the dark halos of galaxies. Hirashita, Takeuchi & Tamura (1998) suggest that this cutoff occurs because dSph galaxies with masses much less than $10^8\,M_\odot$ are not able to retain their interstellar gas. If gas is blown away by supernovae or stellar winds, such low-mass dwarfs may not be able to form more than a single generation of stars. As a result, globs which mostly consist of dark matter with masses $\ll 1 \times 10^8\,M_\odot$, might be difficult or impossible to discover at optical wavelengths.

The data also exhibit a loose correlation between luminosity and metallicity, in the sense that spheroidals, such as NGC 147, NGC 185, and NGC 205, have $M_V \sim -16$ and [Fe/H] -1.0, whereas the faintest dwarf spheroidals, such as Draco and Ursa Minor, have $M_V \approx -8.5$ and [Fe/H] ≈ -2.2 (the latter value being similar to that of

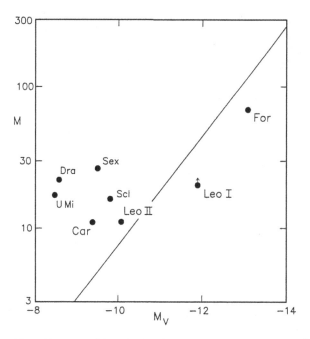

Fig. 15.3. Mass of dwarf spheroidal galaxies, in units of $1 \times 10^6 \, M_\odot$, versus absolute magnitude. The line has $M/L_V = 10$ in solar units. Note the high mass-to-light ratios of the faintest dSph galaxies. It is not yet clear if the lack of dwarf spheroidals with $M < 1 \times 10^7 \, M_\odot$ is an observational selection effect.

the most metal-deficient globular clusters). For those dSph galaxies that are fainter than $M_V = -12$, there appears to be a loose anti correlation between the metallicity and Galactocentric distance. The Ursa Minor and Draco systems, which are closest to the Milky Way system, have [Fe/H] $= -2.2$, whereas the more distant Leo II and Tucanae systems have [Fe/H] $= -1.6$ and [Fe/H] $= -1.8$, respectively. Such a correlation might have originated from early termination of star formation in low-luminosity dwarfs near the Galaxy. Possible support for this hypothesis is provided by the observation (E. K. Grebel, private communication) that the Milky Way dwarf spheroidals with $D > 100$ kpc appear to have a larger intermediate-age population than do those with $D < 100$ kpc.

The fact that the mean periods of Bailey's type ab RR Lyrae stars in dwarf spheroidals are intermediate between those of Galactic globular clusters of Oosterhoff's Types I and II shows that the evolutionary history of dwarf spheroidals must, in some way, differ from that of Galactic globular clusters. Dwarf spheroidal galaxies also differ from Galactic globular clusters (ω Centauri excepted) in that metallicity variations (e.g., in Car, Dra, Scl, and Sex) appear to be ubiquitous. Perhaps the most puzzling feature of dwarf spheroidals is that star formation in them has extended over such a long period of time. The dominant stellar population in Draco and Ursa Minor is old, with ages $\sim$15 Gyr. In Leo II, Carina, and Leo I the main bursts of star formation appear to have taken place, respectively, $\sim$9 Gyr, $\sim$7.5 Gyr, and $\sim$4 Gyr ago. The star forming history of Carina seems to have been particularly complex with bursts of star creation having taken place $\sim$15 Gyr, $\sim$7.5 Gyr, and $\sim$3 Gyr ago. In Sculptor (which still contains H I) some star formation may have taken place as recently as ~1 Gyr ago. Finally, the Phoenix galaxy and Pisces (= LGS 3), which are morphologically rather similar to dwarf spheroidals, are still forming stars at the present time. All of the dwarf galaxies in the Local Group that have been searched to faint enough limits are found to contain RR Lyrae stars, that is, stars with ages >10 Gyr. This would appear to indicate that they started to form stars ab initio. Both this observation and the high mass-to-light ratios of dSph galaxies argue against the suggestion (Dottori, Mirabel & Duc 1994) that many, or most of, such objects formed from gas in tidal tails. This is so because the dark matter density in tidally interacting ancestral giant galaxies is much lower than it is in most dwarf spheroidals, and because tidal arms stretch, thereby reducing the dark matter density even more. As a result of their low mass-to-light ratios, a large fraction of dwarfs formed in tidal tails are expected to suffer tidal disruption (Kroupa 1998). It is presently not yet understood why (see Table 15.8) the specific frequency of anomalous Cepheids in dwarf spheroidal galaxies decreases with increasing parent galaxy luminosity.

The existence of transitional objects, such as Pisces (LGS 3) and Tucana, suggests that dIr galaxies gradually fade into dSph galaxies as their rate of star formation decreases. However, there may be a fundamental difference (Freeman 1999) between the evolution of dIr galaxies that are supported by rotation and dSph galaxies that are not. Since dSph galaxies generally appear to exhibit little systemic rotation, they might be descendants of a nonrotating variety of dIr galaxies.

It is not yet clear if location and gas content of dwarf galaxies are physically related. Phoenix and Pisces, which both still contain gas, are situated far from M31 and the Galaxy. However, Sculptor (Carignan et al. 1998) contains some gas, even though it is situated relatively close to the Milky Way system. Nor does the rate of star formation per unit area (Aparicio 1999) in dwarfs appear to be closely correlated with position. Mateo (1998), Mateo et al. (1998), and Grebel (1997, 1999) have plotted graphs showing the schematic star formation history of dwarf galaxies. These data also show no obvious correlation between the history of star formation in dwarf galaxies and their distance from M31 or the Galaxy. However, as was already noted by Einasto et al. (1974), dwarf irregulars fainter than the SMC do not appear to occur close to the Andromeda Galaxy or the Milky Way system. It is presently not clear if Leo I is a satellite of the Galaxy. If it is bound, then $M(\text{Galaxy}) > 12 \times 10^{11} M_\odot$; however, $M(\text{Galaxy})$ is only $4 \times 10^{11} M_\odot$ if Leo I is unbound.

A comparison between the distribution of ellipticities for three spheroidals and 13 dwarf spheroidals in the Local Group, with the frequency distribution of the ellipticities of E and S0 galaxies (Sandage, Freeman & Stokes 1970), is given in Table 15.10 and is plotted in Figure 15.4. Even though the sample of Local Group spheroidal (NGC 147, NGC 185, and NGC 205) and dwarf spheroidal (Sculptor, Fornax, etc.) galaxies is small, it is clear that it contains fewer nearly circular objects than does the Sandage et al. sample of ellipticals. However, the Local Group Sph + dSph sample clearly contains fewer very highly flattened objects than does the S0 sample of Sandage, Freeman & Stokes. A Kolmogorov–Smirnov test shows that there is only an 8% probability that the Local Group Sph + dSph sample and the Sandage et al. sample of S0 galaxies were drawn from a parent population with the same distribution of ellipticities. It is tentatively concluded that the distribution of the ellipticities of Sph + dSph galaxies is intermediate between those of elliptical and lenticular galaxies.

Table 15.10. *Distribution of Ellipticities*

ϵ	Sph + dSph	E	S0
0.00–0.09	1	40	15
0.10–0.17	2	32	17
0.18–0.26	2	29	24
0.27–0.35	7	24	22
0.36–0.44	1	24	13
0.45–0.53	0	14	24
0.54–0.63	3	4	19
0.64–0.72	0	1	27
0.73–0.81	0	0	23
0.82–0.90	0	0	2
Total	16	168	186

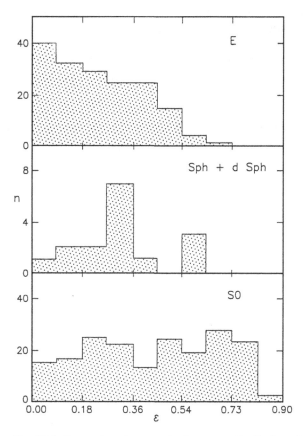

Fig. 15.4. Comparison between the frequency distribution of ellipticities $\epsilon = 1 - b/a$, for Local Group spheroidals and dwarf spheroidals, with that of elliptical and lenticular galaxies. This comparison suggests that the ϵ values of Sph + dSph galaxies are intermediate between those of E and S0 galaxies.

Table 15.11. *Number of C stars in dwarf galaxies (Azzopardi 1999)*

Galaxy	N (carbon stars)	Normalized frequency[a]
Fornax	104	6
Leo I	19	3:
Leo II	8	8
Sculptor	8	10
Draco	4	14
Ursa Minor	1	4
Sextans	0	0
Sagittarius	31?	?
Galactic bulge	>34	?

[a]Normalized to $M_V = -10$.

A tabulation of the number of carbon stars in Local Group galaxies has been given by Azzopardi (1999) and is reproduced in Table 15.11. The data show that dSph galaxies typically have ~6 carbon stars per $M_V = -10$ of parent galaxy luminosity. Some of the C stars in the Galactic halo might be debris from dSph galaxies that were captured, and tidally destroyed, by the Galaxy. However, the observation (Shetrone, Bolte & Stetson 1998) that the abundance pattern in Draco resembles that in globular clusters, more than it does than that in halo field stars, would (if confirmed) appear to militate against the hypothesis that much (or most) of the stellar halo of the Galaxy was built up from disintegrated dSph galaxies (Da Costa 1999).

16

The outer fringes of the Local Group

16.1 Introduction

A detailed discussion of galaxies that may be located along the outer fringes of the Local Group has been given by van den Bergh (1994b). In general three criteria can be used to assess the probability that a galaxy might be associated with the Local Group: (1) The distance to that galaxy should be $\lesssim 1.5$ Mpc (see Section 18.1), (2) it should lie close to the relation between radial velocity and distance from the solar apex (V_r versus $\cos \theta$ diagram) for well-established Local Group members,[1] and (3) it should *not* appear to be associated with a group of galaxies that is known to be located well beyond the limits of the Local Group. On the basis of these criteria van den Bergh (1994b) concluded that it was safe to exclude the following galaxies from membership in the Local Group: (1) Sculptor irregular = UKS 2323-326, (2) Maffei 1 and its companions, (3) UGC-A86[2] = A0355 + 66, (4) NGC 1560, (5) NGC 1569, (6) NGC 5237, and (7) DDO 187 (Hoessel, Saha & Danielson 1998). A particularly strong concentration of Local Group suspects, which includes (2), (3), (4), and (5) listed above, occurs in the IC 342/Maffei group (van den Bergh 1971, Krismer, Tully & Gioia 1995). Krismer et al. place this group at a distance of (3.6 ± 0.5) Mpc. Cassiopeia 1, which was once regarded as a Local Group suspect (Tikhonov 1996), also appears to be a member of the IC 342/Maffei group. Van den Bergh & Racine (1981) failed to resolve Local Group Suspect No. 2 (= LGS 2) on a large reflector plate. They conclude that this object is either Galactic foreground nebulosity or an unresolved stellar system at a distance much greater than that of M31 and M33. Another long-time Local Group suspect is DDO 155 = GR 8. However, observations of this object by Tolstoy et al. (1995) have resulted in the discovery of a single probable Cepheid, which yields a distance modulus of $(m - M)_0 = 26.75 \pm 0.35$, corresponding to a distance of 2.2 Mpc. A distance determination based on the tip of the red giant branch (Dohm-Palmer et al. 1998) is in excellent agreement with the Cepheid value. This places GR 8 beyond the usually accepted boundary of the Local Group. Furthermore, the radial velocity of GR 8 is larger than would be expected for a Local Group member at an apex distance of 107°. The galaxy NGC 55 has recently been listed as a Local Group member by Mateo (1998). However, it appears preferable to follow in the footsteps of de Vaucouleurs

---

[1] The well-established Local Group member Leo I appears to have an above-average deviation from the mean radial velocity versus apex distance relation.

[2] The UGC-A numbers refer to the *Catalogue of Selected Non-UGC Galaxies* (Nilson 1974).

(1975) and Sandage & Tammann (1981, p. 10), who assigned this galaxy to the Sculptor group, which is sometimes also referred to as the South Polar group. The other luminous members of this group are NGC 247, NGC 253, NGC 300, and NGC 7793. Sandage & Bedke (1994, panel 318) write "NGC 55 is very highly resolved into individual stars, about equally well as other galaxies in the South Polar group such as NGC 247 and NGC 300. . . . Evidently, NGC 55 is just beyond the Local Group." Côté, Freeman & Carignan (1994) find that NGC 55 is located close to the center of the distribution of dwarf galaxies associated with the South Polar group. From photometry in J, H, and K, Davidge (1998) concludes that NGC 55, NGC 300, and NGC 7793 are at comparable distances. However, Jerjen, Freeman & Binggeli (1998) derive distances of 1.66 ± 0.20 Mpc, 2.10 ± 0.10 Mpc, and 3.27 ± 0.08 Mpc, respectively, for NGC 55, NGC 300, and NGC 7793. In other words, Jerjen et al. find that NGC 55 is situated on the near side of the Sculptor group.

The galaxy IC 5152, which Sandage & Tammann (1981) classify as Sdm IV–V, is located at a distance of 28° from NGC 7793 and at a distance of 33° from NGC 300. This suggests that it might be an outlying member of the South Polar group. This speculation is supported by a Cepheid distance modulus, $(m - M)_0 \sim 26$, corresponding to a distance $D \approx 1.6$ Mpc, obtained by Caldwell & Schommer (1988). Previously Sérsic & Cerruti (1979) had used a variety of distance indicators to estimate $D \approx 4.6$ Mpc. They concluded that "IC 5152 is too distant to belong to the Local Group." From a color–magnitude diagram in the V and I bands, Zijlstra & Minniti (1999) find that $(m - M)_0 = 26.15 \pm 0.2$. The corresponding distance of 1.70 ± 0.16 Mpc places IC 5152 slightly beyond the outer limits of the Local Group.

The dwarf irregular galaxy UKS 2323-326 was discovered on the plates of the *ESO/SRC Southern Sky Survey* by Longmore et al. (1978). These authors opined that this dwarf might be a member of the Local Group. More detailed observations of this object were subsequently obtained by Capaccioli, Ortolani & Piotto (1987). From CCD observations of the brightest stars in this galaxy in B and V these authors estimate that $(m - M)_0 = 26 \pm 1$, corresponding to a distance $D = 1.6^{+0.9}_{-0.6}$ Mpc. Such a distance is consistent with the hypothesis that UKS 2323-326 is located near the outer fringes of the Local Group. However, the observation that this object lies only $6°6$ from the South Polar group galaxy NGC 7793 strongly suggests that UKS 2323-326 is itself a member of the South Polar group.

The Sculptor dwarf irregular galaxy (Laustsen et al. 1977, Heisler et al. 1997) is another example of a relatively nearby galaxy that appears projected on, and is probably a member of, the South Polar group. This object is located at an angular distance of $4°8$ from NGC 55 and only $2°9$ from NGC 7793. The South Polar group has been discussed in detail by Jerjen, Freeman & Binggeli (1998). After excluding the probable background galaxies NGC 45 and NGC 59, the remaining seven galaxies yield $\langle V_r \rangle = +183 \pm 30$ km s^{-1} and $\sigma = 79$ km s^{-1}. This velocity dispersion is quite similar to that observed in the Local Group (see Section 19.2). From surface brightness fluctuations Jerjen et al. estimate that the nearest Sculptor galaxy, ESO 249-010, has a distance of 1.71 ± 0.07 Mpc.

It should be emphasized that the Local Group is a clumping of galaxies surrounded by other nearby clumps, such as the M81 group, the IC342/Maffei group, and the Sculptor group. As a result it is perhaps not always meaningful to assign objects near the outer fringes of these clumpings to individual groups.

16.2   The dwarf irregular galaxy NGC 3109

This is a highly resolved galaxy, which is classified Sm IV by Sandage & Tammann (1981). Fine images of it are reproduced in Sandage & Bedke (1988, p. 58). Since this galaxy does not appear to show a nucleus, a classification of Ir IV appears to be indicated. From observations of 24 Cepheids Musella, Piotto & Capaccioli (1998) derive a distance modulus of $(m - M)_0 = 25.67 \pm 0.16$, corresponding to a distance of 1.36 ± 0.10 Mpc. This Cepheid distance is consistent with the value $(m - M)_0 = 25.45$ that Richer & McCall (1992) derived from planetary nebulae and that Lee (1993) obtained from the magnitude of the tip of the red giant branch in NGC 3109. For these relatively old stars Lee finds [Fe/H] $= -1.6$. However, it seems probable that younger stars will have a higher metallicity. Adopting the Cepheid distance gives $M_V = -15.8 \pm 0.2$ for NGC 3109. Additional data on this galaxy are collected in Table 16.1. From 21-cm observations Carignan (1985) finds that the rotation curve of this almost edge-on galaxy requires the presence of a substantial dark halo. Huchtmeier (1973) finds that $\sim 20\%$ of the mass in NGC 3109 is in the form of neutral hydrogen. The color–magnitude diagrams for various regions in this galaxy have been studied by Bresolin, Capaccioli & Piotto (1993), who find a plume of blue main-sequence stars ($M \approx 25\,M_\odot$) that extend up to $V \sim 18.5$. Sandage & Carlson (1988) have found the ten brightest red supergiants to be variable. Bresolin et al. group the brightest stars into 18 associations, many of which contain emission nebulosity. Some of these H II regions have a ringlike morphology that is reminiscent of that of the bright Strömgren spheres in the northern part of NGC 6822 (which has a luminosity close to that of NGC 3109). Bresolin et al. failed to detect CO in NGC 3109, which probably indicates that this galaxy is dust poor and hence deficient in H_2. Greggio et al. (1993) find that the rate of star formation has been declining slightly over the past billion years and is now $240\,M_\odot$ per Myr. Davidge (1995) has observed bright asymptotic branch stars in this galaxy, which shows that NGC 3109 also contains an intermediate-age component. Demers, Kunkel & Irwin (1985) have identified 10 objects as globular cluster candidates. If we adopt the Cepheid distance to NGC 3109 these clusters have $\langle M_V \rangle = -5.6$, which is significantly fainter than the mean luminosity of Galactic globulars.

On the basis of its distance and position in the V_r versus $\cos\theta$ diagram, it is concluded in Chapter 18 that NGC 3109 is probably not a member of the Local Group.

16.3   Antlia

The Antlia system is a galaxy of type dSph with no H II or obvious evidence for recent star formation. It was cataloged by Corwin, de Vaucouleurs & de Vaucouleurs (1985) and was noted as a possible nearby galaxy by Feitzinger & Galinski (1985) and by Arp & Madore (1987). This suspicion was strengthened by Fouqué et al. (1990) who found a small radial velocity $V_r = +361 \pm 3$ km s^{-1}. The true nature of this galaxy was revealed by the photometric observations of individual stars by Aparicio et al. (1997), by Sarajedini, Claver & Ostheimer (1997), and by Whiting, Irwin & Hau (1997). A good image of the Antlia dwarf was published by Sarajedini et al. From their V and I photometry these authors find that Antlia has a well-defined red giant branch that possibly shows some evidence for differential internal dust absorption. If Antlia contains a non-negligible amount of dust, then their data yield $(m - M)_0 = 25.47 \pm 0.13$,

Table 16.1. *Data on NGC 3109*

$\alpha(2000) = 10^h\,03^m\,06\!\!.\!\!^s7$		$\delta(2000) = -26°\,09'\,07''$ (1)
$\ell = 262°\!.10$		$b = +23°\!.07$
$V_r = +404 \pm 1$ km s^{-1} (1)		$i = 80° \pm 2°$
$V = 9.87 \pm 0.08$ (1)	$B - V = 0.00$ (2)	$U - B = -0.11 \pm 0.08$ (1)
$(m - M)_o = 25.67 \pm 0.16$ (2)	$E(B - V) = 0.00$ (2)	$A_V = 0.0$ mag
$M_V = -15.80 \pm 0.18$	$D = 1.36 \pm 0.1$ Mpc	$D_{LG} = 1.75$ Mpc
Type = Ir IV	size = $2\!\!.\!\!'8 \times 13\!\!.\!\!'3$ (1.1 kpc	
	$\times$ 5.3 kpc) (2,3)	
$M = 2.3 \times 10^9\,M_\odot$	$M_{HI} = 8.4 \times 10^8\,M_\odot$ (2,7)	M_{H_2} = low (4)
No. globulars = 10? (6)	No. planetaries = 7 (6)	
[Fe/H] $= -1.8 \pm 0.2$ (5)		

(1) de Vaucouleurs et al. (1991).
(2) Musella et al. (1997).
(3) Carignan (1985).
(4) Bresolin et al. (1993).
(5) Minniti et al. (1999).
(6) Richer & McCall (1992).
(7) Huchtmeier (1973).

corresponding to $D = 1.24 \pm 0.10$ Mpc, and [Fe/H] $= -2.41 \pm 0.18$. However, these data yield $(m - M)_o = 25.62 \pm 0.12$, corresponding to $D = 1.33 \pm 0.10$ Mpc, and [Fe/H] $= -1.90 \pm 0.13$ if it is assumed that there is, in fact, no dust absorption. Since dust requires the existence of metals, it seems physically unreasonable to assume that the presence of dust implies such a very low metallicity. In the subsequent discussion, the results that assume no internal dust absorption will therefore be adopted.

From their color–magnitude diagram, Sarajedini et al. conclude that the inner region of Antlia contains a small young, or intermediate-age, population component. They find that the young stars are more centrally concentrated and that the intermediate-age asymptotic branch stars have a more extended distribution than the old stars. Sarajedini et al. conclude that the youngest stars in Antlia have an age of only 0.1 Gyr. In view of the presence of $(8 \pm 2) \times 10^5\,M_\odot$ of H I in Antlia (Fouqué et al. 1990, Whiting et al. 1997) the existence of such relatively young stars is not entirely unexpected. Aparicio et al. find that the surface-brightness distribution of Antlia cannot be represented by a single exponential. For $r < 40''$ they obtain a scale-length of $64''$ (0.4 kpc) and at $r > 40''$ the scale-length is $16\!\!.\!\!''5$ (0.1 kpc). For the inner region of Antlia these authors find a mean rate of star formation of $150 - 400\,M_\odot$ per million years, during the past 1 Gyr. However, in the outer region the rate of star formation has been $<30\,M_\odot$ per million years, during the past 1 Gyr. The presently available data on Antlia are summarized in Table 16.2.

Antlia and NGC 3109 are separated by only $1°\!.18$ on the sky. Furthermore, their distances are, within their errors, identical with $D(\text{N3109}) = 1.36 \pm 0.1$ Mpc and $D(\text{Antlia}) = 1.33 \pm 0.1$ Mpc. (The projected linear separation of these two galaxies is only 28 kpc.) In addition, their radial velocities differ by only 43 ± 2 km s^{-1}. Substituting this value into Eq. 12.2 shows that the total combined mass of the NGC 3109 + Antlia binary system would have to be $\gtrsim 1.6 \times 10^{10}\,M_\odot$ for it to be stable. The corresponding value of M/L_V is $\gtrsim 90$ in solar units. For the Antlia system itself, Whiting et al. (1997) find

Table 16.2. *The Antlia dwarf spheroidal*

$\alpha(2000) = 10^{\rm h}03^{\rm m}34\overset{s}{.}3$		$\delta(2000) = -27°19'48''$ (1)
$\ell = 263\overset{\circ}{.}01$		$b = +22\overset{\circ}{.}24$
$V_r = +361 \pm 2$ km s^{-1}		$\epsilon = 0.35 \pm 0.03$ (1)
$V = 14.8 \pm 0.2$ (2)		
$(m - M)_0 = 25.62 \pm 0.12$ (3)	$E(B - V) = 0.045$ (1)	$A_V = 0.14$ mag
$M_V = -10.96 \pm 0.2$	$D = 1.33 \pm 0.1$ Mpc (3)	$D_{\rm LG} = 1.72$ Mpc
Type = dSph	scale-length = $64''$ (0.41 kpc)	$r < 40''$ (4)
	scale-length = $16\overset{''}{.}5$ (0.11 kpc)	$r > 40''$ (4)
$M = 4 \times 10^7 M_\odot$ (1)	$M_{\rm HI} = (8 \pm 2) \times 10^5 M_\odot$ (1,2)	
No. globulars = 0?		
[Fe/H] = -1.90 ± 0.13		

(1) Whiting et al. (1997).
(2) Fouqué et al. (1990).
(3) Sarajedini et al. (1997).
(4) Aparicio et al. (1997).

$M/L_V = 20 \pm 8$ in solar units. The position of Antlia in a plot of velocity versus apex distance is consistent with its being a gravitationally bound member of the Local Group (although this would not be a requirement). A more detailed discussion of this point is given at the end of Sections 16.5 and 18.

16.4   The dwarf irregular Sextans B

The irregular galaxy Sextans B (= DDO 70) is of morphological type Ir IV–V. Holmberg (1958) attributes the discovery of Sex B to Wilson (1955). However, Wilson's paper, in fact, only refers to the distant globular cluster Palomar 3 in Sextans. A good photograph of Sex B has been published by Sandage & Carlson (1985a), who also studied the Cepheids in this galaxy. More recently additional Cepheids have been discovered in Sex B by Piotto, Capaccioli & Pellegrini (1994). An analysis of the B, V, R, and I photometry of Cepheids in Sex B by Sakai, Madore & Freedman (1997) yields a distance modulus of $(m - M)_0 = 25.69 \pm 0.27$ and an extinction of $A_V = 0.05$ mag. From the I magnitude of the tip of the giant branch in this galaxy, Sakai et al. derive an independent distance modulus of $(m - M)_0 = 25.56 \pm 0.10$ (random) ± 0.16 (systematic). The fact that the Population I Cepheids and Population II red giants give consistent distance estimates strengthens confidence in this result. In the subsequent discussion $(m - M)_0 = 25.6 \pm 0.2$ will be adopted. The corresponding distance is 1.32 ± 0.12 Mpc, which places Sex B near the outer fringes of the Local Group. The CCD color-magnitude diagram of Sex B in V and I (Sakai et al. 1997) is dominated by red giants. A prominent population of intermediate-age asymptotic branch stars is also present. The discovery of a small number of Cepheids attests to the existence of some young stars in Sextans B. Tosi et al. (1991) estimate the average star formation rate over the past 1 Gyr to have been $\sim$2,000 $M_\odot$ per million years. Sandage & Carlson (1985a) find that the brightest red supergiants in Sex B are small-amplitude variables. Hodge (1974) has located four emission regions in Sextans B. From observations of a single H II region in this object, Moles, Aparicio & Masegosa (1990) find that the oxygen abundance is only 0.16 times that in the Sun.

The observed distance of Sex B indicates that this object is situated near the outer fringes of the Local Group. The location of Sex B in a V_r versus $\cos\theta$ plot is consistent with Sextans B being a physical member of the Local Group (although this does not prove it to be). A more detailed discussion of this point is given in Sections 16.5 and 18. Data on Sextans B are compiled in Table 16.3.

16.5   The dwarf irregular Sextans A

Sextans A (= DDO 75) is perhaps the most beautiful known example of a dwarf irregular galaxy.[3] It is classified as being of morphological type Ir V by van den Bergh (1966a). Sex A was discovered by Zwicky (1942), who used its existence to argue (correctly) that the luminosity function of galaxies rises steeply toward faint absolute magnitudes and is not Gaussian, as Hubble (1936, pp. 152–181) had claimed. A nice photograph of this diamond-shaped object, which contains one huge association, is shown in Zwicky (1957, p. 223). Sex A was listed as a "doubtful member" of the Local Group by Humason & Wahlquist (1955). From B, V, R, and I photometry of a small number of Cepheids, Sakai, Madore & Freedman (1996) obtain a true distance modulus of $(m - M)_0 = 25.85 \pm 0.15$. These authors also find $(m - M)_0 = 25.74 \pm 0.13$ from V and I photometry of the tip of the red giant branch of Sex A. The good agreement between these Population I and Population II distance moduli gives confidence in the accuracy of this distance determination. In the subsequent discussion a value of $(m - M)_0 = 25.8 \pm 0.1$, corresponding to $D = 1.45 \pm 0.07$ Mpc, will be adopted. Ables (1971) measured an integrated magnitude $V = 11.63$, which, with a reddening $E(B - V) = 0.015$ (Burstein & Heiles 1984), yields $M_V = -14.2$. For $r > 2'.0$ Ables finds that the surface brightness of Sex A may be represented by an exponential disk with a scale-length of $0'.65$ (275 pc).

The color–magnitude diagram of Sextans A has been studied in great detail by Dohm-Palmer et al. (1997a,b) in B, V, and I with the *Hubble Space Telescope*. These authors find that this galaxy contains a prominent red giant branch, which points to the existence of a strong intermediate-age, or old, population with an age >1 Gyr. Furthermore, Sex A exhibits a well-defined blue main sequence that is populated by stars with ages <10 Myr. Finally, the color–magnitude diagram shows a prominent population of helium core-burning (blue loop) stars that are situated just to the red of the main sequence, with ages of ~10 Myr to ~1 Gyr. These "blue loop" stars allow one to trace the history of star formation in Sex A over the most recent ~700-Myr period. From the distribution of these stars in the color–magnitude diagram, Dohm-Palmer et al. conclude that the rate of star formation in Sex A was nearly constant between 600 Myr and 100 Myr ago. It then increased by a factor of ~4 during the most recent 100 Myr. At present this star formation is strongly concentrated in a single association, with a diameter of 25 pc, that is associated with the brightest H II regions (Hodge, Kennicutt & Strobel 1994) and with the region of greatest H I column density. Dohm-Palmer et al. find that the present rate of star formation, per unit area, is ~20 times greater in Sex A than its average rate over the lifetime of this galaxy. They also note that the regions of most intense star formation have migrated over the surface of this dwarf during the most recent 0.6-Gyr period. (This

---

[3] A color photograph of this object is shown on the front cover of the September 1997 issue of *the Publications of the Astronomical Society of the Pacific*.

Table 16.3. *Data on Sextans B (= DDO 70)*

$\alpha(2000) = 09^h 59^m 59\overset{s}{.}9$		$\delta(2000) = +5°19'42''$ (1)
$\ell = 233\overset{\circ}{.}20$		$b = +43\overset{\circ}{.}78$
$V_r = +301 \pm 4$ km s^{-1} (2)		$i = 35° \pm 15°$ (6)
$V = 11.33 \pm 0.14$ (2)	$B - V = 0.52 \pm 0.03$ (2)	$U - B = -0.13 \pm 0.05$ (2)
$(m - M)_o = 25.6 \pm 0.2$ (1)	$E(B - V) = 0.02$	$A_V = 0.05$ mag (1)
$M_V = -14.3 \pm 0.14$	$D = 1.32 \pm 0.12$ Mpc	$D_{LG} = 1.60$ Mpc
Type = Ir IV–V	size = $5\overset{'}{.}9 \times 7\overset{'}{.}7$ (2.3 kpc	
	$\times$ 3.0 kpc) (3)	
$M = 8.85 \times 10^8 M_\odot$ (6)	$M_{HI} = 4.2 \times 10^7 M_\odot$ (1, 4)	$M_{H_2} < 400 M_\odot$ (6)
No. globulars = 0?		
	$12 + \log(O/H) = 7.56$ (5)	

(1) Sakai et al. (1997).
(2) de Vaucouleurs et al. (1991).
(3) Holmberg (1958).
(4) Fisher & Tully (1975).
(5) Skillman et al. (1989).
(6) Mateo (1998).

behavior is reminiscent of that which Stetson, Hesser and Smecker-Hane 1998 found in the Fornax dwarf.) Possible reasons for the burst of star formation that started in Sex A ∼100 Myr ago, along with a detailed discussion of the migration of star formation, are given by Van Dyk, Puche & Wong (1998). These authors tentatively conclude that the most likely scenario is one in which the current burst of star formation occurs in cold gas that has piled up (and become compressed) along the rim of a hole in the Sex A neutral hydrogen distribution. This hole might have been produced by supernova explosions and stellar winds. From observations of the H II regions in Sex A, Skillman, Kennicutt & Hodge (1989) find that the oxygen abundance in Sextans A is ∼1/30 of that in the Sun.

On the basis of its distance and position in the V_r versus $\cos\theta$ diagram (see Figure 16.1), Sex A is located along the outer fringes of the Local Group, and it might possibly be bound to it. Sex A and Sex B have similar distances and are separated on the sky by only $10\overset{\circ}{.}4$. Their velocity difference is 23 ± 6 km s^{-1}, and their linear separation is 280 kpc. It is quite possible that these objects were formed close together and subsequently drifted apart over a Hubble time, at a relative velocity of 20–30 km s^{-1}. Sex A and NGC 3109 are separated on the sky by $22\overset{\circ}{.}7$. Their linear separation amounts to 560 kpc. These results suggest that NGC 3109, Antlia, Sex A, and Sex B may have formed a loose physical grouping. All four of these galaxies fall above the mean radial velocity versus apex distance solutions of Yahil, Tammann & Sandage (1977). These authors refer to Sex A, Sex B, and NGC 3109 as "possible, but unlikely, members" of the Local Group. The mean distance of the four members of the Antlia–Sextans group from the barycenter of the Local Group (see Chapter 18) is 1.7 Mpc. This places these objects well beyond the zero-velocity surface of the Local Group, for which a radius $R_o = 1.2$ Mpc is found in Chapter 18. In other words, the Antlia–Sextans galaxies constitute the nearest external group of galaxies. A deep search for additional very faint members of this loose grouping might prove rewarding.

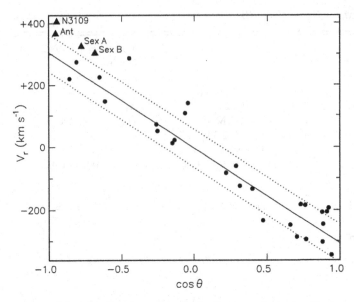

Fig. 16.1. Positions of the Antlia–Sextans galaxies in a V_r versus $\cos\theta$ plot. The figure shows that these objects appear to form a small grouping that is slightly redshifted relative to the Local Group members. The dotted lines are displaced by 1σ from the mean V_r versus $\cos\theta$ relation for Local Group members.

Matthews, Gallagher & Littleton (1995) have found a very small, low-luminosity, object AM 1013-394A at $\alpha = 10^h\ 16^m\ 06\overset{s}{.}0$, $\delta = -39\ 59'\ 17''$ (J 2000) with $V_r = +263$ km s^{-1}. This galaxy is only $14\overset{\circ}{.}1$ from NGC 3109. Based on its 21-cm profile, and on the fact that no high-velocity clouds are known at this position, Matthews et al. believe that this object is a Local Group candidate. Alternatively, it might also be associated with the small nearby Antlia–Sextans group. Finally, it might be a foreground H I cloud superposed on a background galaxy with $V_r = +2982$ km s^{-1}. A summary of the data on Sextans A is given in Table 16.4.

16.6   The dwarf irregular galaxy EGB 0419 + 72
During a search for low surface-brightness planetary nebulae, Ellis, Grayson & Bond (1984) found a number of low surface-brightness objects. One of these turned out to be a nearby dwarf galaxy. An image of EGB 0419 + 72, obtained by Hoessel, Saha & Danielson (1988) with the Hale 5-m telescope, shows it to be a dwarf irregular. (The S0/a classification listed by Huchtmeier & Richter 1986 is incorrect.) According to Hodge (1994b) this late-type dwarf contains 25 H II regions. Depending on the assumed nature of the brightest stars, Hoessel et al. obtain distance moduli that range from 24.5 to 32. The morphology of their image supports a value near the low end of this range. From CCD photometry with the Soviet 6-m telescope in the B, V, and R bands, Karachentsev, Tikhonov & Sazonova (1994) obtained $E(B-V) = 0.95$ and $A_V = 3.04$ mag. From these values, and tentative identification of the two brightest red supergiants, Karachentsev et al. derived a very uncertain distance modulus of $(m-M)_0 = 26.72$. According to Huchtmeier & Richter (1986) the heliocentric velocity of EGB 0419 + 72

Table 16.4. *Data on Sextans A (= DDO 75)*

$\alpha(2000) = 10^{\rm h}11^{\rm m}01\overset{\rm m}{.}3$		$\delta(2000) = -04°42'48''$ (1)
$\ell = 246\overset{\circ}{.}17$		$b = +39\overset{\circ}{.}86$
$V_{\rm r} = +324 \pm 4 \text{ km s}^{-1}$ (1)		$i = 33° \pm 3°$ (2)
$V = 11.63$ (3)	$B - V = 0.30$ (3)	$U - B = -0.5$ (3)
$(m - M)_{\rm o} = 25.8 \pm 0.1$ (4)	$E(B - V) = 0.015$	$A_V = 0.045$ mag (5)
$M_V = -14.2 \pm 0.1$	$D = 1.45 \pm 0.07$ Mpc	$D_{\rm LG} = 1.79$ Mpc
Type = Ir V	size = $8\overset{.}{.}6 \times 9\overset{.}{.}3 (3.6 \text{ kpc}$	
	$\times$ 3.9 kpc) (6)	
	disk scale-length	
	$= 0\overset{.}{.}65$ (275 pc) (3)	
$M > 1.9 \times 10^8 M_\odot$ (2,4)	$M_{\rm HI} = (7.0 \pm 0.2) \times 10^7 M_\odot$	$M_{\rm H_2} = 2.7 \times 10^7 M_\odot$ (7)
		$12 + \log(\text{O/H}) = 7.49$ (8)

(1) de Vaucouleurs et al. (1991).
(2) Skillman et al. (1988).
(3) Ables (1971).
(4) Sakai et al. (1996).
(5) Burstein & Heiles (1984).
(6) Holmberg (1958).
(7) Mateo (1998).
(8) Skillman et al. (1989).

Table 16.5. *Data on EGB 0427 + 63 (= UGC − A92)*

$\alpha(2000) = 04^{\rm h} 32^{\rm m} 01\overset{\rm s}{.}7$		$\delta(2000) = +63° 36' 25''$ (1)
$\ell = 144\overset{\circ}{.}71$		$b = +10\overset{\circ}{.}51$
$V_{\rm r} = -87 \pm 5 \text{ km s}^{-1}$ (2)		
$V = 13.88$ (3)	$B - V = 1.34$ (3)	
$(m - M)_{\rm o} = 26.72$: (4)	$E(B - V) = 0.95$ (4)	$A_V \approx 3.08$ mag (4)
$M_V = -15.9$:		$D = 2.2$: Mpc
Type = Ir	size = $1\overset{.}{.}5 \times 2\overset{.}{.}0$ (1.0	
	$\times$ 1.3 kpc) (1)	

(1) Ellis et al. (1984).
(2) Huchtmeier & Richter (1986).
(3) Karachentseva et al. (1996).
(4) Karachentsev et al. (1994).

is $-87 \pm 5 \text{ km s}^{-1}$. That this object is separated from IC 342 by only $6\overset{\circ}{.}4$ suggests that it is, in fact, an outlying member of the IC 342/Maffei 1 group. This opinion is shared by Kraan-Korteweg & Tammann, who assign EGB 0427 + 63 to their B1 group, which includes Maffei 1 and IC 342. Data on this galaxy are collected in Table 16.5.

16.7 Summary and conclusions

A rather large number of suspected members are found near the outer fringes of the Local Group. Some of these objects may be physically associated with it, while others are, no doubt, situated in the more distant clusters on which they appear projected. It is suggested that NGC 1560, NGC 1569, EGB 0427 + 63, UGC-A86, and Cassiopeia 1 are, in fact, members of the extended IC 342/Maffei group. Furthermore, it appears likely that NGC 55, IC 5152, and UKS 2323-326 are associated with the large nearby South Polar group. Finally, NGC 3109, Antlia, Sextans A, Sextans B, and perhaps AM 1013-394A may form a physical association of galaxies that lies beyond the zero-velocity surface that forms the physical boundary of the Local Group.

17

Intergalactic matter in the Local Group

17.1   Introduction

Gaseous material is expected to flow into galaxies by the capture of high-velocity clouds (Oort 1966, 1970). Oort estimated that the mass of the Milky Way system might be increasing by ~1% per Gyr as a result of such inflow. However, tidal interactions between galaxies will result in the return of some interstellar gas in galaxies to intergalactic space (Morris & van den Bergh 1994). In addition, some gas may be ejected from galaxies by supernova-induced fountains (Heiles 1987). Perhaps the best example of inflow into the Galactic Disk is provided by "Complex C" (Wakker & van Woerden 1997). From the nondetection of absorption features in stars of known distance, these authors conclude that $D > 2.4$ kpc and that Complex C is falling toward us with a velocity >100 km s^{-1}. From spectroscopic observations of S II absorption lines (which are not affected by depletion onto dust) with the *Hubble Space Telescope*, Wakker et al. (1999) find that Complex C has a metallicity of 0.094 ± 0.020 times solar. This very low metallicity rules out the possibility that the gas in Complex C was ejected in a Galactic fountain. Furthermore, the data on the age–metallicity relation of the Small Magellanic Cloud (Mighell, Sarajedini & French 1998) show that the gas in Complex C is so metal poor that it could only have been stripped from the SMC $\gtrsim 5$ Gyr ago (i.e., long before the Magellanic Stream was formed). For a more detailed discussion of the ~1.5 Gyr old Magellanic Stream, and of the ~0.2 Gyr old Bridge between the LMC and SMC, the reader is referred to Sections 6.12 and 7.9.

17.2   Tidal debris

Perhaps the most striking example of the importance of tidal debris is provided by the M81 group (Appleton, Davies & Stephenson 1981), in which $1.4 \times 10^9\ M_\odot$ of neutral hydrogen is found to be located in tidal structures that lie beyond the Holmberg radii of the interacting galaxies NGC 3031 (=M81), NGC 3034 (=M82), and NGC 3077. Within the Local Group the most spectacular tidal features are the Magellanic Stream (Mathewson & Ford 1984) and the more recently formed Magellanic Bridge. Some high-velocity clouds may also represent the remnants of tidal features.

17.3   High-velocity clouds

Bland-Hawthorn (1998) and Tufte, Reynolds & Haffner (1999) have shown that high-velocity clouds can be observed in Hα. This provides a powerful new technique for the study of these objects. Maloney & Bland-Hawthorn (1999) show that ionization

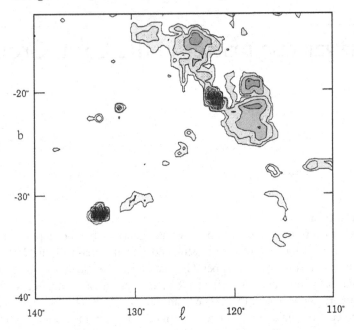

Fig. 17.1. Distribution of neutral hydrogen (with velocities between -140 km s^{-1} and -85 km s^{-1} relative to the local standard of rest) near M31 and M33. Note the large hydrogen complex surrounding the Andromeda galaxy ($\ell = 121°$, $b = -22°$) and the broken chain of clouds between M31 and M33 ($\ell = 134°$, $b = -31°$). (Adapted from Blitz et al. 1998.)

in the Local Group will be dominated by the cosmic background. Wakker, van Woerden & Gibson (1999) point out that observations of the strength of Hα in the compact clouds discussed by Blitz et al. (1998) should be able to establish if these objects are truly intracluster objects, or if they are high-velocity clouds associated with the most massive Local Group members. In the latter case one would expect them to strongly emit Hα because they are close to the OB stars in such galaxies. However, truly intergalactic objects would typically be quite far removed from luminous Local Group members and hence dim in Hα. This suggests that Hα emission from individual galaxies will only dominate the Hα emission of high-velocity clouds in a few percent of the volume of the Local Group. The remaining high-velocity clouds will only be illuminated by the cosmic background.

Blitz et al. (1998) postulate that most high-velocity clouds constitute the surviving building blocks of the Local Group. They find that the kinematics of the entire ensemble of high-velocity clouds is inconsistent with a Galactic origin. However, in their view, it is consistent with a scenario in which such high-velocity clouds mainly consist of intercluster gas that is falling into the Local Group. Such continuing accretion of gas and dark matter is an inevitable consequence of hierarchical merger models of structure in the Universe. Blitz et al. regard the more distant high-velocity clouds as dark matter "minihalos" that are moving along filaments toward the Local Group. These authors suggest that such high-velocity clouds may be analogs to the Lyman-limit absorbing clouds seen in the spectra of distant quasars. According to Blitz et al. a typical high-velocity cloud has a mass of $\sim 3 \times 10^8 \, M_\odot$, a hydrogen mass of $\sim 3 \times 10^7 \, M_\odot$, a diameter of 30 kpc, a

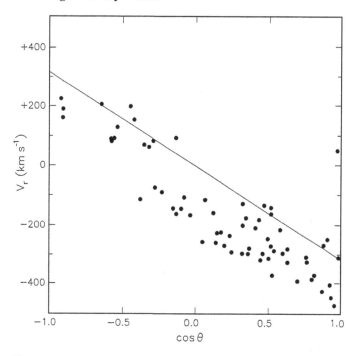

Fig. 17.2. Radial velocity versus $\cos\theta$ plot, in which θ is the distance from the solar apex, for compact, isolated, high-velocity clouds. Note that many of these clouds are redshifted relative to the plotted relation for Local Group galaxies that is derived in Section 19; that is, these clouds are falling into the Local Group, whereas galaxies are not. Figure adapted from Burton & Braun (1998). The deviant point is a galaxy at an estimated distance of 6 Mpc (Burton et al. 1999).

distance of ~ 1 Mpc, and a mean gas density of $\sim 1 \times 10^{-4}$ cm^{-3}. Of particular interest is a complex of large H I structures, with dimensions of $5° \times 15°$, surrounding M31, which is shown in Figure 17.1. Furthermore, there appears to be a broken chain of smaller H I features that stretches from M31 to M33. Finally, Wright (1974, 1979) has found a $3° \times 5°$ hydrogen structure near M33 itself. If it is at the same distance as the Triangulum galaxy, it has a mass of $\sim 1 \times 10^8 \, M_\odot$.

Burton & Braun (1998) find a solar motion of $V = 293$ km s^{-1} toward $\ell = 88°$, $b = -19°$ from compact, isolated, high-velocity clouds. This is essentially indistinguishable from $V = 306 \pm 18$ km s^{-1} toward $\ell = 99° \pm 5°$, $b = -4° \pm 4°$, which Courteau & van den Bergh (1999) find from the radial velocities of Local Group galaxies. It should, however, be noted that many high-velocity clouds (see Figure 17.2) are observed to exhibit significant blueshifts, relative to the V_r versus $\cos\theta$ relation of Local Group galaxies. Such blueshifts are not found among Local Group galaxies; that is, gas clouds still appear to be falling into the Local Group, whereas galaxies are not.

Blitz et al. find that the one-dimensional velocity dispersion of the high-velocity clouds in the Local Group is 106 km s^{-1}. Assuming that these clouds are virialized, and occupy a volume with a radius of 1.5 Mpc, they find a total Local Group mass of $\sim 2 \times 10^{12} \, M_\odot$, which is close to the mass that Kahn & Woltjer (1959) derived from their M31/Galaxy timing argument. A similar mass of $(2.3 \pm 0.6) \times 10^{12} \, M_\odot$ is derived by Courteau & van den Bergh (1999) from the radial velocities, and half-mass radius of the Local Group.

18

Dynamical and physical evolution

18.1   Missing mass

The centers of M31 and the Galaxy, which are the dominant galaxies in the Local Group, are approaching each other with a velocity of 125 ± 25 km s^{-1}. If one assumes that these two objects were formed close to each other (Kahn & Woltjer 1959), then they must already have performed the larger part of one orbit around their center of gravity during a Hubble time ($\sim$15 Gyr). This requires that the effective mass at their center of gravity must be $\gtrsim 2 \times 10^{12}$ $M_\odot$. Since this value is significantly greater than the combined baryonic mass of Andromeda and the Milky Way system, Kahn & Woltjer concluded that a major fraction of the mass required to stabilize the Local Group was "missing." These authors estimated that at least 1.5×10^{12} $M_\odot$ of intracluster material was required to stabilize the Local Group. Interestingly, Kahn & Woltjer did not reference Zwicky (1933), in which a similar "missing mass" problem had been noted in the Coma cluster. Furthermore, Smith (1936) had already found that "a great mass of internebular material" would be required to stabilize the Virgo cluster. The total mass of the Local Group can also be estimated from the velocity dispersion of Local Group members (Courteau & van den Bergh 1999). If one assumes the Local Group to be in virial equilibrium, and that its velocity dispersion is isotropic, then (Spitzer 1969 and Binney & Tremaine 1987, p. 214) one has

$$M \text{ (Local Group)} \simeq \frac{7.5}{G}\langle\sigma_r^2\rangle R_h = 1.74 \times 10^6 \langle\sigma_r^2\rangle R_h M_\odot, \qquad (18.1)$$

in which $\langle\sigma_r\rangle = 61 \pm 8$ km s^{-1} (Courteau & van den Bergh 1999), and R_h was assumed to be 350 kpc, which is the distance of M31 from the center of mass of the Local Group. With these values the dynamical mass of the Local Group is found to be M (Local Group) $=$ $(2.3 \pm 0.6) \times 10^{12}$ $M_\odot$. Lynden-Bell (1981) finds that the radius R_0 of the zero-velocity surface of the Local Group is related to its mass M and age T_0 by the relation

$$GMT_0^2 = \pi^2(R_0/2)^3. \qquad (18.2)$$

Substituting the Local Group mass derived above yields

$$R_0(\text{Mpc}) = (1.23 \pm 0.16) \times (T_0(\text{Gyr})/15)^{2/3}. \qquad (18.3)$$

The reason for the rather small mean error of this result is, of course, that the zero-velocity radius is proportional to the cube root the Local Group mass. It seems probable that a significant fraction of the total Local Group mass is located in the halos of M31 and of the Galaxy. From the radial velocities of galaxies within the Andromeda subgroup of the Local Group, van den Bergh (1981a) found a mass in the range $(4-11) \times 10^{11} M_\odot$. More recently Courteau & van den Bergh (1999) have derived a mass of $11.5 \times 10^{11} M_\odot$ by application of the virial theorem to the Andromeda subgroup and $M = 15 \times 10^{11} M_\odot$ by using the projected mass method. It is of interest to note that the latter value is slightly larger than the mass[1] of $(4.6-12.5) \times 10^{11} M_\odot$ that Zaritsky et al. (1989) found for the Galaxy and its closest companions. These data suggest that the combined masses of Andromeda and Galactic subgroups is comparable to the total mass required to stabilize the Local Group.

Assuming the Sun to have $M_V = +4.82 \pm 0.02$ (Hayes 1985), one finds that the total luminosity of the galaxies of the Local Group is $5.2 \times 10^{10} L_\odot$, corresponding to $M_V = -22.0$. Of this luminosity 86% is provided by M31 and the Galaxy. The main uncertainty in this value is attributed to the poorly determined value for the total luminosity of the Milky Way system. In conjunction with a total Local Group mass of $(2.3 \pm 0.6) \times 10^{12} M_\odot$ this yields a Local Group mass-to-light ratio of $M/L_V = 44 \pm 12$ in solar units.

18.2 Orbits of individual galaxies

Valtonen et al. (1993) and Peebles (1994) have emphasized the fact that the Local Group cannot be viewed in isolation and that interactions with other nearby structures, such as the IC 342/Maffei group, have to be taken into account when studying the orbits of individual Local Group members. Such tidal interactions may have imparted so much angular momentum to M31 and the Galaxy that these two objects will not collide, and merge, during the next $\sim 1 \times 10^{10}$ yr. Dunn & Laflamme (1993) have stressed that tidal forces will be dominated by galaxies beyond the present Local Group membership during the era when $z > 1$. Furthermore, Byrd et al. (1994) have pointed out that the large angular momentum of the Magellanic Clouds, relative to the Milky Way system, suggests that they might be orbiting within the Local Group, rather than just around the Galaxy. (A similar argument can be made for Leo I.) From their modeling of the interactions between Local Group members, these authors conclude that the Magellanic Clouds left the neighborhood of M31 ~ 10 Gyr ago and were subsequently captured by the Galaxy ~ 6 Gyr ago. Simulations by Sawa (1999) suggest that the LMC and SMC have remained a bound pair for ~ 15 Gyr. However, detailed modeling by Li and Thronson (1999) indicates that the SMC lost so much mass during its two most recent close encounters with the LMC that it cannot have survived many such interactions. They therefore conclude that the LMC and SMC must have captured each other rather recently. These two points of view might be reconciled by assuming that the separation between the LMC and SMC has decreased with time. Much more accurate proper motions of the Clouds, relative to an inertial frame provided by quasars and active galaxy nuclei, will be required to provide additional constraints on the orbits of the Large and the Small Magellanic Cloud.

---

[1] The lower value assumes that Leo I is not bound to the Galactic subgroup.

18.3   Luminosity evolution

To understand the high frequency of very faint blue galaxies it is necessary to assume that small low-mass galaxies experienced short bursts of star formation during the era that is presently viewed at redshifts $0.5 \lesssim z \lesssim 1.0$ (Babul & Ferguson 1996). Biggs (1997) finds that there is no evidence for an undiscovered new population of faint hydrogen-rich objects that might be associated with such starburst galaxies. This suggests that some present members of the Local Group might, in their distant past, have been starburst galaxies similar to the "boojums" that are seen at intermediate redshifts. There is presently no evidence that any of the Local Group irregular galaxies experienced an enormous burst of star formation in its recent past. (The only well-documented, relatively ancient, starburst is the one that started in the LMC $\sim$4 Gyr ago.) Although some dwarf spheroidal galaxies experienced higher rates of star formation in their past, it is not clear that these starbursts were violent enough to make them appear like "boojums." For the Carina dSph galaxy, Hurley-Keller, Mateo & Nemec (1998) conclude that its luminosity, during the starburst that occurred 7 Gyr ago, was only $M_V \sim -16$. In this connection Biggs (1997) has raised the possibility that the local volume of space is atypical. Such a non-Copernican view is unattractive, but it cannot yet be ruled out with certainty.

19

Properties of the Local Group

19.1   Introduction

A summary of derived properties for the probable members of the Local Group is given in Table 19.1. In the following sections these data will be used to derive various global properties of Local Group galaxies.

19.2   The motions of Local Group galaxies

Table 19.1 contains information on 35 probable Local Group members. Giving equal weight to each object, one may use these data to derive both the size of the solar motion relative to the centroid of the Local Group and the position of the apex toward which the Sun is moving. Originally Humason, Mayall & Sandage (1956) adopted a solar motion of $V_{\odot} = 300$ km s^{-1}. More recently Yahil, Tammann & Sandage (1977) derived $V_{\odot} = 308$ km s^{-1} toward $\ell = 105°$, $b = -7°$. Their study included NGC 404, NGC 6949, IC 342, IC 5152, Maffei 1, Maffei 2, and DDO 187 (Aparicio, García-Pelayo & Moles 1988), all of which are presently considered to be nonmembers of the Local Group, and NGC 3109, Sextans A, and Sextans B, which are probable nonmembers (see Chapter 16). An unweighted least-squares solution for the 27 Local Group galaxies with known radial velocities that are listed in Table 2.1 yields a solar motion of 306 ± 18 km s^{-1} toward an apex at $\ell = 99° \pm 5°$ and $b = -4° \pm 4°$, in which the quoted mean errors were determined using a "bootstrap" technique (Courteau & van den Bergh 1999). Figure 19.1 shows a plot of the heliocentric radial velocities of individual Local Group members versus angular distance from the solar apex. Perhaps the most surprising feature of Figure 19.1 is that the dispersion around the adopted regression relation is only $\sigma_{\rm r} = 61 \pm 8$ km s^{-1}. It might be argued that this value is too low since galaxies with large redshifts were censored. This is so because radial velocity was one of the three criteria used to select candidate Local Group members. However, the absence of galaxies with large blueshifts in the sample suggests that such censorship was probably not a major factor affecting the Local Group velocity dispersion. Substituting the observed radial velocity dispersion into the equation (Binney & Tremaine 1987, p. 214)

$$\langle \sigma^2 \rangle \simeq 0.4 \, GM/R_{\rm h}, \qquad (19.1)$$

with $R_{\rm h} = 350$ kpc, yields a total Local Group mass of $(2.3 \pm 0.6) \times 10^{12} \, M_{\odot}$. In deriving this value it was assumed that (1) the Virial theorem applies (which may not be true for the more distant Local Group members) and that (2) the motions are isotropic, so that $\langle \sigma^2 \rangle = 3 \langle \sigma_{\rm r}^2 \rangle$.

Table 19.1. *Derived data on Local Group members*

Name	Alias	DDO Type	$(m - M)_0$	M_V	ℓ	b	D(kpc)	$\cos\theta$
M31	N 224	Sb I–II	24.4	−21.2	121.17	−21.57	760	0.880
Milky Way	Galaxy	S(B)bc I–II:	14.5	−20.9	000.00	00.00	8	−0.150
M33	N 598	Sc II–III	24.5	−18.9	133.61	−31.33	795	0.729
LMC	...	Ir III–IV	18.5	−18.5	280.19	−33.29	50	−0.801
SMC	...	Ir IV/IV–V	18.85	−17.1	302.81	−44.33	59	−0.609
M32	NGC 221	E2	24.4	−16.5	121.15	−21.98	760	0.878
NGC 205	...	Sph	24.4	−16.4	120.72	−21.14	760	0.884
IC 10	...	Ir IV:	24.1	−16.3	118.97	−03.34	660	0.938
NGC 6822	...	Ir IV–V	23.5	−16.0	025.34	−18.39	500	0.292
NGC 185	...	Sph	24.1	−15.6	120.79	−14.48	660	0.910
IC 1613	...	Ir V	23.3	−15.3	129.73	−60.56	725	0.473
NGC 147	...	Sph	24.1	−15.1	119.82	−14.25	660	0.917
WLM	DDO 221	Ir IV–V	24.85	−14.4	075.85	−73.63	925	0.318
Sagittarius	...	dSph(t)	17.0	−13.8	005.61	−14.09	24	−0.036
Fornax	...	dSph	20.7	−13.1	237.24	−65.66	138	−0.253
Pegasus	DDO 216	Ir V	24.4	−12.3	094.77	−43.55	760	0.764
Leo I	Regulus	dSph	22.0	−11.9	225.98	+49.11	250	−0.443
And I	...	dSph	24.55	−11.8	121.69	−24.85	810	0.859
And II		dSph	24.2	−11.8	128.91	−29.15	700	0.782
Leo A	DDO 69	Ir V	24.2	−11.5	196.90	+52.41	690	−0.136
Aquarius[a]	DDO 210	V	25.05	−11.3	034.04	−31.35	1025	0.398
SagDIG[a]	...	Ir V	25.7	−10.7	021.13	−16.23	1300:	0.224
Pegasus II	And VI	dSph	24.45	−10.6	106.01	−36.30	830	0.834
Pisces	LGS 3	dIr/dSph	24.55	−10.4	126.77	−40.88	810	0.705
And V	...	dSph	24.55	−10.2	126.22	−15.12	810	0.870
And III	...	dSph	24.4	−10.2	119.31	−26.25	760	0.864
Leo II	...	dSph	21.6	−10.1	220.14	+67.23	210	−0.258
Phoenix	...	dIr/dSph	23.0	−9.8	272.19	−68.95	395	−0.299
Sculptor		dSph	19.7	−9.8	287.69	−83.16	87	−0.057
Cassiopeia	And VII	dSph	24.2	−9.5	109.46	−09.95	690	0.976
Tucana[a]	...	dSph	24.7	−9.6	322.91	−47.37	870	−0.439
Sextans	...	dSph	19.7	−9.5	243.50	+42.27	86	−0.645
Carina	...	dSph	20.0	−9.4	260.11	−22.22	100	−0.853
Draco	...	dSph	19.5	−8.6	086.37	+34.71	79	0.767
Ursa Minor	...	dSph	19.0	−8.5	104.88	+44.90	63	0.660

[a]Membership in the Local Group not yet firmly established.

19.3   The luminosity distribution for Local Group galaxies

The data in Table 19.1 may be used to construct the luminosity distribution[1] for the Local Group, which is shown in Figure 19.2. It should be emphasized that the data on Local Group membership are probably quite incomplete for objects fainter than $M_V \approx -11$.[2] Irwin (1994) is, however, quite optimistic on this point because he found

[1] The term "luminosity function" is not used because it is defined as the number of galaxies per magnitude interval *per unit volume*.

[2] That the luminosity distribution for distant Local Group members is incomplete at dim luminosities is attested to by the fact that only one (And VII) galaxy fainter than $M_V = -10.0$ is known in the M31 subgroup, versus five such objects (Car, Dra, Scl, Sex, and UMi) near the Galaxy.

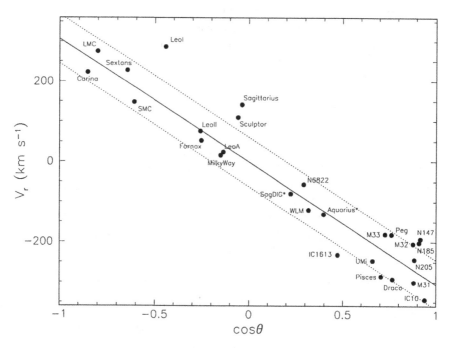

Fig. 19.1. Heliocentric radial velocity V_r versus the cosine of the solar apex distance for all Local Group members with known radial velocities. The line has $V_r = -306$ km s^{-1} at $\cos\theta = 1.0$.

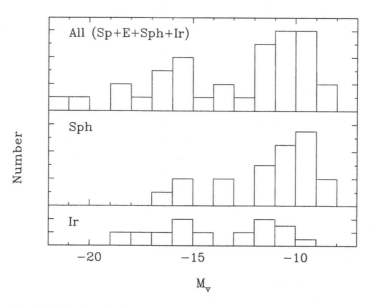

Fig. 19.2. Luminosity distribution of Local Group members. The data are probably quite incomplete fainter than $M_V \approx -11$. The luminosity distribution of dIr galaxies appears flat, whereas that for dSph galaxies rises toward fainter luminosities.

only one new dwarf galaxy (Sextans, which has $M_V = -9.5$), in a deep Schmidt survey of a 20,000 square degree area at high Galactic latitude. The fact that IC 1613 ($M_V = -15.3$) has been known for almost a century indicates that the sample (except perhaps at very low Galactic latitude) is quite complete at intermediate luminosities. The data in Figure 19.2 may be fit with a Schechter (1976) function of the form

$$\phi(L_V)\,dL_V \propto (L_V/L_V^*)^\alpha \exp(-L_V/L_V^*)\,d(L_V/L_V^*). \qquad (19.2)$$

Pritchet & van den Bergh (1999) find that the number of luminous Local Group galaxies is too small to determine L_V^* with any degree of confidence. However, they do find that the fainter part of the Local Group Luminosity function may be fit with a Schechter function having $\alpha = -1.1 \pm 0.1$. It is noted in passing that this value of the slope is less steep than has been found in some (Trentham 1998a), but not in all (Trentham 1998b), rich clusters of galaxies. From a study of small compact groups of galaxies, Hunsberger, Charlton & Zaritsky (1998) conclude that their data are best represented by a superposition of two Schechter functions. However, the number of galaxies in the Local Group is too small to justify such a fit, which has four free parameters.

The luminosity distributions for Local Group irregular and dwarf spheroidal galaxies are shown separately in the two bottom panels of Figure 19.2. This figure shows that the luminosity distribution of irregulars is flat, whereas that for dwarf spheroidals appears to rise toward faint luminosities. From a deep, blind H I survey, Schneider, Spitzak & Rosenberg (1998) find possible evidence for a faint-end steepening of the mass function of field galaxies. In fact Klypin et al. (1999) suggest that the faint end of the Local Group luminosity function is incomplete because (1) many of the least massive objects have been called high-velocity clouds, rather than galaxies (Blitz et al. 1999), (2) because many are not observable because they failed to accrete gas that could form stars, or (3) because this gas was ejected by supernova-driven winds, by heating resulting from intergalactic ionizing background radiation or at a very early epoch by photoionization during recombination (Barkana & Loeb 1999).

The luminosity function of the Local Group may be compared with that of the nearby M81 group that is given by van Driel et al. (1998). Assuming an apparent M81 distance modulus of $(m - M)_B = 28.0$, and $\langle B - V \rangle = 0.5$, one can compare the B magnitudes of galaxies in the M81 group with the M_V values listed in Table 19.1. Under the assumption that both samples are complete to $M_V = -10.0$, it is possible to compare the luminosity distributions of 27 Local Group galaxies more luminous than $M_V = -10.0$ with that of the 38 probable M81 group members having $B < 17.5$. A nonparametric Kolmogorov–Smirnov two-sample test shows that the luminosity distributions of the M81 group and of the Local Group do not differ at a respectable level of statistical significance.

19.4   Local Group statistics

19.4.1   *Metallicity versus luminosity*

Figure 19.3 shows a plot of the metallicity [Fe/H] versus parent galaxy luminosity M_V, for Local Group galaxies. The plot illustrates the well-known tendency for luminous galaxies, which have deep potential wells, to be more metal rich than less luminous systems. Most of the irregular galaxies (plotted as filled squares) are seen to

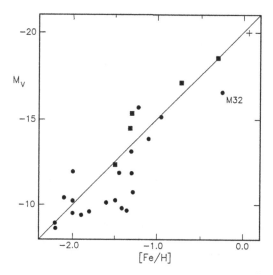

Fig. 19.3. Plot of parent galaxy luminosity M_V versus galaxy metallicity [Fe/H]. Irregular galaxies, which are plotted as filled squares, are mostly seen to lie above the line defined by Eq. 19.3, while the majority of early-type Local Group galaxies fall below it. In other words irregular galaxies are presently too bright for their metallicity. They will, however, fall among the early-type galaxies after they have evolved and faded by 1–2 mag. The cross in the upper right corner marks the position of the solar neighborhood.

fall above the relation

$$M_V = -20 - 5[\text{Fe/H}], \tag{19.3}$$

whereas the majority of early-type Local Group members fall below it. After they have evolved, and faded by one or two magnitudes, the irregular galaxies will fall among the early-type objects.

Figure 19.4 is a plot of the oxygen abundance parameter $12 + 5 \log (\text{O/H})$ versus parent galaxy luminosity for Local Group members. The plot shows that the oxygen abundance increases by a factor of ~100, as parent galaxy luminosity increases by four orders of magnitude. It is noteworthy that the scatter in Figure 9.4 is large. It is particularly puzzling why the oxygen abundance in the spheroidal galaxy NGC 205 is four times as high as that in the spheroidal NGC 185, even though these objects differ in luminosity by only 0.8 mag.

Note that the tidal dwarfs near NGC 5291, which were studied by Duc & Mirabel (1998), have much higher metallicities than do the faint Local Group dwarfs. This, no doubt, occurs because tidal tails are drawn from the outer regions of luminous, and hence relatively metal rich, spirals. Inspection of the prints of the *Palomar Sky Survey* shows that such tidal tails are almost exclusively associated with interacting (metal rich) giant spirals.

19.4.2 *Luminosity versus Disk scale-length*

Figure 19.5 shows a plot of disk scale-length versus parent galaxy absolute magnitude M_V. As expected the disk scale-length grows with increasing galaxy luminosity.

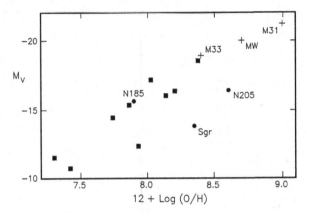

Fig. 19.4. Plot of M_V versus oxygen abundance parameter, $12 + \log(\mathrm{O/H})$, for Local Group galaxies. There is a clear trend for the oxygen abundance to increase as parent galaxy luminosity increases. However, this relationship is seen to exhibit a surprisingly large scatter. Spirals are shown as plus signs, and irregulars as filled squares.

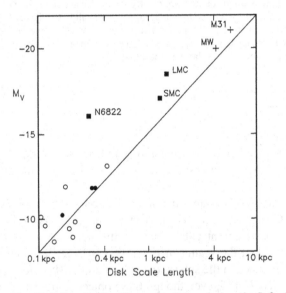

Fig. 19.5. Luminosity M_V versus disk scale-length, S, for Local Group galaxies. Irregulars are plotted as squares, spirals as plus signs, dSph galaxies near M31 as filled circles, and other Local Group dSph galaxies as open circles. Dwarf irregulars are seen to be overluminous for their size.

The irregular galaxies (filled squares) are seen to lie well above the relation

$$M_V = -15 - 7 \log S, \tag{19.5}$$

in which S is the disk scale-length. The dwarf spheroidal galaxies near M31 (filled circles) are scattered among the dSph galaxies located near the Milky Way (open circles).

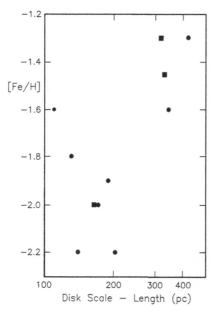

Fig. 19.6. [Fe/H] versus disk scale-length for dwarf spheroidal galaxies. Metallicity is seen to increase as the scale-length grows. Filled squares represent objects associated with M31.

19.4.3 Metallicity versus disk scale-length

A plot of metallicity [Fe/H] versus disk scale-length for Local Group dwarf spheroidal galaxies is shown in Figure 19.6. The figure shows that galaxy luminosity increases with scale-length. The M31 dwarf spheroidals (filled squares) are seen to fall among the other Local Group dwarf spheroidals (filled circles); that is, the satellites of the Andromeda galaxy and of the Milky Way appear to obey similar metallicity versus scale-length relationships.

19.5 Distribution of Local Group members

Courteau & van den Bergh (1999) have calculated the distances of individual Local Group galaxies from the centroid of the Local Group. They assumed this centroid to be located in the direction of the Andromeda galaxy, and at a distance of 0.6 times the separation between M31 and the Galaxy. Figure 19.7 shows a plot of the distribution of the distances of Local Group members from this assumed centroid. The observation that relatively few galaxies are situated within 0.4 Mpc of the adopted centroid is, of course, due to the fact that this centroid is located in a density minimum between M31 and the Galaxy. The Local Group is seen to be rather compact with few members located at more than 0.9 Mpc from its center. The normalized cumulative distribution of distances from the Local Group center of Courteau and van den Bergh is plotted in Figure 19.8. This figure shows that half of the members of the Local Group are situated within ∼0.45 Mpc of its center. Note that this value is significantly smaller than the 0.35-Mpc half-mass radius adopted in Section 19.2. The reason for this difference is, of course, that the massive Andromeda and Milky Way galaxies are located at only 0.30 Mpc and 0.46 Mpc, respectively, from the adopted centroid of the Local Group.

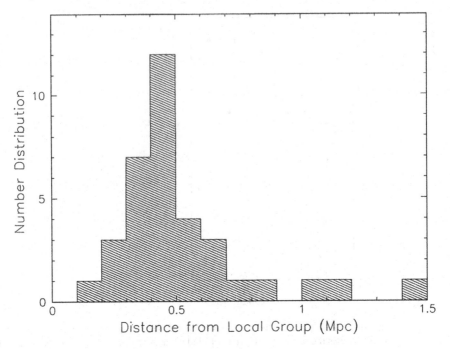

Fig. 19.7. Distribution of distances of Local Group members from the adopted centroid. The figure shows that the Local Group has few members beyond a distance of 0.9 Mpc.

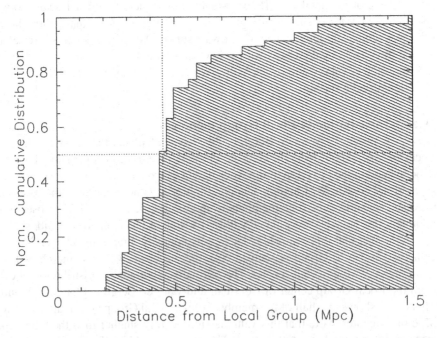

Fig. 19.8. Cumulative distribution of distances from the centroid of the Local Group. The figure shows that half of all Group members are situated within 0.45 Mpc of the adopted centroid.

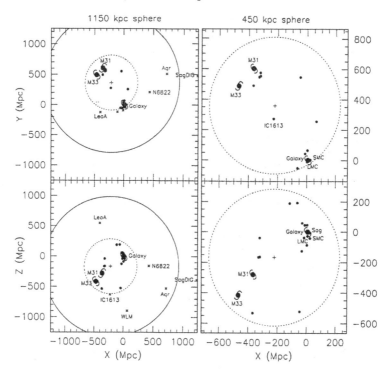

Fig. 19.9. Positions of Local Group galaxies in Cartesian Galactocentric coordinates, as viewed from two orthogonal directions. The solid circle in the left-hand panels shows a sphere of radius 1150 kpc, corresponding to the zero-velocity surface of the Local Group. The dotted circle has a radius of 450 kpc. Both spheres are centered on the Local Group barycenter at $X = +220$, $Y = +361$, and $Z = -166$ kpc. Points outside of a radius of 450 kpc are plotted as an x.

A plot of the distribution of Local Group members projected onto the X, Y and the X, Z planes is shown in Figure 19.9. The figure clearly shows that the galaxies of the Local Group are strongly clustered into the Galactic and M31 subclusters. The real clustering may, however, be less pronounced than this figure suggests. This is so because (1) the area of the sky near M31 has been more thoroughly searched for dwarf galaxies than have other regions of the Local Group and because (2) the census of very faint dwarf galaxies near the Milky Way is more complete than it is at larger Galactocentric distances. Table 19.2 contains information on the distribution of Local Group members with $M_V > -16.5$ (i.e., objects fainter than M32). Since IC 10 and Pisces ($=$ LGS 3)

Table 19.2. *Classification type versus environment*[a]

Environment	Sph + dSph	dSph/dIr	Ir
Isolated	1	2	6
?	1	1	1
In subcluster	15	0	0

[a] All Local Group galaxies with $M_V > -16.5$.

Table 19.3. *Classification type versus environment*[a]

Environment	Sph + dSph	dSph/Ir	Ir
Isolated	0	1	5
?	1	0	1
In subcluster	7	0	0

[a]Local Group galaxies with $-16.5 < M_V < -11.0$.

are *possible* members of the Andromeda subgroup of the Local Group, their subgroup membership has been listed as "?" in the Table. The morphological classification of the Aquarius system (=DDO 210) is uncertain, and so it has been included in the dSph/dIr column of the Table. Leo I is so distant, and has such a high radial velocity, that its physical association with the Galactic subgroup of the Local Group may be questioned. The data in Table 19.2 show a strong correlation between morphological type and environment, in the sense that Sph and dSph galaxies are strongly clustered, whereas dIr galaxies are mostly not associated with the two main subclusterings within the Local Group. A χ^2 test shows that the observed correlation between environment and morphological type in Table 19.2 is significant at >99.9%. However, the observed correlation might have been artificially strengthened because faint dSph galaxies are more likely to have been discovered close to the Galaxy and M31. This problem may be avoided (or greatly reduced) by only considering the relationship between environment and morphological type for the more luminous galaxies in the range $-18.5 < M_V < -11.0$. The data for Local Group galaxies in this restricted luminosity range are given in Table 19.3. This table again shows that the locations, and morphological types, of Local Group dwarf galaxies are strongly correlated. A χ^2 test shows that the observed correlation is significant at the 99% level. This effect is probably related to that discovered by Einasto et al. (1974), who found that early-type companions tend to occur closer to their parents than do late-type companions. These results may, however, have been affected by a bias that creeps in because (see Figure 19.2) there are more faint dSph than dIr galaxies. As a result the discovery of dwarf spheroidals will be enhanced in the most deeply searched regions, that is, in the vicinity of M31 and the Galaxy.

Presently available data are not yet sufficiently numerous to be able to confirm the suggestion by van den Bergh (1995b) that star formation in dSph galaxies near M31 and the Milky Way started out earlier than it did in more remote dSph galaxies. However, the observation (see Section 15.9) that intrinsically faint dSph galaxies within 100 kpc of the Galaxy are more metal poor than those at greater distances is consistent with the hypothesis that star formation was truncated early in dim dwarfs near the Milky Way system. The observation (E. K. Grebel, private communication) that distant Galactic dSph companions contain a larger fraction of intermediate-age stars than do nearer dSph galaxies also appears consistent with this notion.

Van Driel et al. (1998) have noted that galaxies in the M81 group show a clear morphological segregation. Most of the early-type galaxies are found in the dense core of the M81 group, while the dwarf irregulars spread out over the entire area that they surveyed. In this respect the behavior of the dwarfs in the M81 group is similar to that of the dwarfs in the Local Group. This suggests that morphological segregation may be a common evolutionary feature of small groups and clusters of galaxies.

20

Conclusions

20.1 Introduction

Since its existence was first noted by Edwin Hubble, the Local Group has been the subject of continuing exploration. Hubble (1936, p. 125) refers to it as "a typical, small group of nebulae which is isolated in the general field." The elongated core of the Local Group, consisting of the Andromeda galaxy, the Milky Way, and their close companions does, indeed, form a rather well-isolated group. However, the low-density outer envelope of the Local Group may mingle with the coronae of other nearby clusterings, such as the South Polar group and the Mafei/IC 342 group. As was already noted in Section 18.1 the velocity dispersion of Local Group galaxies establishes a rather well-determined radius of 1.2 Mpc for the zero-velocity surface of the Local Group. If we adopt M_V (Sun) $= +4.82 \pm 0.02$ (Hayes 1985), the total luminosity of the Local Group galaxies listed in Table 19.1 is found to be $4.2 \times 10^{10} L_\odot(V)$. Of this amount 86% is provided by M31 and the Galaxy. The discovery of additional faint Local Group members would not increase the estimated total luminosity of the Local Group significantly. Comparison with the Local Group mass of $M = (2.3 \pm 0.6) \times 10^{12} M_\odot$ yields $M/L_V = 44 \pm 12$. This shows that dark matter in the Local Group outweighs visible matter by about an order of magnitude. Data on the mass-to-light ratios of the Galactic and Andromeda subgroups of the Local Group are given in Table 20.1.

20.2 Local Group calibrators

Modern research has vindicated Hubble's expectation that Local Group members ("these neighboring systems") would furnish a small sample collection of nebulae, from which criteria might be developed to explore the remoter regions of space. In particular it has been possible to determine the extragalactic distance scale by calibrating the Cepheid period–luminosity relation in individual Local Group galaxies. Novae, planetary nebulae, and the tip of the red giant branch of Population II have also proved to be useful distance calibrators. Studies of the evolution of stellar populations in individual Local Group members should, eventually, help us to understand the nature of distant galaxies that are presently being observed at significant look-back times. Furthermore, observations of the motions (and compositions) of globular clusters and old metal-poor individual stars have provided us with a wealth of new information on conditions that prevailed in the early Universe.

Table 20.1. *Local Group properties*

Radius of zero-velocity surface	$R_o = 1.18 \pm 0.16\,\mathrm{Mpc}^a$ (1)
Half-mass radius	$R_h \approx 350$ kpc
Total Local Group luminosity	$L_V = 4.2 \times 10^{10}\,L_\odot$ (2,4)
Radial velocity dispersion	$\sigma = 61 \pm 8\,\mathrm{km\,s}^{-1}$ (1)
Total Local Group mass	$M = (2.3 \pm 0.6) \times 10^{12}\,M_\odot$ (1)
Local Group mass-to-light ratio	$M/L_V = 44 \pm 14$ (in solar units) (2)
Mass of Andromeda subgroup	$M(A) = (1.15-1.5) \times 10^{12}\,M_\odot$ (3)
Luminosity of Andromeda subgroup	$L_V = 3.0 \times 10^{10}\,L_\odot$ (4)
Andromeda subgroup mass-to-light ratio	$M(A)/L_V = 38-50$
Mass of Milky Way subgroup[b]	$M(G) = (0.46-1.25) \times 10^{12}\,M_\odot$ (5)
Luminosity of Milky Way subgroup	$L_V = 1.1 \times 10^{10}\,L_\odot$ (4)
Milky Way subgroup mass-to-light ratio[b]	$M(G)/L_V = 42-114$
Bound multiple systems	M31 + M32 + NGC 205, LMC + SMC, NGC 147 + NGC 185

[a]$T_o = 14 \pm 2$ Gyr assumed.
[b]Lower value assumes Leo I is *not* a satellite of the Galaxy.
(1) van den Bergh & Courteau (1999).
(2) See Section 20.1.
(3) See Section 18.1.
(4) From Table 19.1.
(5) Zaritsky et al. (1989), Zaritsky (1998).

20.3   Evolution of Local Group galaxies

One of the most interesting results of such population studies in individual Local Group galaxies is that all of these objects except Leo I, and with the possible exception of Leo A, appear to contain very old stars with ages >10 Gyr. In other words, star formation started out with a bang – although it ended (at least in early-type galaxies) with a whimper. It is puzzling that globular clusters (which are diagnostic of the oldest stellar populations) first appear almost simultaneously (and with rather similar masses) in the main body of the Galaxy, the outer Galactic halo (NGC 2419), and in the Large Magellanic Cloud. In contrast, globular cluster formation appears to have ramped up more slowly in M33, and, perhaps, also in the Small Cloud.

In those galaxies for which detailed population studies have been published, it is usually found that the star forming region contracts with time (i.e., galaxies mostly shrink as they evolve). The observation that many (old) globular clusters occur in galactic halos is consistent with such an evolutionary scenario. It is not yet clear how the conclusion that Local Group galaxies shrank with time is to be reconciled with the observation (Bouwens, Broadhurst & Silk 1998; Pascarelle et al. 1996; Pascarelle, Windhorst & Keel 1998) that distant galaxies in the Hubble Deep Field typically appear smaller than galaxies at lower redshifts. Perhaps we only see the nuclear bulges of young galaxies at large redshifts (Friaça & Terlevich 1999). Other evidence for the evolution of Local Group galaxies is provided by the great burst of star formation that took place in the Large Cloud during the past 3–5 Gyr. None of the Local Group galaxies appears to have preserved any fossil evidence for a great burst of star formation 5–10 Gyr ago, such as would have been expected if they had once been "boojums."

20.4   Mergers and galaxy destruction

The present disruption of the Sagittarius dwarf spheroidal by Galactic tidal forces provides the most direct evidence for the destruction of galaxies in the Local Group. That no other dSph galaxies presently exist within 60 kpc may also be because any such objects that once existed have long since been destroyed by tidal forces. The existence of the Magellanic Stream, and of the Magellanic Bridge, shows that the LMC and SMC have suffered strong interactions during the past 2 Gyr. Such violent interactions place the long-term existence of the Small Cloud in doubt. The E2 galaxy M32 may also have been severely truncated by the tides of M31. Finally, the fact that the Andromeda galaxy exhibits an $R^{1/4}$ profile out to large radii may indicate that this object was, itself, formed by the merger of two massive ancestral galaxies.

20.5   Historical perspective

This book has reviewed the enormous progress made since Edwin Hubble first drew attention to the existence of the Local Group in 1936. These researches have shown that the Local Group is a quite typical small cluster, which provides an environment similar to that in which about half of all galaxies are located at the present time. Furthermore, these studies have confirmed Hubble's hope that our companion galaxies in the Local Group would furnish us with the criteria that would be needed to explore the more remote regions of the Universe. As Hubble (1936, p. 125) pointed out "The fact that the [G]alactic system is a member of a group is a very fortunate accident."

Starting with the work by Baade (1944a,b), the studies of stellar populations in Local Group galaxies have produced deep insights into the evolution of stellar populations, and of their parent galaxies. However, this new work has also introduced us to a number of new, and previously unsuspected, mysteries. Among the most insistent questions that present themselves are, Why did the Large Cloud suddenly enter a phase of enhanced activity 3–5 Gyr ago? and Why did the rate of cluster formation in the LMC increase much more rapidly than the rate of field star creation? A possibly related question is Why are luminous carbon stars (which have ages ~0.5 Gyr) only observed in the LMC field, and not in LMC clusters? That is, was there a rather recent phase of star formation in the Large Cloud, during which cluster formation was suppressed? Was the burst of star and cluster creation in the LMC triggered by the formation of the Large Cloud Bar? This also raises the question whether the formation of the LMC Bar was part of a general increase with time, in the fraction of barred galaxies in the Universe. It would also be interesting to know if any of the Local Group galaxies were once active blue galaxies (boojums). Moreover, little work has yet been done on the possible effects of a hypothetical (mini) quasarlike phase during the evolution of M31 and the Galaxy, which resulted from the assembly of their massive central black holes. It would also be interesting to follow up Freeman's idea that M31 might have formed from the early merger of two ancestral galaxies, whereas the Milky Way more likely evolved by collapse of a single protogalaxy. In situ studies of the motions of Galactic halo stars should also provide new insights into the formation, and early evolution, of the Galaxy.

The presence of RR Lyrae stars in all Local Group galaxies (except Leo I, and perhaps Leo A), in which deep searches have been made, suggests that star formation started early (and nearly simultaneously) in most galaxies. Furthermore, available evidence indicates that the earliest star formation took place in a larger volume than it does at the present time; that is, Local Group galaxies shrank as they evolved. This is unexpected because recent studies of galaxies at high redshifts (large look-back times) appear to show such objects to be *smaller* than similar nearby galaxies at low redshifts. Hopefully, research during the first decades of the new millennium will provide answers to many of these questions.

On completing the manuscript of this book I was reminded of the lines of Ovid, who wrote: *"Evening has overtaken me, and the Sun has dipped below the horizon of the ocean, yet I have not had time to tell you of all the things that have evolved into new forms."*

Glossary

Color system see Table G1

Core radius see Eq. (14) of King (1962)

Effective radius r_e radius within which is emitted half of the light of a galaxy (or cluster) in projection

Ellipticity ϵ $1 - b/a$, where a and b are the major and minor axis diameters, respectively

Early-type stars of types O, B, A, or galaxies of types E, S0, Sa

Globular cluster old (usually metal poor) cluster associated with the halo, bulge, or thick Disk

Half-light radius r_h radius within which is emitted half of the light of a galaxy in projection

Holmberg radius radius of the isophote with $m_{pg} = 26.5$ mag arcsec^{-2} (Holmberg 1958, p. 6)

Inclination i angle between the fundamental plane of a galaxy and the plane of the sky

Late-type stars of types K and M, or galaxies of types Sc, Sd, Im

Open cluster star cluster (usually young or of intermediate age) exhibiting disk kinematics

Population I member of a young metal-rich population (Baade 1944)

Population II member of an old metal-poor population (Baade 1944)

Populous cluster rich young or intermediate-age star cluster

Specific cluster frequency S number of clusters normalized to $M_V(\text{total}) = -15$ (see Harris & van den Bergh 1981)

Tidal radius r_t see Eq. (14) of King (1962)

Tilt angle $90° - i$

Table G1. *Color system*

Color passband	$A/E(B-V)^a$	Effective wavelength (μm)
U	5.43	0.36
B	4.10	0.44
V	3.10	0.55
R	2.32	0.64
I	1.50	0.79
J	0.88	1.23
H	0.54	1.66
K'	...	2.12
K	0.36	2.22

[a]Mostly from Fitzpatrick (1999). See Schlegel et al. (1998) for additional information.

Bibliography

Aaronson, M. 1983, *Astrophys. J.*, 166, L11.

Aaronson, M., Gordon, G., Mould, J., Olszewski, E. & Suntzeff, N. 1985, *Astrophys. J.*, 296, L7.

Aaronson, M., Olszewski, E.W. & Hodge, P.W. 1983, *Astrophys. J.*, 267, 271.

Ables, H.D. 1971, *Publ. U.S. Naval Obs.*, Second Series, 20, Part 4.

Ables, H.D. & Ables, P.G. 1977, *Astrophys. J. Suppl.*, 34, 245.

Abraham, R.G., Merrifield, M.R., Ellis, R.S., Tanvir, N.R. & Binchmann, J. 1999, *Mon. Not. R. Astron. Soc.*, in press.

Afonso, C. et al. 1998, *Astron. Astrophys.*, 337, L17.

Alard, C. 1996, *Astrophys. J.*, 458, L17.

Alcock, C. et al. 1996, *Astrophys. J.*, 470, 583.

Alcock, C. et al. 1997a, *Astrophys. J.*, 474, 217.

Alcock, C. et al. 1997b, *Astrophys. J.*, 482, 89.

Alcock, C. et al. 1997c, *Astrophys. J.*, 490, L59.

Alcock, C. et al. 1998a, *Astron. J.*, 115, 1921.

Alcock, C. et al. 1998b, *Astrophys. J.*, 492, 190.

Alcock, C. et al. 1998c, *Astrophys. J.*, 499, L9.

Alcock, C. et al. 1999, *Astron. J.*, 117, p. 920.

Allen, R.J., Le Bourlot, J., Lequeux, J., Pineau des Forêts, G. & Roueff, E. 1995, *Astrophys. J.*, 444, 147.

Allen, W.H., et al. 1993, *Astrophys. J.*, 403, 239.

Allsopp, N.J. 1978, *Mon. Not. R. Astron. Soc.*, 184, 397.

Alvarez, H., Aparicio, J. & May, J. 1989, *Astron. Astrophys.*, 213, 13.

Alves, D.R. et al.1997, in *Planetary Nebulae (IAU Symposium No. 180)*, ed. H.J. Habing and H.J.G.L.M. Lamers, Dordrecht: Kluwer, p. 468.

Alves, D. et al. 1998, *Astrophys. J.*, 511, 225.

Alves, D.R. & Sarajedini, A. 1998, *Astrophys. J.*, 511, 225.

Ambartsumian, V.A. 1949, *Astron. J. USSR*, 26, 3.

Ambartsumian, V.A. 1962, *IAU Trans.*, 11B, 145.

Amy, S.W. & Ball, L. 1993, *Astrophys. J.*, 411, 761.

Anders, E. & Grevesse, N. 1989, *Geochimie et Cosmochimie Acta*, 53, 197.

Anguita, C. 1999, in *New Views of the Magellanic Clouds (IAU Symposium No. 190)*, ed. Y.-H. Chu et al., San Francisco: ASP, in press.

Ansari, R. et al. 1999, *Astron. Astrophys.*, in press.

Aparicio, A. 1999, astro-ph/9811186.

Aparicio, A., Dalcanton, J.J., Gallart, C. & Martínez-Delgado, D. 1997, *Astron. J.*, 114, 1447.

Aparicio, A. & Gallart, C. 1995, *Astron. J.*, 110, 2105.

Aparicio, A., Gallart, C. & Bertelli, G. 1997a, *Astron. J.*, 114, 669.

Aparicio, A., Gallart, C. & Bertelli, G. 1997b, *Astron. J.*, 114, 680.

Aparicio, A., García-Pelayo, J.M. & Moles, M. 1988, *Astron. Astrophys. Suppl.*, 74, 367.

Aparicio, A., Herrero, A. & Sanchez, F. 1998, *Stellar Astrophysics for the Local Group*, Cambridge: Cambridge University Press.

Appleton, P.N., Davies, R.D. & Stephenson, R.J. 1981, *Mon. Not. R. Astron. Soc.*, 195, 327.

Ardeberg, A., Gustafsson, B., Linde, P. & Nissen, P.-E. 1997, *Astron. Astrophys.*, 322, L13.

Argelander, F. 1903, *Bonner Durchmusterung*, Vol. II, Bonn: Marcus & Weber, p. 344.

Armandroff, T.E. 1998, Colloquium given in Victoria, Canada, 1998 May 25.

Armandroff, T.E., Da Costa, G.S., Caldwell, N. & Seitzer, P. 1993, *Astron. J.*, 106, 986.

Armandroff, T.E., Davies, J.E. & Jacoby, G.H. 1998, *Astron. J.*, 116, 2287.

Armandroff, T.E. & Massey, P. 1985, *Astrophys. J.*, 291, 685.

Armandroff, T.E., Olszewski, E.W. & Pryor, C. 1995, *Astron. J.*, 110, 2131.

Arp, H.C. 1956, *Astron. J.*, 61, 15.

Arp, H.C. 1962, in *Symposium on Stellar Evolution*, ed. J. Sahade, La Plata: Observatorio Astronomico, p. 87.

Arp, H.C. 1964, *Astrophys. J.*, 139, 1045.

Arp, H. & Brueckel, F. 1973, *Astrophys. J.*, 179, 445.

Arp, H.C. & Madore, B.F. 1987, *Catalogue of Southern Peculiar Galaxies and Associations*, Cambridge: Cambridge Univ. Press.

Ashman, K.M. & Bird, C.M. 1993, *Astron. J.*, 106, 228.

Aurière, M., Coupinot, G. & Hecquet, J. 1992, *Astron. Astrophys.*, 256, 95.

Azzopardi, M. 1999, in *Stellar Content of the Local Group*, (*IAU Symposium No. 192*), ed. P. Whitelock and R.D. Cannon, San Francisco: Astron. Soc. Pac., p. 144.

Azzopardi, M., Lequeux, J., Rebeirot, E. & Westerlund, B.E. 1991, *Astron. Astrophys. Suppl.*, 88, 265.

Baade, W. 1935, in *Annu. Rep. Mt. Wilson Observatory*, 1934–35, p. 185.

Baade, W. 1944a, *Astrophys. J.*, 100, 137.

Baade, W. 1944b, *Astrophys. J.*, 100, 147.

Baade, W. 1951, *Publ. Obs. U. Michigan*, 10, 7.

Baade, W. 1954, *IAU Trans.* 8, 398.

Baade, W. 1963, *Evolution of Stars and Galaxies*, Cambridge: Harvard University Press.

Baade, W. & Arp, A. 1964, *Astrophys. J.*, 139, 1027.

Baade, W. & Hubble, E. 1939, *Publ. Astron. Soc. Pac.*, 51, 40.

Baade, W. & Swope, H.H. 1961, *Astron. J.*, 66, 300.

Baade, W. & Swope, H.H. 1963, *Astron. J.*, 68, 435.

Baade, W. & Swope, H.H. 1965, *Astron. J.*, 70, 212.

Babcock, H.W. 1939, *Lick Obs. Bull.*, 19, 41 (No. 498).

Babul, A. & Ferguson, H.C. 1996, *Astrophys. J.*, 458, 100.

Backer, D.C. & Sramek, R.A. 1982, *Astrophys. J.*, 260, 512.

Bacon, R., Emsellem, E., Monnet, G. & Nieto, J.L. 1994, *Astron. Astrophys.*, 281, 691.

Bahcall, J.N. & Tremaine, S. 1981, *Astrophys. J.*, 244, 805.

Balona, L.A. 1992, in *Variable Stars and Galaxies* (*ASP Conference Series No. 30*), ed. B. Warner, San Francisco: Astron. Soc. Pac., p. 155.

Barbá, R.H., Niemela, V.S., Baume, G. & Vazquez, R.A. 1995, *Astrophys. J.*, 446, L23.

Barbuy, B., Bica, E. & Ortolani, S. 1998, *Astron. Astrophys.*, 333, 117.

Barbuy, B., de Freitas Pacheco, J.-A. & Castro, S. 1994, *Astron. Astrophys.*, 283, 32.

Barkana, R. & Loeb, A. 1999, astro-ph/9901114.

Barlow, M.J. 1995, in *Highlights of Astronomy*, ed. I. Appenzeller, Dordrecht: Kluwer, p. 476.

Barnes, J.E., Moffett, T.J. & Gieren, W.P. 1993, *Astrophys. J.*, 405, L51.

Barnes, T.G. 1996, in *Formation of the Galactic Halo...Inside and Out* (*ASP Conference Series No. 92*), ed. H. Morrison and A. Sarajedini, San Francisco: Astron. Soc. Pac., p. 415.

Battinelli, P. & Demers, S. 1992, *Astron. J.*, 104, 1458.

Battistini, P., Bònoli, F., Bracessi, A., Frederici, L., Fusi Pecci, F., Marano, B. & Börgen, F. 1987, *Astron. Astrophys. Suppl.*, 67, 447.

Battistini, P., Bònoli, F., Bracessi, A., Fusi Pecci, F., Malagnini, M. & Marano, B. 1980, *Astron. Astrophys. Suppl.*, 42, 357.

Battistini, P.L., Bònoli, F., Casavecchia, M., Ciotti, L., Frederici, L. & Fusi Pecci, F. 1993, *Astron. Astrophys.*, 272, 77.

Bauer, F. et al. 1998, *Astron. Astrophys.*, (= astro-ph/9807094).

Baum, W. & Schwarzschild, M. 1955, *Astron. J.*, 60, 247.

Baumgardt, H., Dettbarn, C., Fuchs, B., Rockmann, J. & Wielen, R. 1999, astro-ph/9812437.

Beauchamp, D., Hardy, E., Suntzeff, N.B. & Zinn, R. 1995, *Astron. J.*, 109, 1628.

Beaulieu, J.-P. & Sackett, P.D. 1998, *Astron. J.*, 116, 209.

Beaulieu, J-P., Lamers, H.J.G.L.M. & de Wit, W.J. 1999, astro-ph/9811363.

Bell, S.A., Hill, G., Hilditch, R.W., Clausen, J.V. & Reynolds, A.P. 1993, *Mon. Not. R. Astron. Soc.*, 265, 1047.

Bellazzini, M., Ferraro, F.R. & Buonanno R. 1999 *Mon. Not. R. Astron. Soc.*, 304, 633.

Bellazzini, M., Fusi Pecci, F. & Ferraro, F.R. 1996, *Mon. Not. R. Astron. Soc.*, 278, 952.

Bender, R., Kormendy, J. & Dehnen, W. 1996, *Astrophys. J.*, 464, L123.

Bender, R., Paquet, A. & Nieto, J.-L. 1991, *Astron. Astrophys.*, 246, 349.

Benvenuti, P., D'Odorico, S. & Dumontel, M. 1979, *Astrophys. Space Sci.*, 66, 39.

Bergeat, A., Knapik, A. & Rutily, B. 1998, *Astron. Astrophys.*, 332, L53.

Berman, B.G. & Suchkov, A.A. 1991, *Astrophys. Space Sci.*, 184, 169.

Bertelli, G., Mateo, M., Chiosi, C. & Bressan, A. 1992, *Astrophys. J.*, 388, 400.

Bertola, F., Bressan, A., Burstein, D., Buson, L.M., Chiosi, C. & Di Serego Alighieri, S. 1995, *Astrophys. J.*, 438, 680.

Bessell, M.S. 1991, *Astron. Astrophys.*, 242, L17.

Bianchi, L., Clayton, G.C., Bohlin, R.C., Hutchings, J.B. & Massey, P. 1996, *Astrophys. J.*, 471, 203.

Bica, E., Alloin, D. & Schmidt, A.A. 1990, *Astron. Astrophys.*, 228, 23.

Bica, E., Claría, J.J., Dottori, H., Santos, J.F.C. & Piatti, A.E. 1996, *Astrophys. J. Suppl.*, 102, 57.

Bica, E., Geisler, D., Dottori, H., Clariá, J.-J., Piatti, A.E. & Santos, J.F.C. 1998, *Astron. J.*, 116, 723.

Bica, E.L.D. & Schmitt, H.R. 1995, *Astrophys. J. Suppl.*, 101, 41.

Bica, E.L.D., Schmitt, H.R., Dutra, C.M. & Oliveira, H.L. 1999, *Astron. J.*, 117, 238.

Biggs, F.H. 1997, *Astrophys. J.*, 484, L29.

Binney, J. & Tremaine, S. 1987, *Galactic Dynamics*, Princeton: Princeton Univ. Press.

Blaauw, A. 1958, in *Stellar Populations*, ed. D.J.K. O'Connell, Amsterdam: North Holland Publ. Co., p. 105.

Blair, W.P. & Davidsen, A.F. 1993, *Publ. Astron. Soc. Pac.*, 105, 494.

Blair, W.P., Kirshner, R.P. & Chevalier, R.A. 1981, *Astrophys. J.*, 247, 879.

Blair, W.P., Kirshner, R.P. & Chevalier, R.A. 1982, *Astrophys. J.*, 254, 50.

Blanco, V.M. & McCarthy, M.F. 1983, *Astron. J.*, 88, 1442.

Bland-Hawthorn, J. & Maloney, P.R. 1998, *Astrophys. J.*, 510, L33.

Bland-Hawthorne, J., Veilleux, S., Cecil, G.N., Putman, M.E., Gibson, B.K. & Maloney, P.R., 1998, *Mon. Not. R. Astron. Soc.*, 299, 611.

Blitz, L. & Spergel, D.N. 1991, *Astrophys. J.*, 379, 631.

Blitz, L., Spergel, D.N., Teuben, P.J. & Hartmann, D. 1999, astro-ph/9901307.

Blitz, L., Spergel, D.N., Teuben, P.J., Hartman, D. & Burton, W.B. 1998, *Astrophys. J.*, 514, 818.

Blitz, L. & Teuben, P. 1996, *Unsolved Problems of the Milky Way*, San Francisco: Astron. Soc. Pac.

Blom, J.J., Paglione, A.D. & Carramiñana, A. 1999, astro-ph/9811389.

Bolte, M. & Hogan, C.J. 1995, *Nature*, 376, 399.

Bomans, D.J., Dennerl, K. & Kürster, M. 1994, *Astron. Astrophys.*, 283, L21.

Bothun, G.D. 1991, *Astron. J.*, 103, 104.

Bothun, G.D. & Thompson, I.B. 1988, *Astron. J.*, 96, 877.

Boulesteix, J., Colin, J., Athenassoula, E. & Monnet, G. 1979, in *Photometry, Kinematics and Dynamics of Galaxies*, ed. D.S. Evans, Austin: Univ. Texas Press, p. 271.

Boulesteix, J., Courtès, G., Laval, A., Monnet, G. & Petit, H. 1974, *Astron. Astrophys.*, 37, 33.

Boulesteix, J. & Monnet, G. 1970, *Astron. Astrophys.*, 9, 350.

Bouwens, R., Broadhurst, T. & Silk, J. 1998, *Astrophys. J.*, 506, 557.

Bowen, D.V., Tolstoy, E., Ferrara, A., Blades, J.C., & Brinks, E. 1997, *Astrophys. J.*, 478, 530.

Bowen, J. & Monnet, G. 1970, *Astron. Astrophys.*, 9, 350.

Brandt, W.N., Ward, M.J., Fabian, A.C. & Hodge, P.W. 1997, *Mon. Not. R. Astron. Soc.*, 291, 709.

Braun, R. & Walterbos, R.A.M. 1993, *Astron. Astrophys. Suppl.*, 98, 327.

Bresolin, F., Capaccioli, M. & Piotto, G. 1993, *Astron. J.*, 105, 1779.

Brodie, J.P. & Huchra, J.P. 1990, *Astrophys. J.*, 362, 503.

Brodie, J.P. & Huchra, J.P. 1991, *Astrophys. J.*, 379, 157.

Brosche, N., Shara, M., MacKenty, J., Zurek, D. & McLean, B. 1999, *Astron. J.*, 117, 206.

Brosche, P., Einasto, J. & Rümmel, U. 1974, *Heidelberg Veröff. No. 26*.

Brown, T.M., Ferguson, H.C., Stanford, S.A. & Deharveng, J.-M. 1998, *Astrophys. J.*, 504, 113.

Brown, T.M., Ferguson, H.C., Stanford, S.A., Deharveng J.-M. & Davidsen, A.F. 1998, *Astrophys. J.*, 504, 113.

Buchler, J.R. & Moskalik, P. 1994, *Astron. Astrophys.*, 292, 450.

Buonanno, R., Corsi, C.E., Battistini, P., Bònoli, F. & Fusi Pecci, F. 1982, *Astron. Astrophys. Suppl.*, 47, 451.

Buonanno, R., Corsi, C., Ferraro, F.R., Fusi Pecci, F. & Bellazzini, M. 1996, in *Formation of the Galactic Halo (ASP Conf. Series No. 92)*, ed. H. Morrison and A. Sarajedini, San Francisco: Astron. Soc. Pac., p. 520.

Buonanno, R., Corsi, C.E., Fusi Pecci, F., Richer, H.B. & Fahlman, G.C. 1993, *Astron. J.*, 105, 184.

Buonanno, R., Corsi, C.E., Fusi Pecci, F., Fahlman, G.G. & Richer, H.B. 1994, *Astrophys. J.*, 430, L121.

Buonanno, R., Corsi, C.E., Zinn, R., Fusi Pecci, F., Hardy, E. & Suntzeff, N.B. 1998, *Astrophys. J.*, 501, L33.

Burke, C.J. & Mighell, K. 1998, *Bull. Am. Astron. Soc.*, 30, 1998.

Burleigh, M.R. & Barstow, M.A. 1998, *Mon. Not. R. Astron. Soc.*, 295, L15.

Burrows, A. 1998, preprint.

Burstein, D., Faber, S.M., Gaskell, C.M. & Krumm, N. 1984, *Astrophys. J.*, 287, 586.

Burstein, D. & Heiles, C. 1984, *Astrophys. J. Suppl.*, 54, 33.

Burton, W.B. & Braun, R. 1998, preprint.

Burton, W.B., Braun, R., Walterbos, R.A.M. & Hoopes, C.G. 1999, *Astron. J.*, 117, 194.

Butcher, H. 1977, *Astrophys. J.*, 216, 372.

Butler, D., Demarque, P. & Smith, H.A. 1982, *Astrophys. J.*, 257, 592.

Byrd, G., Valtonen, M., McCall, M. & Innanen, K. 1994, *Astron. J.*, 107, 2055.

Caldwell, J.A.R. & Coulson, I.M. 1986, *Mon. Not. R. Astron. Soc.*, 218, 223.

Caldwell, J.A.R. & Laney, C.D. 1991, in *The Magellanic Clouds (IAU Symposium No. 148)*, ed. R. Haynes and D. Milne, Dordrecht: Kluwer, p. 249.

Caldwell, J.A.R. & Maeder, D.L. 1992, in *Variable Stars and Galaxies (ASP Conference Series No. 30)*, ed. B. Warner, San Francisco: Astron. Soc. Pac., p. 173.

Caldwell, N., Armandroff, T.E., Da Costa, G.S. & Seitzer, P. 1998, *Astron. J.*, 115, 535.

Caldwell, N., Armandroff, T.E., Seitzer, P. & Da Costa, G.S. 1992, *Astron. J.*, 103, 840.

Caldwell, N. & Schommer, R. 1988, in *The Extragalactic Distance Scale (ASP Conference Series No. 4)*, ed. S. van den Bergh and C.J. Pritchet, San Francisco: Astron. Soc. Pac., p. 77.

Calzetti, D., Kinney, A.L., Ford, H., Doggett, J. & Long, K.S. 1995, *Astron. J.*, 110, 2739.

Cannon, R.D., Hawarden, T.G. & Tritton, S.B. 1977, *Mon. Not. R. Astron. Soc.*, 880, 81p.

Cannon, R.D., Niss, B. & Nørgaard-Nielsen, H.U. 1981, *Mon. Not. R. Astron. Soc.*, 196, 1p.

Canterna, R. & Flower, P.J. 1977, *Astrophys. J.*, 212, L57.

Capaccioli, M., Della Valle, M., D'Onofrio, M. & Rosino, L. 1989, *Astron. J.*, 97, 1622.

Capaccioli, M., Ortolani, S. & Piotto, G. 1987, in *Stellar Evolution and Dynamics in the Outer Halo of the Galaxy*, ed. M. Azzopardi and F. Matteucci, Garching: ESO, p. 281.

Cappellari, M., Bertola, F., Burstein, D., Buson, L.M., Greggio, L. & Renzini, A. 1999, *Astrophys. J.*, 515, L17.

Caputo, F., Cassisi, S., Castellani, M., Marconi, G. & Santolamazza, P. 1999, *Astron. J.*, 117, 2199.

Capuzzo-Dolcetta, R. & Vignola, L. 1997, *Astron. Astrophys.*, 327, 130.

Caraveo, P.A., Magnani, R. & Bignami, G.F. 1998, astro-ph/9810016.

Carignan, C. 1985, *Astrophys. J.*, 299, 59.

Carignan, C., Beaulieu, S., Côté, S., Demers, S. & Mateo, M. 1998, *Astron. J.*, 116, 1690.

Carignan, C., Demers, S. & Côté, S. 1991, *Astrophys. J.*, 381, L13.

Carlson, G. & Sandage, A. 1990, *Astrophys. J.*, 352, 587.

Carney, B.W., Laird, J.B., Latham, D.W. & Aguilar, L.A. 1996, *Astron. J.*, 112, 668.

Carney, B.W., Latham, D.W. & Laird, J.B. 1990, *Astron. J.*, 99, 572.

Carney, B.W., Lee, J.-W. & Habgood, M.J. 1998, *Astron. J.*, 116, 424.

Carraro, G., Ng, Y.K. & Portinari, L. 1998, *Mon. Not. R. Astron. Soc.*, 296, 1045.

Carraro, G., Vallenari, A., Girardi, L. & Richichi, A. 1999, *Astron. Astrophys.*, in press.

Carroll, L. 1876, *The Hunting of the Snark*, Berkeley: Univ. California Press.

Carter, D. & Jenkins, C.R. 1993, *Mon. Not. R. Astron. Soc.*, 263, 1049.

Carter, D. & Sadler, E.M. 1990, *Mon. Not. R. Astron. Soc.*, 245, 12p.

Castellani, M., Marconi, G. & Buonanno, R. 1996a, *Astron. Astrophys.*, 310, 715.

Castellani, M., Marconi, G. & Buonanno, R. 1996b, in *Formation of the Galactic Halo (ASP Conference Series No. 92)*, ed. H. Morrison and A. Sarajedini, San Francisco: Astron. Soc. Pac., p. 511.

Castro, S., Rich, S.M., McWilliam, A., Ho, L.C., Spinrad, H., Filippenko, A.V. & Bell, R.A. 1996, *Astron. J.*, 111, 2439.

Catelan, M. 1998, *Astrophys. J.*, 495, L81.

Cepa, J. & Beckman, J.E. 1988, *Astron. Astrophys.*, 200, 21.

Cesarsky, D.A., Laustsen, S., Lequeux, J., Schuster, H.-E. & West, E.M. 1977, *Astron. Astrophys.*, 61, L31.

Cesarsky, D.A., Lequeux, J., Pagani, L., Ryter, C., Loinard, L. & Sauvage, M. 1998, *Astron. Astrophys.*, 337, L35.

Chaboyer, B., Demarque, P. & Sarajedini, A. 1996, *Astrophys. J.*, 459, 558.

Chaboyer, B., Green, E.M. & Liebert, J. 1999, *Astron. J.*, 117, 1360.

Chandar, R. 1998, *Bull. Am. Astron. Soc.*, 30, 1336.

Chamcham, K. & Hendry, M.A. 1996, *Mon. Not. R. Astron. Soc.*, 279, 1083.

Chappell, D. & Scalo, J. 1997, preprint.

Chen, B. 1998, *Astrophys. J.*, 495, L1.

Chin, Y.-N., Henkel, C., Langer, N. & Mauersberger, R. 1999, *Astrophys. J.*, 512, L143.

Chin, Y.-N., Henkel, C., Millar, T.J., Whiteoak, J.B. & Marx-Zimmer, M. 1998, *Astron. Astrophys.*, 330, 901.

Christian, C. 1993, in *The Globular Cluster–Galaxy Connection (ASP Conference Series Vol. 48)*, ed. G.H. Smith and J.P. Brodie, San Francisco: Astron. Soc. Pac., p. 448.

Christian, C.A. & Heasley, J.N. 1991, *Astron. J.*, 101, 848.

Christian, C.A. & Schommer, R.A. 1982, *Astrophys. J. Suppl.*, 49, 405.

Christian, C.A. & Schommer, R.A. 1983, *Astrophys. J.*, 275, 92.

Christian, C.A. & Schommer, R.A. 1987, *Astron. J.*, 93, 557.

Christian, C.A. & Schommer, R.A. 1988, *Astron. J.*, 95, 704.

Christianson, G.E. 1995, *Edwin Hubble: Mariner of the Nebulae*, New York: Farrer, Straus & Geiroux.

Christodoulou, D.M., Tohline, J.E. & Keenan, F.P. 1997, *Astrophys. J.*, 486, 810.

Chu, Y.-H. 1997, *Astron. J.*, 113, 1815.

Chu, Y.-H., Caulet, A., Dickel, J.R., Williams, R. Arias-Montaño, L., Rosado, M., Ambrocio-Cruz, P., Laval, A. & Bomans, D. 1998, *Bull. Am. Astron. Soc.*, 30, 1365.

Chu, Y.-H., Hesser, J.E., Suntzeff, N.B. & Bohlender, D. 1999, *New Views of the Magellanic Clouds (IAU Symposium No. 190)*, San Francisco: Astron. Soc. Pac., in press.

Chu, Y.-H. & Mac Low, M.-M 1990, *Astrophys. J.*, 365, 510.

Chu, Y.-H. & Kennicutt, R.C. 1994, *Astrophys. J.*, 425, 720.

Ciani, A., D'Odorico, S. & Benvenuti, P. 1984, *Astron. Astrophys.*, 137, 223.

Ciardullo, R., Ford, H.C., Neill, J.D., Jacoby, G.H. & Shafter, A.W. 1987, *Astrophys. J.*, 318, 521.

Ciardullo, R., Ford, H.C., Williams, R.E., Tamblyn, P. & Jacoby, G.H. 1990, *Astron. J.*, 99, 1079.

Ciardullo, R. & Jacoby, G.H. 1992, *Astrophys. J.*, 388, 268.

Ciardullo, R., Jacoby, G.H., Ford, H.C. & Neill, J.D. 1989, *Astrophys. J.*, 339, 53.

Ciardullo, R., Rubin, V.C., Jacoby, G.H., Ford, H.C. & Ford, W.K. 1988, *Astron. J.*, 95, 438.

Ciardullo, R., Shafter, A.W., Ford, H.C., Neill, J.D., Shara, M.M. & Tomaney, A.B. 1990, *Astrophys. J.*, 356, 472.

Clampin, M., Nota, A., Golimowski, D.A., Leitherer, C. & Durrance. S.T. 1993, *Astrophys. J.*, 410, L35.

Clark, D.H. & Stephenson, F.R. 1977, *The Historical Supernovae*, Oxford: Pergamon.

Clayton, G.C., Kilkenny, D., Welch, D.L. 1999, in *New Views of the Magellanic Clouds (IAU Symposium No. 190)*, ed. Y.-H. Chu et al., San Francisco: Astron. Soc. Pac., in press.

Code, A.D. & Houck, T.E. 1955, *Astrophys. J.*, 121, 553.

Cohen, J.G. 1993, in *The Globular Cluster–Galaxy Connection (ASP Conference Series Vol. 48)*, ed. G.H. Smith and J.P. Brodie, San Francisco: Astron. Soc. Pac., p. 438.

Cohen, J.G. & Blakeslee, J.P. 1998, *Astron. J.*, 115, 2356.

Cohen, J.G., Persson, S.E. & Searle, L. 1984, *Astrophys. J.*, 218, 141.

Cohen, R.S., Dame, T.M., Garay, G., Montani, J., Rubio, M. & Thaddeus, P. 1988, *Astrophys. J.*, 331, L95.

Cole, A.A. 1998, *Astrophys. J.*, 500, L137.

Cole, A.A. et al. 1998, *Astrophys. J.*, 505, 230.

Collier, J., Hodge, P.W. & Kennicutt, R.C. 1995, *Publ. Astron. Soc. Pac.*, 107, 361.

Combes, F., Debbasch, F., Friedli, D. & Pfenniger, D. 1990, *Astron. Astrophys.*, 233, 82.

Cook, K.H. 1987, Univ. Arizona PhD thesis.

Cook, K.H., Aaronson, M. & Norris, J. 1986, *Astrophys. J.*, 305, 634.

Cook, K.H., et al. 1997, in *Variable Stars and Astrophysical Returns of Microlensing Surveys (12th IAP Colloquium)*, ed. R. Ferlet, J.-P. Milliard and B. Raban, Paris: Editions Frontières, p. 17.

Cook, K.H. & Olszewski, E. 1989, *Bull. Am. Astron. Soc.*, 21, 775.

Corbelli, E. & Schneider, S.E. 1997, *Astrophys. J.*, 479, 244.

Corbelli, E., Schneider, S.E. & Salpeter, E.E. 1989, *Astron. J.*, 97, 390.

Corwin, H.G., de Vaucouleurs, G. & de Vaucouleurs, A. 1985, *Southern Galaxy Catalog*, Austin: Univ. Texas Press.

Côté, S., Freeman, K. & Carignan, C. 1994, in *Dwarf Galaxies*, ed. G. Meylan and P. Prugniel, Garching: ESO, p. 101.

Courteau, S. & van den Bergh, S. 1999, in *Astron. J.*, 118, 337.

Courtès, G., Petit, H., Sivan, J.-P., Dodonov, S. & Petit, M. 1987, *Astron. Astrophys.*, 174, 28.

Courtès, G., Viton, M., Bowyer, S., Lampton, M., Sasseen, T.P. & Wu, X.-Y. 1995, *Astron. Astrophys.*, 297, 338.

Cowley, A.P. & Hartwick, F.D.A. 1991, *Astrophys. J.*, 373, 80.

Cowley, A.P., Schmidtke, P.C., Hutchings, J.B., Crampton, D. & McGrath, T.K. 1993, *Astrophys. J.*, 418, L63.

Cowley, A.P., Schmidtke, P.C., McGrath, T.K., Ponder, A.L., Fertig, M.R., Hutchings, J.B. & Crampton, D. 1997, *Publ. Astron. Soc. Pac.*, 109, 21.

Cram, T.R., Roberts, M.S. & Whitehurst, R.N. 1980, *Astron. Astrophys. Suppl.*, 40, 215.

Crampton, D., Cowley, A.P., Schade, D. & Chayer, P. 1985, *Astrophys. J.*, 288, 494.

Crampton, D., Gussie, G., Cowley, A.P. & Schmidtke, P.C. 1997, *Astron. J.*, 114, 2353.

Crane, P.C., Cowan, J.J., Dickel, J.R. & Roberts, D.A. 1993, *Astrophys. J.*, 417, L61.

Crane, P.C., Dickel, J.R. & Cowan, J.J. 1992, *Astrophys. J.*, 390, L9.

Crotts, A.P.S. 1997, in *Dark and Visible Matter in Galaxies (ASP Conference Series No. 117)*, ed. M. Persic and P. Salucci, San Francisco: Astron. Soc. Pac., p. 289.

Crowther, P.A. & Dessart, L.1998, *Mon. Not. R. Astron. Soc.*, 296, 622.

Da Costa, G.S. 1991, in *The Magellanic Clouds (IAU Symposium No. 148)*, ed. R. Haynes and D. Milne, Dordrecht: Kluwer, p. 183.

Da Costa, G.S. 1994, in *Dwarf Galaxies*, ed. G. Meylan and P. Prugniel, Garching: ESO, p. 221.

Da Costa, G.S. 1998, in *Stellar Astrophysics for the Local Group*, Cambridge: Univ. Cambridge Press, p. 351.

Da Costa, G.S. 1999, in *The Galactic Halo: Bright Stars and Dark Matter*, ed. B.K. Gibson, T.S. Axelrod, and M.E. Putman, San Francisco: Astron. Soc. Pac., p. 153.

Da Costa, G.S. & Armandroff, T.E. 1995, *Astron. J.*, 109, 2533.

Da Costa, G.S. & Hatzidimitriou, D. 1998, astro-ph/9802008.

Da Costa, G.S. & Mould, J.R. 1988, *Astrophys. J.*, 334, 159.

Dame, T.M., Koper, E., Israel, F.P. & Thaddeus, P. 1993, *Astrophys. J.*, 418, 730.

Danziger, I.J., Dopita, M.A., Hawarden, T.G. & Webster, B.L. 1978, *Astrophys. J.*, 220, 466.

Danziger, J. 1987, *ESO Workshop on the SN 1987A*, ed. I.J. Danziger, Garching: ESO.

Davidge, T.J. 1992, *Astrophys. J.*, 397, 457.

Davidge, T.J. 1994, *Astron. J.*, 108, 2123.

Davidge, T.J. 1995, *Astron. J.*, 105, 1392.

Davidge, T.J. 1998, *Astrophys. J.*, 497, 650.

Davidge, T.J. & Jones, J.H. 1992, *Astron. J.*, 104, 1365.

Davies, M.B., Blackwell, R., Bailey, V.C. & Sigurdsson, S. 1998, *Mon. Not. R. Astron. Soc.*, 301, 745.

Davies, R.D., Elliot, K.H. & Meaburn, J. 1976, *Mem. R. Astron. Soc.*, 81, 89.

Davis, D.S., Bird, C.M., Mushotzky, R.F. & Odewahn, S.C. 1995, *Astrophys. J.*, 440, 48.

de Boer, K.S., Braun, J.-M., Vallenari, A. & Mebold, U. 1998a, *Astron. Astrophys.*, 329, L49.

de Boer, K.S., Richter, P., Bomans, D.J., Heithausen, A. & Koornneef, J. 1998b, *Astron. Astrophys.*, 338, L5.

de Boer, K.S. & Savage, B.D. 1980, *Astrophys. J.*, 238, 86.

de Freitas Pacheco, J.A. 1998, *Astron. J.*, 116, 1701.

de Freitas Pacheco, J.A., Barbuy, B. & Idiart, T. 1998, *Astron. Astrophys.*, 332, 19.

Deharveng, J.-M., Joubert, M., Monnet, G. & Donas, J. 1982, *Astron. Astrophys.*, 106, 16 de Laverny, p. et. al. 1998, *Astron. Astrophys.*, 335, L93.

Della Valle, M. 1988, in *The Extragalactic Distance Scale (ASP Conference Series Vol. 4)*, ed. S. van den Bergh and J.C. Pritchet, Provo: Astron. Soc. Pac., p. 73.

Della Valle, M. & Duerbeck, H.W. 1993, *Astron. Astrophys.*, 271, 175.

Della Valle, M. & Livio, M. 1994, *Astron. Astrophys.*, 286, 786.

Della Valle, M. & Livio, M. 1995, *Astrophys. J.*, 452, 704.

Della Valle, M. & Livio, M. 1998, *Astrophys. J.*, 506, 818.

Della Valle, M., Rosino, L., Bianchini, A. & Livio, M. 1994, *Astron. Astrophys.*, 287, 403.

Demers, S. & Battinelli, P. 1998, *Astron. J.*, 115, 154.

Demers, S., Battinelli, P., Irwin, M.J. & Kunkel, W.E. 1995, *Mon. Not. R. Astron. Soc.*, 274, 491.

Demers, S., Beland, S. & Kunkel, W.E. 1983, *Publ. Astron. Soc. Pac.*, 95, 354.

Demers, S. & Harris, W.E. 1983, *Astron. J.*, 88, 329.

Demers, S. & Irwin, M.J. 1993, *Mon. Not. R. Astron. Soc.*, 261, 657.

Demers, S., Irwin, M.J. & Gambu, I. 1994 *Mon. Not. R. Astron. Soc.*, 266, 7.

Demers, S., Irwin, M.J. & Kunkel, W.E. 1994, *Astron. J.*, 108, 1648.

Demers, S., Kunkel, W.E. & Irwin, M.J. 1985, *Astron. J.*, 90, 1967.

Deul, E.R. & den Hartog, R.H. 1990, *Astron. Astrophys.*, 229, 362.

Deul, E.R. & van der Hulst, J.M. 1987, *Astron. Astrophys. Suppl.*, 67, 509.

de Vaucouleurs, G. 1954a, *Australian J. Sci.*, 17, 4.

de Vaucouleurs, G. 1954b, *Observatory*, 74, 158.

de Vaucouleurs, G. 1955a, *Astron. J.*, 60, 126.

de Vaucouleurs, G. 1955b, *Astron. J.*, 60, 219.

de Vaucouleurs, G. 1958, *Astrophys. J.*, 128, 465.

de Vaucouleurs, G. 1959a, *Astrophys. J.*, 130, 728.

de Vaucouleurs, G. 1959b, in *Handb. d. Physik*, Vol. 53 (Stellar Systems), ed. S. Flügge, Berlin: Springer-Verlag, p. 311.

de Vaucouleurs, G. 1961, *Astrophys. J. Suppl.*, 5, 233.

de Vaucouleurs, G. 1975, in *Galaxies and the Universe*, ed. A. Sandage, M. Sandage, and J. Kristian, Chicago: Univ. Chicago Press, p. 557.

de Vaucouleurs, G. & Ables, H. 1965, *Publ. Astron. Soc. Pac.*, 77, 272.

de Vaucouleurs, G. & Corwin, H.G. 1985, *Astrophys. J.*, 295, 287.

de Vaucouleurs, G., de Vaucouleurs, A., Corwin, H.G., Buta, R.J., Paturel, G. & Fouqué, P. 1991, *Third Reference Catalog of Bright Galaxies*, Berlin: Springer-Verlag.

de Vaucouleurs, G. & Freeman, K.C. 1972, *Vistas in Astronomy*, 14, 163.

de Vaucouleurs, G. & Leach, R.W. 1981, *Publ. Astron. Soc. Pac.*, 93, 190.

Devereux, N.A., Price, R., Wells, L.A. & Duric, N. 1994, *Astron. J.*, 108, 1667.

Dieball, A. & Grebel, E.K. 1998, *Astron. Astrophys.*, 339, 773.

Di Nella-Courtois, H., Lanoix, P. & Paturel, G. 1998, astro-ph/9810211.

Djorgovski, S. 1993, in *Structure and Dynamics of Globular Clusters* (*ASP Conference Series No. 50*), ed. S.G. Djorgovski and G. Meylan, San Francisco: Astron. Soc. Pac., p. 373.

Dohm-Palmer, R. Skillman, E.D., Gallagher, J., Tolstoy, E., Mateo, M., Dufour, R.J., Saha, A., Hoessel, J. & Chiosi, C. 1998, *Astron. J.*, 116, 1227.

Dohm-Palmer, R.C., Skillman, E.D., Saha, A., Tolstoy, E., Mateo, M., Gallagher, J., Hoessel, J., Chiosi, C & Dufour, R.J. 1997a, *Astron. J.*, 114, 2514.

Dohm-Palmer, R.C., Skillman, E.D., Saha, A., Tolstoy, E., Mateo, M., Gallagher, J., Hoessel, J., Chiosi, C. & Dufour, R.J. 1997b, *Astron. J.*, 114, 2527.

D'Odorico, S., Dopita, M.A. & Benvenuti, P. 1980, *Astron. Astrophys. Suppl.*, 40, 67.

D'Onofrio, M., Capaccioli, M., Wagner, S.J. & Hopp, U. 1994, *Mem. Soc. Astron. Italiana*, 65, 731.

Dopita, M.A. 1984, in *Structure and Evolution of the Magellanic Clouds* (*IAU Symposium No. 108*), ed. S. van den Bergh and K.S. de Boer, Dordrecht: Reidel, p. 271.

Dopita, M.A., Ford, H.C., Lawrence, C.J. & Webster, B.L. 1985, *Astrophys. J.*, 296, 390.

Dopita, M.A., Wood, P.R., Meatheringham, S.J., Vacillates, E., Bohlin, R.C., Ford, H.C., Harrington, J.P., Stacker, T.P. & Maran, SP. 1997, in *Planetary Nebulae* (*IAU Symposium No. 180*), ed. H.J. Habing and H.J.G.L.M. Lamers, Dordrecht: Kluwer, p. 417.

Dostoevsky, F. 1980, *The Brothers Karamazov*, translated by R. Pevear & L. Volokhonsky, New York: Random House.

Dottori, H., Bica, E., Claría, J.J., Puerari, I. 1996, *Astrophys. J.*, 461, 742.

Dottori, H., Mirabel, F. & Duc, P.-A. 1994, in *Dwarf Galaxies*, ed. G. Meylan and P. Prugniel, Garching: ESO, p. 393.

Downes, D. & Solomon, P.M. 1998, *Astrophys. J.*, 507, 615.

Dressler, A. & Richstone, D.O. 1988, *Astrophys. J.*, 324, 701.

Dreyer, J.L.E. 1908, *Mem. R. Astron. Soc.*, 59, 105.

Drissen, L., Moffat, A.F.J. & Shara, M.M. 1993, *Astron. J.*, 105, 1400.

Dubath, P. & Grillmair, C.J. 1997, *Astron. Astrophys.*, 321, 379.

Dubath, P., Meylan, G. & Mayor, M. 1992, *Astrophys. J.*, 400, 510.

Dubus, G., Charles, P.A., Long, K.S., Hakala, P.J. & Kuulkers, E. 1998, astroph/9810050.

Duc, P.-A. & Mirabel, I.F. 1998, *Astron. Astrophys.*, 333, 813.

Dufour, R.J. 1984, in *Structure and Evolution of the Magellanic Clouds (IAU Symposium No. 108)*, ed. S. van den Bergh and K.S. de Boer, Dordrecht: Reidel, p. 353.

Dufour, R.J. & Talent, D.L. 1980, *Astrophys. J.*, 235, 22.

Duncan, J.C. 1922, *Publ. Astron. Soc. Pac.*, 34, 290.

Dunn, A.M. & Laflamme, R. 1993, *Mon. Not. R. Astron. Soc.*, 264, 865.

Durand, S., Dejonghe, H. & Acker, A. 1996, *Astron. Astrophys.*, 310, 97.

Duric, N., Viallefond, F., Goss, W.M. & van der Hulst, J.M. 1993, *Astron. Astrophys. Suppl.*, 99, 217.

Durrell, P.R., Harris, W.E. & Pritchet, C.J. 1994, *Astron. J.*, 108, 2114.

Dyer, J. 1895, *Mon. Not. R. Astron. Soc.*, 51, 185.

Eckart, A. & Genzel, R. 1997, *Mon. Not. R. Astron. Soc.*, 284, 576.

Edvardsson, B., Andersen, J., Gustafsson, B., Lambert, D.L., Nissen, P.E. & Tomkin, J. 1993, *Astron. Astrophys.*, 275, 101.

Efremov, Y.N. 1991, *Sov. Astron. Lett.*, 17, 173.

Efremov, Y.N. 1995, *Astron. J.*, 110, 2757.

Efremov, Y.N. & Elmegreen, B.C. 1998, *Mon. Not. R. Astron. Soc.*, 299, 643.

Efremov, Y.N., Elmegreen, B.C. & Hodge, P.W. 1998, *Astrophys. J.*, 501, L163.

Eggen, O.J., Lynden-Bell, D. & Sandage, A.R. 1962, *Astrophys. J.*, 136, 748.

Einasto, J., Saar, E., Kaasik, A. & Chernin, A.D. 1974, *Nature*, 252, 111.

Eisenhauer, F., Quirrenbach, A., Zinnecker, H. & Genzel, R. 1998, *Astrophys. J.*, 498, 278.

Ekers, R.D., van Gorkom, J.H., Schwartz, U.J. & Goss, W.M. 1983, *Astron. Astrophys.*, 122, 143.

Ellis, G.L., Grayson, E.T. & Bond, H.E. 1984, *Publ. Astron. Soc. Pac.*, 96, 283.

Elsässer, H. 1959, *Z. Astrophys.*, 47, 1.

Elson, R.A., Gilmore, G.F. & Santiago, B.X. 1997, *Mon. Not. R. Astron. Soc.*, 289, 157.

Elson, R.A.W., Sigurdsson, S., Davies, M., Hurley, J. & Gilmore, G. 1998a, *Mon. Not. R. Astron. Soc.*, 300, 857.

Elson, R.A.W., Sigurdsson, S., Hurley, J., Davies, M.B. & Gilmore, G.F. 1998b, *Astrophys. J.*, 499, L53.

Emerson, D.T. 1974, *Mon. Not. R. Astron. Soc.*, 169, 607.

Eskridge, P.B. 1988a, *Astron. J.*, 95, 1706.

Eskridge, P.B. 1988b, *Astron. J.*, 96, 1614.

Eskridge, P.B. 1995, *Publ. Astron. Soc. Pac.*, 107, 561.

Eskridge, P.B. & White, R.E. 1997, *Astron. J.*, 114, 988.

Faber, S.M. 1973, *Astrophys. J.*, 179, 731.

Fahlman, G.G., Mandushev, G., Richer, H.B., Thompson, I.B. & Sivaramakrishnan, A. 1996, *Astrophys. J.*, 459, L65.

Feast, M.W. & Catchpole, R.M. 1997, *Mon. Not. R. Astron. Soc.*, 186, L1.

Feast, M.W., Thackeray, A.D. & Wesselink, A.J. 1960, *Mon. Not. R. Astron. Soc.*, 121, 337.

Feast, M.W. & Whitelock, P. 1997, *Mon. Not. R. Astron. Soc.*, 291, 683.

Fehrenbach, C., Duflot, M. & Petit, M. 1970, *Astron. Astrophys. Special Supplement No. 1.*

Feitzinger, J.V. & Galinski, T. 1985, *Astron. Astrophys. Suppl.*, 61, 503.

Ferguson, H.C. & Davidsen, A.F. 1993, *Astrophys. J.*, 408, 92.

Fernley, J., Barnes, T.G., Skillen, I., Hawley, S.L., Hanley, C.J., Evans, D.W., Solano, E. & Garrido, R. 1998a, *Astron. Astrophys.*, 330, 515.

Fernley, J., Carney, B.W., Skillen, I., Cacciari, C. & Jones, K. 1998b, *Mon. Not. R. Astron. Soc.*, 293, L61.

Ferraro, F.R., Fusi Pecci, F., Tosi, M. & Buonanno, R. 1989, *Mon. Not. R. Astron. Soc.*, 241, 433.

Ferraro, I., Ferraro, F.R., Fusi Pecci, F., Corsi, C.E. & Buonanno, R. 1995, *Mon. Not. R. Astron. Soc.*, 275, 1057.

Fesen, R.A., Gerardy, C.L., McLin, K.M. & Hamilton, J.S. 1999, *Astrophys. J.*, 514, 195.

Fesen, R.A., Hamilton, A.J.S. & Saken, J.M. 1989, *Astrophys. J.*, 341, L55.

Fich, M. & Hodge, P. 1991, *Astrophys. J.*, 374, L17.

Filipović, M.D. et al. 1998, *Astron. Astrophys. Suppl.*, 127, 119.

Fischer, P., Pryor, C., Murray, S., Mateo, M. & Richtler, T. 1998, *Astron. J.*, 115, 592.

Fisher, J.R. & Tully, R.B. 1975, *Astron. Astrophys.*, 44, 151.

Fisher, J.R. & Tully, R.B. 1979, *Astron. J.*, 84, 62.

Fitzpatrick, E.L. 1999, *Publ. Astron. Soc. Pac.*, 111, 63.

Florsch, A. 1972, *Strasbourg Obs.*, Publ. 2. fasc. 1.

Forbes, D.A., Brodie, J.P. & Grillmair, C.J. 1997, *Astron. J.*, 113, 1652.

Ford, H.C. 1983, in *Planetary Nebulae* (*IAU Symposium No. 103*), ed. D.R. Flower, Dordrecht: Reidel, p. 443.

Ford, H.C. & Jacoby, G.H. 1978a, *Astrophys. J.*, 219, 437.

Ford, H.C. & Jacoby, G.H. 1978b, *Astrophys. J. Suppl.*, 38, 351.

Ford, H.C., Jacoby, G. & Jenner, D.C. 1977, *Astrophys. J.*, 213, 18.

Forest, T.A., Spenny, D.L. & Johnson, R.W. 1988, *Publ. Astron. Soc. Pac.*, 100, 683.

Fouqué, R., Bottinelli, L., Durand, N., Gougenheim, L. & Paturel, G. 1990, *Astron. Astrophys. Suppl.*, 86, 473.

Frederici, L., Bònoli, F., Ciotti, L., Fusi Pecci, F., Marano, B., Lipovetsky, V.A., Neizvestny, S.I. & Spassova, N. 1993, *Astron. Astrophys.*, 274, 87.

Frederici, L., Bònoli, F., Fusi Pecci, F., Marano, B., Börgen, F. & Meusinger, H. 1994, in *Astronomy from Wide Field Imaging* (*IAU Symposium No. 161*), ed. H.T. MacGillivray et al., Dordrecht: Reidel, p. 583.

Freedman, W.L. 1985, *Astrophys. J.*, 299, 74.

Freedman, W.L. 1988a, in *The Extragalactic Distance Scale* (*ASP Conference Series Vol. 4*), ed. S. van den Bergh and C.J. Pritchet, Provo: Astron. Soc. Pac., p. 24.

Freedman, W.L. 1988b, *Astrophys. J.*, 326, 691.

Freedman, W.L. 1988c, *Astrophys. J.*, 96, 1248.

Freedman, W.L. 1992, *Astron. J.*, 104, 1349.

Freedman, W.L. & Madore, B.F. 1990, *Astrophys. J.*, 365, 186.

Freedman, W.L., Wilson, C.D. & Madore, B.F. 1991, *Astrophys. J.*, 372, 455.

Freeman, K.C. 1999, in *Stellar Content of the Local Group*, (*IAU Symposium No. 192*), ed. P. Whitelock and R.D. Cannon, San Francisco: Astron. Soc. Pac., p. 383.

Friaça, A.C.S. & Terlevich, R.J. 1999, *Mon. Not. R. Astron. Soc.*, in press.

Friel, E.D. 1995, *Annu. Rev. Astron. Astrophys.*, 33, 381.

Frogel, J.A., Terndrup, D.M., Blanco, V.M. & Whitford, A.E. 1990, *Astrophys. J.*, 353, 494.

Fusi Pecci, F. et al. 1994, *Astron. Astrophys.*, 284, 349.

Fusi Pecci, F., Buonanno, R., Cacciari, C., Corsi, C.E., Djorgovski, S.G., Frederici, L., Ferraro, F.R., Parmeggiano, G. & Rich, R.M. 1996, *Astron. J.*, 112, 1461.

Fusi Pecci, F., Cacciari, C., Frederici, L. & Pasquali, A. 1993, in *The Globular Cluster–Galaxy Connection* (*ASP Conference Series Vol. 48*), ed. G.H. Smith and J.P. Brodie, San Francisco: Astron. Soc. Pac., p. 410.

Gallagher, J.S. et al. 1996, *Astrophys. J.*, 466, 732.

Gallagher, J.S., Tolstoy, E., Dohm-Palmer, R.C., Skillman, E.D., Cole, A.A., Hoessel, J.G., Saha, A. & Mateo, M. 1998, *Astron. J.*, 115, 1869.

Gallagher, J.S. & Wyse, R.F.G. 1994, *Publ. Astron. Soc. Pac.*, 106, 1225.

Gallart, C. 1998, *Astrophys. J.*, 495, L43.

Gallart, C. et al. 1999, *Astrophys. J.*, 514, 665.

Gallart, C., Aparicio, A., Bertelli, G. & Chiosi, C. 1996, *Astron. J.*, 112, 1950.

Gallart, C., Aparicio, A., Freedman, W., Bertelli, G. & Chiosi, C. 1998, in *The Magellanic Clouds and Other Dwarf Galaxies*, ed. T. Richtler and J.M. Braun, Aachen: Shaker Verlag, p. 147.

Gallart, C., Aparicio, A. & Vílchez, J.M. 1996, *Astron. J.*, 112, 1929.

Gallouet, L., Heidmann, N. & Dampierre, F. 1975, *Astron. Astrophys. Suppl.*, 19, 1.

Gaposchkin, C.P. & Gaposchkin, S. 1966, *Smithsonian Center for Astrophys.*, 9, 1.

Gaposchkin, S. 1943, *Astrophys. J.*, 97, 166.

Gardiner, L.T. & Hatzidimitriou, D. 1992, *Mon. Not. R. Astron. Soc.*, 257, 195.

Gardiner, L.T. & Hawkins, M.R.S. 1991, *Mon. Not. R. Astron. Soc.*, 251, 174.

Gardiner, L.T. & Noguchi, M. 1996, *Mon. Not. R. Astron. Soc.*, 278, 191.

Garnavich, P.M., VandenBerg, D.A., Zurek, D.R. & Hesser, J.E. 1993, *Astron. J.*, 107, 1097.

Garnett, D.R., Shields, G.A., Skillman, E.D., Sagan, S.P. & Dufour, R.J. 1997, *Astrophys. J.*, 489, 63.

Gascoigne, S.C.B. & Kron, G.E. 1952, *Publ. Astron. Soc. Pac.*, 64, 196.

Gascoigne, S.C.B. & Kron, G.E. 1965, *Mon. Not. R. Astron. Soc.*, 130, 333.

Gay, P.L. 1998, *Bull. Am. Astron. Soc.*, 30, 1353.

Geha, M.C. et al. 1998, *Astron. J.*, 115, 1045.

Gehrz, R.D., Truran, J.W., Williams, R.E. & Starrfield, S. 1998, *Publ. Astron. Soc. Pac.*, 110, 3.

Geisler, D., Bica, E., Dottori, H., Claría, J.J., Piatti, A.E. & Santos, J.F.C. 1997, *Astron. J.*, 114, 1920.

Geisler, D. & Hodge, P. 1980, *Astrophys. J.*, 242, 66.

Geisler, D. & Sarajedini, A. 1996, in *Formation of the Galactic Halo* (*ASP Conference Series Vol. 92*), ed. H. Morrison and A. Sarajedini, San Francisco: Astron. Soc. Pac., p. 524.

Geisler, D., Armandroff, T.E., Da Costa, G., Lee, M.G. & Sarajedini, A., 1999, in *Stellar Content of the Local Group (IAU Symposium No. 192)*, ed. P. Whitelock and R.D. Cannon, San Francisco: Astron. Soc. Pac., p. 231.

Genzel, R., Thatte, N., Krabbe, A., Kroker, H. & Tacconi-Garman, L.E. 1996, *Astrophys. J.*, 472, 153.

Genzel, R. & Townes, C.H. 1987, *Annu. Rev. Astron. Astrophys.*, 25, 377.

Georgiev, L., Borissova, J., Rosado, M., Kurtev, R., Ivanov, G. & Koenigsberger, G. 1999, *Astron. Astrophys. Suppl.*, 134, 21.

Ghez, A.M., Klein, B.L., Morris, M. & Becklin, E.E. 1998, *Astrophys. J.*, 509, 678.

Gieren, W.P., Fouqué, P. & Gómez, M. 1998, *Astrophys. J.*, 496, 17.

Gilmore, G., King, I.R. & van der Kruit, P.C. 1989, *The Milky Way as a Galaxy*, Mill Valley, CA: Univ. Science Books.

Gilmore, G. & Wyse, R.F.G. 1991, *Astrophys. J.*, 367, L55.

Gingrich, O. 1997, private communication.

Girardi, L., Groenewegen, M.A.T., Weiss, A. & Salaris, M. 1998, *Mon. Not. R. Astron. Soc.*, 301, 149.

Gizis, J.E., Mould, J.R. & Djorgovski, S. 1993, *Publ. Astron. Soc. Pac.*, 105, 871.

Gordon, K.G. 1969, *Q. J. R. Astron. Soc.*, 10, 293.

Gordon, K.D., Calzetti, D. & Witt, A.N. 1997, *Astrophys. J.*, 487, 625.

Gordon, K.D. & Clayton, G.C. 1998, *Astrophys. J.*, 500, 816.

Gordon, K.D., Hanson, M.M., Clayton, G.C., Rieke, G.H. & Misselt, K.A. 1999, *Astrophys. J.*, 519, 165.

Gordon, S.M., Kirshner, R.P., Duric, N. & Long, K.S. 1993, *Astrophys. J.*, 418, 743.

Goss, W.M. & Lozinskaya, T.A. 1995, *Astrophys. J.*, 439, 637.

Gottesman, S.T., Davies, R.D. & Reddish, V.C. 1966, *Mon. Not. R. Astron. Soc.*, 133, 359.

Gottesman, S.T. & Weliachew, L. 1977, *Astron. Astrophys.*, 61, 523.

Gould, A. 1998, *Astrophys. J.*, 499, 728.

Gould, A., Guhathakurta, P., Richstone, D. & Flynn, C. 1992, *Astrophys. J.*, 388, 345.

Gould, A. & Popovski, P. 1998, *Astrophys. J.*, 508, 844.

Gould, A. & Uza, O. 1998, *Astrophys. J.*, 494, L118.

Graham, J.A. 1975, *Publ. Astron. Soc. Pac.*, 87, 641.

Graham, J.A. & Araya, G. 1971, *Astron. J.*, 76, 768.

Gratton, R.G. 1998, *Mon. Not. R. Astron. Soc.*, 296, 739.

Grebel, E.K. 1997, *Rev. Mod. Astron.*, 10, 29.

Grebel, E.K. 1999, in *Stellar Content of the Local Group (IAU Symposium No. 192)*, ed. P. Whitelock and R.D. Cannon, San Francisco: Astron. Soc. Pac., p. 17.

Grebel, E.K. & Brandner, W. 1998, in *The Magellanic Clouds and Other Dwarf Galaxies*, ed. T. Richtler and M. Braun, Aachen: Shaker Verlag, p. 151.

Grebel, E.K. & Guthakurta, P. 1999, *Astrophys. J.*, 511, L101.

Greenhill, L.J., Moran, J.M., Reid, M.J., Menten, K.M. & Hirabayashi, H. 1993, *Astrophys. J.*, 406, 482.

Greggio, L., Marconi, G., Tosi, M. & Focardi, P. 1993, *Astron. J.*, 105, 894.

Grillmair, C.J. 1998 in *Galectic Halos (ASP Conference Series No. 136)*, ed. D. Zaritsky, San Francisco: Astron. Soc. Pac., p. 45.

Grillmair, C.J. et al. 1996, *Astron. J.*, 112, 1996.

Grillmair, C.J. et al. 1997, in *The Nature of Elliptical Galaxies (ASP Conference Series Vol. 116)*, ed. M. Arnaboldi, G.S. Da Costa, and P. Saha, San Francisco: Astron. Soc. Pac., p. 308.

Grillmair, C.J. et al. 1998, *Astron. J.*, 115, 144.

Grillmair, C.J. & Irwin, M. 1998, *Bull. Am. Astron. Soc.*, 30, 1257.

Guhathakurta, P. & Reitzel, D.B. 1998 in *Galactic Halos (ASP Conference Series Vol. 136)*, ed. D. Zaritsky, San Francisco: Astron. Soc. Pac., p. 22.

Guidoni, U., Messi, R. & Natali, G. 1981, *Astron. Astrophys.*, 96, 215.

Guinan, E.F., Fitzpatrick, E.L., DeWarf, L.E., Maloney, F.P., Maurone, P.A., Ribas, I., Pritchard, J.D., Broadstreet, D.H. & Giménez, A. 1999, *Astrophys. J.*, 509, L21.

Gurwell, M. & Hodge, P. 1990, *Publ. Astron. Soc. Pac.*, 102, 849.

Gurzadyan, V.G., Kocharyan, A.A. & Petrosian, A.R. 1993, *Astrophys. Space Sci.*, 201, 243.

Haas, M. 1998, *Astron. Astrophys.*, 337, L1.

Haas, M., Lemke, D., Stickel, M., Hippelein, H., Kunkel, M., Herbstmeier, U. & Mattila, K. 1998, *Astron. Astrophys.*, 338, L33.

Habing, H.J., Olnon, F.M., Chester, T., Gillett, F., Rowan-Robinson, M. & Neugebauer, G. 1985, *Astron. Astrophys.*, 152, L1.

Hambly, N.C., Dufton, P.L., Keenan, F.P., Rolleston, W.R.J., Howarth, I.D. & Irwin, M.J. 1994, *Astron. Astrophys.*, 285, 716.

Hambly, N.C., Fitzsimmons, A., Keenan, F.P., Dufton, P.L., Brown, P.J.F., Irwin, M.J. & Rolleston, W.R.J. 1995, *Astrophys. J.*, 448, 628.

Han, C. & Ryden, B.S. 1994, *Astrophys. J.*, 433, 80.

Han, J.L., Beck, R. & Berkhuijsen, E.M. 1998, *Astron. Astrophys.*, 335, 1117.

Hansen, B.M.S. 1999, *Astrophys. J.*, 512, L120.

Hardy, E., Buonanno, R., Corsi, C.E., Janes, K.A. & Schommer, R.A,. 1984, *Astrophys. J.*, 278, 592.

Hardy, E., Couture, J., Couture, C. & Joncas, G. 1994, *Astron. J.*, 107, 195.

Hardy, E., Suntzeff, N.B. & Azzopardi, M. 1989, *Astrophys. J.*, 344, 210.

Hargreaves, J.C., Gilmore, G., Irwin, M.J. & Carter, D. 1994b, *Mon. Not. R. Astron. Soc.*, 271, 693.

Hargreaves, J.C., Gilmore, G., Irwin, M.J. & Carter, D. 1994c, in *Dwarf Galaxies*, ed. G. Meylan and P. Prugniel, Garching: ESO, p. 253.

Hargreaves, J.C., Gilmore, G., Irwin, M.J. & Carter, D. 1996, *Mon. Not. R. Astron. Soc.*, 282, 305.

Harringon, R.G. & Wilson, A.G. 1950, *Publ. Astron. Soc. Pac.*, 62, 118.

Harris, J., Zaritsky, D. & Thompson, I. 1997, *Astron. J.*, 114, 1933.

Harris, W.E. 1991, *Annu. Rev. Astron. Astrophys.*, 29, 543.

Harris, W.E. 1997, http://www.physics.mcmaster.ca/Globular.html.

Harris, W.E. et al. 1997, *Astron. J.*, 114, 1030.

Harris, W.E. & van den Bergh, S. 1981, *Astron. J.*, 86, 1627.

Hartmann, D.H., Brown, L.E. & Schnepf, N. 1993, *Astrophys. J.*, 408, L13.

Hatano, K., Branch, D., Fisher, A. & Starrfield, S. 1997a, *Astrophys. J.*, 487, L45.

Hatano, K., Branch, D., Fisher, A. & Starrfield, S. 1997b, *Mon. Not. R. Astron. Soc.*, 290, 113.

Hatzidimitriou, D., Cannon, R.D. & Hawkins, M.R.S. 1993, *Mon. Not. R. Astron. Soc.*, 261, 873.

Hatzidimitriou, D. & Hawkins, M.R.S. 1989, *Mon. Not. R. Astron. Soc.*, 241, 667.

Hawley, S.L., Jefferys, W.H., Barnes, T.G., & Lai, W. 1986, *Astrophys. J.*, 302, 626.

Hayes, D.S. 1985, in *Calibration of Fundamental Stellar Quantities (IAU Symposium No. 111)*, ed. D.S. Hayes, L.E. Pasinetti, and A.G. Davis Philip, Dordrecht: Reidel, p. 225.

Haynes, R.F., Klein, U., Wielebinski, R. & Murray, J.D. 1986, *Astron. Astrophys.*, 159, 22.

Haynes, R. & Milne, D. 1991, *The Magellanic Clouds (IAU Symposium No. 148)*, Dordrecht: Kluwer.

Heatherington, N.S. 1972, Q.J.R. Astron. Soc., 13, 25.

Heggie, D.C., Griest, K. & Hut, P. 1993, in *Structure and Dynamics of Globular Clusters (ASP Conf. Series No. 50)*, San Francisco: Astron. Soc. Pac., p. 137.

Heikkilä, A., Johansson, L.E.B. & Olofsson, H. 1998, *Astron. Astrophys.*, 332, 493.

Heiles, C. 1987, *Astrophys. J.*, 315, 555.

Heisler, C.A., Hill, T.L., McCall, M.L. & Hunstead, R.W. 1997, *Mon. Not. R. Astron. Soc.*, 285, 374.

Heisler, J., Tremaine, S. & Bahcall, J.N. 1985, *Astrophys. J.*, 298, 8.

Held, E.V., Mould, J.R. & de Zeeuw, P.J. 1990, *Astron. J.*, 100, 415.

Hendry, M.A. 1997, *Observatory*, 117, 329.

Henry, R.B.C. 1990, *Astrophys. J.*, 356, 229.

Henry, R.B.C. 1998, in *Abundance Profiles: Diagnostic Tools for Galaxy History*, ed. D. Friedli et al. (*ASP Conference Series Vol. 147*), San Francisco: Astron. Soc. Pac., p. 59.

Hernandez, X. & Gilmore, G. 1997, 1998 *Mon. Not. R. Astron. Soc.*, 297, 517.

Herschel, J.F.W., 1847, *Results of Astronomical Observations at the Cape of Good Hope*, London: Smith & Elder, p. 143.

Hill, J.K., Isensee, J.E., Bohlin, R.C., O'Connell, R.W., Roberts, M.S., Smith, A.M. & Stecher, T.P. 1993, *Astrophys. J.*, 414, L9.

Hill, V., Barbuy, B. & Spite, M. 1997, *Astron. Astrophys.*, 323, 461.

Hindman, J.V. 1967, *Australian J. Phys.*, 20, 147.

Hirashita, H., Takeuchi, T.T. & Tamura, N. 1998, *Astrophys. J.*, 504, L83.

Hirimoto, N., Maihara, T. & Oda, N. 1983, *Publ. Astron. Soc. Jpn.*, 35, 413.

Hodge, P.W. 1961a, *Astron. J.*, 66, 249.

Hodge, P.W. 1961b, *Astron. J.*, 66, 83.

Hodge, P.W. 1961c, *Astron. J.*, 66, 3984.

Hodge, P.W. 1961d, *Publ. Astron. Soc. Pac.*, 81, 875.

Hodge, P.W. 1963, *Astron. J.*, 68, 691.

Hodge, P.W. 1965, *Astrophys. J.*, 142, 1390.

Hodge, P.W. 1969, *Publ. Astron. Soc. Pac.*, 81, 875.

Hodge, P.W. 1973, *Astrophys. J.*, 182, 671.

Hodge, P.W. 1974, *Astrophys. J. Suppl.*, 27, 113.

Hodge, P.W. 1976, *Astron. J.*, 81, 25.

Hodge, P.W. 1977, *Astrophys. J. Suppl.*, 33, 69.

Hodge, P.W. 1978, *Astrophys. J. Suppl.*, 37, 145.

Hodge, P.W. 1980, *Astrophys. J.*, 241, 125.

Hodge, P.W. 1981, *Atlas of the Andromeda Nebula*, Seattle: Univ. Washington Press.

Hodge, P.W. 1982, *Astron. J.*, 87, 1668.

Hodge, P. 1992, *The Andromeda Galaxy*, Dordrecht: Kluwer.

Hodge, P.W. 1994a, in *Dwarf Galaxies* (*ESO Conference No. 49*), ed. G. Meylan and P. Prugniel, Garching: ESO, p. 501.

Hodge, P.W. 1994b, in *The Formation and Evolution of Galaxies*, ed. C. Muñoz-Tuñón and F. Sánchez, Cambridge: Cambridge Univ. Press, p. 5.

Hodge, P.W. 1998a, Colloquium given in Victoria, Canada, 1998 April 21.

Hodge, P.W. 1998b, private communication.

Hodge, P.W., Kennicutt, R.C. & Lee, M.G. 1988, *Publ. Astron. Soc. Pac.*, 100, 917.

Hodge, P.W., Kennicutt, R.C. & Strobel, N. 1994, *Publ. Astron. Soc. Pac.*, 106, 765.

Hodge, P. & Lee, M.G. 1990, *Publ. Astron. Soc. Pac.*, 102, 26.

Hodge, P.W., Lee, M.G. & Gurwell, M. 1990, *Publ. Astron. Soc. Pac.*, 102, 1245.

Hodge, P. & Miller, B.W. 1995, *Astrophys. J.*, 451, 176.

Hodge, P.W., Smith, T.R., Eskridge, P.B., MacGillivray, H.T. & Beard, S.M. 1991a, *Astrophys. J.*, 369, 372.

Hodge, P.W., Smith, T., Eskridge, P., MacGillivray, H. & Beard, S. 1991b, *Astrophys. J.*, 379, 621.

Hodge, P.W. & Wright, F.W. 1967, *The Large Magellanic Cloud*, Washington, DC: Smithsonian Inst.

Hodge, P.W. & Wright, F.W. 1969, *Astrophys. J. Suppl.*, 17, 467.

Hodge, P.W. & Wright, F.W. 1977, *The Small Magellanic Cloud*, Seattle: Univ. Washington Press.

Hodge, P.W. & Wright, F.W. 1978, *Astron. J.*, 83, 228.

Hoessel, J.G., Abbot, M.J., Saha, A., Mossman, A.E. & Danielson, G.E. 1990, *Astron. J.*, 100, 1151.

Hoessel, J.G. & Mould, J.R. 1982, *Astrophys. J.*, 254, 38.

Hoessel, J.G., Saha, A. & Danielson, G.E. 1988, *Publ. Astron. Soc. Pac.*, 100, 680.

Hoessel, J.G., Saha, A. & Danielson, G.E. 1998, *Astron. J.*, 116, 1679.

Hoessel, J.G., Saha, A., Krist, J. & Danielson, G.E. 1994, *Astron. J.*, 108, 645.

Holland, S. 1998, *Astron. J.*, 115, 1916.

Holland, S., Fahlman, G. & Richer, H.B. 1995, *Astron. J.*, 109, 2061.

Holmberg, E. 1958, *Lund Medd. Ser. II*, No. 136.

Holtzman, J.A. 1997, *Astron. J.*, 113, 656.

Holtzman, J.A., Watson, A.M., Baum, W.A., Grillmair, C.J., Groth, E.J., Light, R.M., Lynds, R. & O'Neil, E.J. 1998, *Astron. J.*, 115, 1946.

Hopp, U., Schulte-Ladbeck, R.E., Greggio, L. & Mehlert, D. 1999, *Astron. Astrophys.*, in press.

Hoskin, M.A. 1976, *J. Hist. Astron.*, 7, 169.

Hubble, E.P. 1925a, *Observatory*, 48, 139.

Hubble, E.P. 1925b, *Popular Astron.*, 33, 252.

Hubble, E. 1925c, *Astrophys. J.*, 62, 409.

Hubble, E. 1926, *Astrophys. J.*, 63, 236.

Hubble, E. 1929, *Astrophys. J.*, 69, 103.

Hubble, E. 1932, *Astrophys. J.*, 76, 44.

Hubble, E. 1936, *The Realm of the Nebulae*, New Haven: Yale University Press.

Hubble, E. & Sandage, A. 1953, *Astrophys. J.*, 118, 353.

Huchra, J.P. 1993, in *The Globular Cluster–Galaxy Connection* (*ASP Conference Series Vol. 48*), ed. G.H. Smith and J.P. Brodie, San Francisco: Astron. Soc. Pac., p. 420.

Huchra, J.P., Brodie, J.P. & Kent, S.M. 1991, *Astrophys. J.*, 370, 495.

Huchtmeier, W. 1973, *Astron. Astrophys.*, 22, 27.

Huchtmeier, W.K. 1979, *Astron. Astrophys.*, 75, 170.

Huchtmeier, W.K. & Richter, O.-G. 1986, *Astron. Astrophys. Suppl.*, 63, 323.

Huchtmeier, W.K., Seiradakis, J.H. & Materne, J. 1981, *Astron. Astrophys.*, 102, 134.

Hughes, J.P., Hayashi, I. & Koyama, K. 1998, *Astron. J.*, 505, 732.

Hughes, J.P. & Smith, R.C. 1994, *Astron. J.*, 107, 1363.

Humason, M.L., Mayall, N.U. & Sandage, A.R. 1956, *Astron. J.*, 61, 97.

Humason, M.L. & Wahlquist, H.D. 1955, *Astron. J.*, 60, 254.

Humphreys, R.M. 1975, *Astrophys. J.*, 200, 426.

Humphreys, R.M. 1978, *Astrophys. J.*, 219, 445.

Humphreys, R.M. 1980, *Astrophys. J.*, 238, 65.

Humphreys, R.M. 1993, *The Minnesota Lectures on the Structure and Dynamics of the Milky Way*, San Francisco: Astron. Soc. Pac.

Humphreys, R.M. & Sandage, A. 1980, *Astrophys. J. Suppl.*, 44, 319.

Hunsberger, S.D., Charlton, J.C. & Zaritsky, D. 1998, *Astrophys. J.*, 505, 536.

Hunter, D.A. 1997, *Publ. Astron. Soc. Pac.*, 109, 937.

Hunter, D.A., Elmegreen, B.C. & Baker, A.L. 1998, *Astrophys. J.*, 493, 595.

Hunter, D.A., Howley, W.N. & Gallagher, J.S. 1993, *Astron. J.*, 106, 1797.

Hurley-Keller, D., Mateo, M. & Grebel, E.K. 1999, *Astrophys. J.*, 523, L25.

Hurley-Keller, D., Mateo, M. & Nemec, J. 1998, *Astron. J.*, 115, 1840.

Hutchings, J.B., Bianchi, L., Lamers, H.J.G.L.M., Massey, P. & Morris, S.C. 1992, *Astrophys. J.*, 400, L35.

Hutchinson, J.L. 1973, *Publ. Astron. Soc. Pac.*, 85, 119.

Ibata, R.A., Gilmore, G. & Irwin, M.J. 1994, *Nature*, 370, 194.

Ibata, R.A., Gilmore, G. & Irwin, M.J. 1995, *Mon. Not. R. Astron. Soc.*, 277, 781.

Ibata, R.W. & Lewis, G.F. 1998, astro-ph/9802212.

Ibata, R.A. & Razoumov, A.O. 1998, *Astron. Astrophys.*, 336, 130.

Ibata, R.A., Wyse, R.F.G., Gilmore, G., Irwin, M.J. & Suntzeff, N.B. 1997, *Astron. J.*, 113, 634.

Idiart, T.P., de Freitas Pacheco, J.A. & Costa, R.D.D. 1996, *Astron. J.*, 113, 1169.

Imbert, M. 1994, *Astron. Astrophys. Suppl.*, 105, 1.

Innanen, K.A., Kamper, K.W., Papp, K.A. & van den Bergh, S. 1982, *Astrophys. J.*, 254, 515.

Innanen, K.A. & Papp, K.A. 1979, *Astron. J.*, 84, 601.

Irwin, M.J. 1991, in *The Magellanic Clouds* (*IAU Symposium No. 148*), ed. R. Haynes and D. Milne, Dordrecht: Kluwer, p. 453.

Irwin, M.J. 1994, in *Dwarf Galaxies*, ed. G. Meylan and P. Prugniel, Garching: ESO, p. 27.

Irwin, M.J. 1999, in *Stellar Content of the Local Group* (*IAU Symposium No. 192*), ed. P. Whitelock and R.D. Cannon, San Francisco: Astron. Soc. Pac., p. 409.

Irwin, M.J., Bunclark, P.S., Bridgeland, M.T. & McMahon, R.G. 1990, *Mon. Not. R. Astron. Soc.*, 244, 16p.

Irwin, M.J., Demers, S. & Kunkel, W.E. 1990, *Astron. J.*, 99, 191.

Irwin, M. & Hatzidimitriou, D. 1993, in *The Globular Cluster–Galaxy Connection* (*ASP Conference Series Vol. 48*), ed. G.S. Smith and J.P. Brodie, San Francisco: Astron. Soc. Pac., p. 322.

Irwin, M. & Hatzidimitriou, D. 1995, *Mon. Not. R. Astron. Soc.*, 277, 1354.

Israel, F.P. 1988, *Astron. Astrophys.*, 194, 24.

Israel, F.P. 1998, *Astron. Astrophys.*, 328, 471.

Israel, F.P. et al. 1993, *Astron. Astrophys.*, 276, 25.

Israel, F.P., Tilanus, R.P.J. & Baas, F. 1998, *Astron. Astrophys.*, 339, 398.

Israel, G.L., Campana, S., Cusumano, G., Frontera, F., Orlandini, M., Santangelo, A. & Stella, L. 1998a, *Astron. Astrophys.*, 334, L65.

Israel, G.L., Stella, L., Campana, S., Covino, S., Ricci, D. & Oosterbrock, T. 1998b, *IAU Circular No. 6999*.

Ivanov, G.R. 1987, Astrophysics & Space Science, 136, 113.

Ivanov, G.R. 1992, *Astron. Lett. Comm.*, 28, 281.

Ivanov, G.R. 1998, *Astron. Astrophys.*, 337, 39.

Ivanov, G.R., Freedman, W.L. & Madore, B.F. 1993, *Astrophys. J. Suppl.*, 89, 85.

Iyudin, A.F. et al. 1998, *Nature*, 396, 142.

Jablonka, P., Bica, E., Bonatto, C., Bridges, T.J., Langlois, M. & Carter, D. 1998, *Astron. Astrophys.*, 335, 867.

Jacobsson, S. 1970, *Astron. Astrophys.*, 5, 413.

Jacoby, G.H. 1997, in *Planetary Nebulae* (*IAU Symp. No. 180*), ed. H.J. Habing and H.J.G.L.M. Lamers, Dordrecht: Reidel, p. 448.

Jacoby, G.H., Branch, D., Ciardullo, R., Davies, R.L., Harris, W.E., Pierce, M.J., Pritchet, C.J., Tonry, J.L. & Welch, D.L. 1992, *Publ. Astron. Soc. Pac.*, 104, 599.

Jacoby, G.H. & Ciardullo, R. 1999, *Astrophys. J.*, 515, 169.

Jacoby, G.H. & Ford, H.C. 1986, *Astrophys. J.*, 304, 490.

Jacoby, G.H. & Lesser, M.P. 1981, *Astron. J.*, 86, 185.

Jerjen, H., Freeman, K.C. & Binggeli, B. 1998, *Astron. J.*, 116, 2873.

Jimenez, R., Flynn, C. & Kotoneva, E. 1998, *Mon. Not. R. Astron. Soc.*, 299, 515

Johnson, D.W. & Gottesman, S.T. 1983, *Astrophys. J.*, 275, 549.

Johnson, H.M. & Hanna, M.M. 1972, *Astrophys. J.*, 174, L71.

Johnson, J.A., Bolte, M., Bond, H.E., Hesser, J.E., de Oliveira, C.M., Richer, H.B., Stetson, P.B. & VandenBerg, D.A. 1999, in *New Views of the Magellanic Clouds (IAU Symposium No. 190)*, ed. Y.-H. Chu et al., San Francisco: Astron. Soc. Pac., in press.

Johnson, S.B. & Joner, M.D. 1987, *Astron. J.*, 94, 324.

Johnston, K.V., Sigurdsson, S. & Hernquist, L. 1998, *Mon. Not. R. Astron. Soc.*, in press.

Johnston, K.V., Spergel, D.N. & Hernquist, L. 1995, *Astrophys. J.*, 451, 598.

Johnston, K.V., Zhao, H.S., Spergel, D.N. & Hernquist, L. 1998, *Bull. AAS*, 30, 1378.

Jones, B.F., Klemola, A.R. & Lin, D.N.C. 1994, *Astron. J.*, 107, 1333.

Jones, D.H. et al. 1996, *Astrophys. J.*, 466, 742.

Jones, J.H. 1993, Astron. J., 105, 933.

Jones, L.A. & Rose, J.A. 1994, *Bull. Am. Astron. Soc.*, 26, 942.

Jørgensen, U.F. & Jiminez, R. 1997, *Astron. Astrophys.*, 317, 54.

Kafatos, M. & Michalitsianos, A. 1988, *Supernova 1987A in the Large Magellanic Cloud*, Cambridge: Cambridge Univ. Press.

Kahabka, P. 1998, *Astron. Astrophys.*, 332, 189.

Kahn, F.D. & Woltjer, L. 1959, *Astrophys. J.*, 130, 705.

Kaler, J.B. 1995, in *Highlights of Astronomy*, Vol. 10, ed. I. Appenzeller, Dordrecht: Kluwer, p. 480.

Kaluzny, J., Krzeminski, W. & Mazur, B. 1995, *Astron. J.*, 110, 2206.

Kaluzny, J., Kubiak, M. Szymański, M., Udalski, A., Krzemiński, W., Mateo, M. & Stanek, K.Z. 1995, *Astron. Astrophys. Suppl.*, 112, 407.

Kaluzny, J., Kubiak, M. Szymański, M., Udalski, A., Krzemiński, W., Mateo, M. & Stanek, K.Z. 1998, *Astron. Astrophys. Suppl.*, 128, 19.

Kaluzny, J. & Rucinski, S.M. 1995, *Astron. Astrophys. Suppl.*, 114, 1.

Karachentsev, I. 1994, *Astron. Astrophys. Trans.*, 6, 1.

Karachentsev, I.D. & Karachentseva, V.E. 1999, *Astron. Astrophys.*, 341, 355.

Karachentsev, I.D., Makarova, B.L.N. & Andersen, M.I. 1999, *Astron. Astrophys.*, in press.

Karachentsev, I.D., Tikhonov, N.A. & Sazonova, L.N. 1994, *Astron. Lett.*, 20, 84.

Karachentseva, V.E. 1976, *Comm. Special Astrophys. Obs.*, 18, 42.

Karachentseva, V.E., Prugniel, P., Vennick, J., Richter, G.M., Thuan, T.X. & Martin, J.M. 1996, *Astron. Astrophys. Suppl.*, 117, 343.

Keller, S.C. & Bessell, M.S. 1998, *Astron. Astrophys.*, 340, 397.

Kenney, J.D.P. 1990, in *The Interstellar Medium in Galaxies,* ed. H.A. Thronson and J.A. Shull, Dordrecht: Kluwer, p. 151.

Kennicutt, R.C. 1983, *Astrophys. J.*, 272, 54.

Kennicutt, R.C., Bresolin, F., Bomans, D.J., Bothun, G.D. & Thompson, I.B. 1995, *Astron. J.*, 109, 594.

Kennicutt, R. et al. 1998, *Astrophys. J.*, 498, 181.

Kent, S.M. 1983, *Astrophys. J.*, 266, 562.

Kent, S.M. 1987, *Astron. J.*, 94, 306.

Kent, S.M., Dame, T.M. & Fazio, G.1991, *Astrophys. J.*, 378, 131.

Kerr, F.J., Hindman, J.V. & Robinson, B.J. 1954, *Australian J. Phys.*, 7, 297.

Kerr, F.J. & Knapp, G.R. 1972, *Astron. J.*, 77, 573.

Killen, R.M. & Dufour, R.J. 1982, *Publ. Astron. Soc. Pac.*, 94, 444.

Kim, S., Staveley-Smith, L., Dopita, M.A., Freeman, K.C., Sault, R.J., Kesteven, M.J. & McConnell, D. 1998, *Astrophys. J.*, 503, 673.

Kim, S.C. & Lee, M.G. 1998, *J. Korean Astron. Soc.*, 31, 51.

King, I.R. 1962, *Astron. J.*, 67, 471.

King, I.R. 1966, *Astron. J.*, 71, 64.

King, I.R. 1993, in *Galactic Bulges (IAU Symposium No. 153)*, ed. H. Dejonghe and H.J. Habing, Dordrecht: Kluwer, p. 3.

King, I.R., Stanford, S.A. & Crane, P. 1995, *Astron. J.*, 109, 164.

King, N.L., Walterbos, R.A.M. & Braun, R. 1998, *Astrophys. J.*, 507, 210.

Kinman, T.D. 1959, *Mon. Not. R. Astron. Soc.*, 119, 538.

Kinman, T.D., Mould, J.R. & Wood, P.R. 1987, *Astron. J.*, 93, 833.

Kinman, T.D., Stryker, L.L., Hesser, J.E., Graham, J.A., Walker, A.R., Hazen, M.L. & Nemec, J.M. 1991, *Publ. Astron. Soc. Pac.*, 103, 1279.

Kleyna, J.T., Geller, M.J., Kenyon, S.J., Kurtz, M.J. & Thorstensen, J.R. 1998, *Astron. J.*, 115, 2359.

Klypin, A.A., Kravtsov, A.V., Venezuela, O. & Prado, F. 1999, astro-ph/9901240.

Knapp, G.R., Kerr, F.J. & Bowers, P.F. 1978, *Astron. J.*, 83, 360.

Kobulnicky, H.A. & Skillman, E.D. 1998, *Astrophys. J.*, 497, 601.

Kochanek, C.S. 1996, *Astrophys. J.*, 457, 229.

König, C.H.B., Nemec, J.M., Mould, J.R. & Fahlman, G.G. 1993, *Astron. J.*, 106, 1819.

Kontizas, M., Morgan, D.H., Hatzidimitriou, D. & Kontizas, E. 1990, *Astron. Astrophys. Suppl.*, 84, 527.

Koribalski, B., Johnston, S. & Otrupcek, R. 1994, *Mon. Not. R. Astron. Soc.*, 270, L43.

Kormendy, J. 1982, in *Morphology and Dynamics of Galaxies*, ed. L. Martinet and M. Mayor, Geneva: Geneva Obs., p. 115.

Kormendy, J. 1985, *Astrophys. J.*, 295, 73.

Kormendy, J. 1988, *Astrophys. J.*, 325, 128.

Kormendy, J. 1998, *Bull. Am. Astron. Soc.*, 30, 1281.

Kormendy, J. & McClure, R.D. 1993, *Astron. J.*, 105, 1793.

Kormendy, J. & Richstone, D. 1995, *Annu. Rev. Astron. Astrophys.*, 33, 581.

Kowal, C., Lo, K.Y. & Sargent, W.L.W. 1978, *IAU Circ. No. 3305*.

Kraan-Korteweg, R.C. & Tammann, G.A. 1979, *Astron. Nachr.*, 300, 181.

Krismer, M., Tully, R.B. & Gioia, I.M. 1995, *Astron. J.*, 110, 1584.

Kron, G.E. 1956, *Publ. Astron. Soc. Pac.*, 68, 125.

Kron, G.E. & Mayall, N.U. 1960, *Astron. J.*, 65, 581.

Kroupa, P. 1998, *Mon. Not. R. Astron. Soc.*, 300, 200.

Kroupa, P. & Bastian, U. 1997, *New Astronomy*, 2, 77.

Kudritzki, R.P., Pauldrach, A. & Puls, J. 1987, *Astron. Astrophys.*, 173, 293.

Kuhn, J.R., Smith, H.A. & Hawley, S.L. 1996, *Astrophys. J.*, 469, L93.

Kuijken, K. 1996, in *Barred Galaxies (ASP Conference Series Vol. 91)*, ed. R. Buta, D.A. Crocker, and B.G. Elmegreen, San Francisco: Astron. Soc. Pac., p. 504.

Kunkel, W.E. 1998, private communication.

Kunkel, W.E. & Demers, S. 1977, *Astrophys. J.*, 214, 21.

Kunkel, W.E., Demers, S., Irwin, M.S. & Albert, L. 1997, *Astrophys. J.*, 488, L129.

Kurt, C.M. & Dufour, R.J. 1998, *Rev. Mex. Astron. Astrophys.*, (Conference Series), 7, 202.

Kwitter, K.B. & Aller, L.H. 1981, *Mon. Not. R. Astron. Soc.*, 195, 939.

Lake, G. & Skillman, E.D. 1989, *Astron. J.*, 98, 1274.

Laney, C.D. & Stobie, R.S. 1994, *Mon. Not. R. Astron. Soc.*, 266, 441.

Lauer, T.R. et al. 1992, *Astron. J.*, 104, 552.

Lauer, T.R. et al. 1993, *Astron. J.*, 106, 1436.

Lauer, T.R. et al. 1996, *Astrophys. J.*, 471, L79.

Lauer, T.R., Faber, S.M., Ajhar, E.A., Grillmair, C.J. & Scowen, P.A. 1998, *Astron. J.*, 116, 2263.

Laustsen, S., Richter, W., van der Lans, J., West, R.M. & Wilson, R.N., 1977, *Astron. Astrophys.*, 54, 639.

Lavery, R.J. 1990, *IAU Circ. No. 5139*.

Lavery, R.J. & Mighell, K.J. 1992, *Astron. J.*, 103, 81.

Lawrie, D.G. 1983, *Astrophys. J.*, 273, 562.

Layden, A.C. 1999, in *Post-Hipparcos Cosmic Candles*, ed. A. Heck and F. Caputo, Dordrecht: Kluwer, p. 37.

Layden, A.C. & Sarajedini, A. 1997, *Astrophys. J.*, 486, L107.

Layden, A., Smith, R.C. & Storm, J. 1994, *The Local Group: Comparative and Global Properties*, Garching: ESO.

le Coarer, E., Rosado, M., Georgelin, Y., Viale, A. & Goldes, G. 1993, *Astron. Astrophys.*, 280, 365.

Leavitt, H.S. 1907, *Harvard Ann.*, 60, 87.

Lee, M.G. 1993, *Astrophys. J.*, 408, 409.

Lee, M.G. 1995a, *J. Korean Astron. Soc.*, 28, 169.

Lee, M.G. 1995b, in *Stellar Populations (IAU Symposium No. 164)*, ed. P.C. van der Kruit and G. Gilmore, Dordrecht: Kluwer, p. 413.

Lee, M.G. 1995c, *Astron. J.*, 110, 1129.

Lee, M.G. 1995d, *Astron. J.*, 110, 1155.

Lee, M.G. 1996, *Astron. J.*, 112, 1483.

Lee, M.G. 1999, in *Stellar Content of the Local Group (IAU Symposium No. 192)*, ed. P. Whitelock and R.D. Cannon, San Francisco: Astron. Soc. Pac., p. 268.

Lee, M.G., Freedman, W.L. & Madore, B.F. 1993a, in *New Perspectives on Stellar Pulsations and Pulsating Variable Stars (IAU Colloquium No. 139)*, ed. J.M. Nemec and J.M. Matthews, Cambridge: Cambridge Univ. Press, p. 92.

Lee, M.G., Freedman, W.L. & Madore, B.F. 1993b, *Astrophys. J.*, 417, 553.

Lee, M.G., Freedman, W.L. & Madore, B.F. 1993c, *Astron. J.*, 106, 964.

Lee, M.G., Freeman, W., Mateo, M., Thompson, I., Roth, M. & Ruiz, M.-T. 1993d, *Astron. J.*, 106, 1420.

Lee, Y.-W. 1992, *Publ. Astron. Soc. Pac.*, 104, 798.

Lee, Y.-W., Demarque, P. & Zinn, R. 1994, *Astrophys. J.*, 423, 248.

Leggett, S.K., Ruiz, M.T. & Bergeron, P. 1998, *Astrophys. J.*, 497, 294.

Lehnert, M.D., Bell, R.A., Hesser, J.E. & Oke, J.B. 1992, *Astrophys. J.*, 395, 466.

Leon, S., Bergond, G. & Vallenari, A. 1999, *Astron. Astrophys.*, in press.

Lequeux, J. 1999, in *Stellar Content of the Local Group (IAU Symposium No. 192)*, ed. P. Whitelock and R.D. Cannon, San Francisco: Astron. Soc. Pac., p. 185.

Lerner, M.S., Sundin, M. & Thomasson, M. 1999, *Astron. Astrophys.*, in press.

Li, P.S. & Thronson, H.A. 1999, in *New Views of the Magellanic Clouds (IAU Symposium No. 190)*, ed. Y.-H. Chu, J.E. Hesser, and N. Suntzeff, San Francisco: Astron. Soc. Pac., in press.

Light, E.S., Danielson, R.E. & Schwarzschild, M. 1974, *Astrophys. J.*, 194, 257.

Liller, W. & Mayer, B. 1987, *Publ. Astron. Soc. Pac.*, 99, 606.

Lindblad, B. 1922, *Astrophys. J.*, 55, 85.

Lindblad, B. 1923, *Mon. Not. R. Astron. Soc.*, 83, 503.

Lindblad, B. 1927, *Mon. Not. R. Astron. Soc.*, 87, 553.

Lindblad, B. & Stenquist, E. 1934, *Stockholm Obs. Ann.*, 11, No. 12.

Lindsay, E.M. 1958, *Mon. Not. R. Astron. Soc.*, 118, 172.

Lisenfeld, U. & Ferrara, A. 1998, *Astrophys. J.*, 496, 145.

Lloyd-Evans, T. 1992, in *Variable Stars and Galaxies (ASP Conference Series Vol. 30)*, ed. B. Warner, San Francisco: *Astron. Soc. Pac.*, p. 169.

Lo, K.Y., Sargent, W.L.W. & Young, K. 1993, *Astron. J.*, 106, 507.

Loeb, A. & Perna, R. 1998, *Astrophys. J.*, 503, L35.

Loewenstein, M., Hayashida, K., Tomeri, T. & Davis, D.S. 1998, *Astrophys. J.*, 497, 681.

Loinard, L. & Allen, R.J. 1998, *Astrophys. J.*, 499, 227.

Long, K.S. 1996, in *Supernovae and Supernova Remnants*, ed. R. McCray and Z. Wang, Cambridge: Cambridge University Press, p. 349.

Long, K.S., Blair, W.P., Kirshner, R.P. & Winkler, P.F. 1990, *Astrophys. J. Suppl.*, 72, 61.

Long, K.S., Helfand, D.J. & Grabelsky, D.A. 1981, *Astrophys. J.*, 248, 925.

Longmore, A.J., Hawarden, T.G., Webster, B.L., Goss, W.M. & Mebold, U. 1978, *Mon. Not. R. Astron. Soc.*, 183, 97p.

Lortet, M.-C. 1986, *Astron. Astrophys. Suppl.*, 64, 325.

Lozinskaya, T.A., Sil'chenko, O.K., Helford, D.J. & Goss, W.M. 1998, *Astron. J.*, 116, 2328.

Lu, L., Savage, B.D., Sembach, K.R., Wakker, B.P., Sargent, W.L.W. & Oosterloo, T.A. 1998, *Astron. J.*, 115, 162.

Luck, R.E., Moffet, T.J., Barnes, T.G., & Gieren, W.P. 1998, *Astron. J.*, 115, 605.

Lucke, P.B. 1974, *Astrophys. J. Suppl.*, 28, 73.

Lucke, P.B. & Hodge, P.W. 1970, *Astron. J.*, 75, 171.

Luks, T. & Rohlfs, K. 1992, *Astron. Astrophys.*, 263, 41.

Lumsden, S.L. & Puxley, P.J. 1995, *Mon. Not. R. Astron. Soc.*, 276, 723.

Lundmark, K. 1923, *Publ. Astron. Soc. Pac.*, 35, 95.

Luri, X., Gómez, A.E., Torra, J., Figueras, F. & Mennessier, M.O. 1998, *Astron. Astrophys.*, 335, L81.

Lynden-Bell, D. 1981, *Observatory*, 101, 111.

Lynden-Bell, D. & Lynden-Bell, R.M. 1995, *Mon. Not. R. Astron. Soc.*, 275, 429.

Lyngå, G. & Westerlund, B.E. 1963, *Mon. Not. R. Astron. Soc.*, 127, 31.

Madden, S.C., Poglitsch, A., Geis, N., Stacey, G.J. & Townes, C.H. 1997, *Astrophys. J.*, 483, 200.

Madore, B.F. 1997, private communication.

Madore, B.F. & Freedman, W.L. 1991, *Publ. Astron. Soc. Pac.*, 103, 933.

Madore, B.F. & Freedman, W.L. 1998, *Astrophys. J.*, 492, 110.

Madore, B.F., McAlary, C.W., McLaren, R.A., Welch, D.L., Neugebauer, G. & Matthews, K. 1985, *Astrophys. J.*, 294, 560.

Madore, B.F., van den Bergh, S. & Rogstad, D.H. 1974, *Astrophys. J.*, 191, 317.

Maeder, A. 1999, in *Stellar Content of the Local Group* (*IAU Symposium No. 192*), ed. P. Whitelock and R.D. Cannon, San Francisco: Astron. Soc. Pac., p. 291.

Magnier, E.A., Battinelli, P., Lewin, W.H.G., Haiman, Z., van Paradijs, J., Hasinger, G., Pietsch, W., Supper, R. & Trümpler, J. 1993, *Astron. Astrophys.*, 278, 36.

Magnier, E.A., Hodge, P., Battinelli, P., Lewin, W.H.G. & van Paradijs, J. 1997a, *Mon. Not. R. Astron. Soc.*, 292, 490.

Magnier, E.A., Prins, S., Augustijn, T., van Paradijs, J. & Lewin, W.H.G. 1997b, *Astron. Astrophys.*, 326, 442.

Magnier, E.A., Prins, S., Augusteijn, T., van Paradijs, J. & Lewin, W.H.G. 1997, *Astron. Astrophys.*, 326, 442.

Magorrian, J. & Tremaine, S. 1999, *Mon. Not. R. Astron. Soc.*, in press.

Mahadevan, R. 1998, *Nature*, 394, 651.

Majewski, S.R. 1992, *Astrophys. J. Suppl.*, 78, 87.

Majewski, S.R. 1993, *Galaxy Evolution: The Milky Way Perspective*, (*ASP Conference Series Vol. 49*), San Francisco: Astron. Soc. Pac.

Majewski, S.R., Munn, J.A. & Hawley, S.L. 1996, *Astrophys. J.*, 459, L73.

Maloney, P.R. & Bland-Hawthorn, J. 1999, *Astrophys. J.*, 522, L81.

Malumuth, E.M. & Heap, S.R. 1994, *Astron. J.*, 107, 1054.

Manchester, R.N., Staveley-Smith, L. & Kesteven, M.J. 1993, *Astrophys. J.*, 411, 756.

Maran, S. P., Gull, T.R., Stacker, T.P., Aller, L.H. & Keyes, C.D. 1984, *Astrophys. J.*, 280, 615.

Marconi, G., Buonanno, R., Castellani, M., Iannicola, G., Molaro, P., Pasquini, L. & Pulone, L. 1998, *Astron. Astrophys.*, 330, 453.

Marconi, G., Buonanno, R., Corsi, C.E. & Zinn, R. 1999, in *The Stellar Content of Local Group Galaxies* (*IAU Symposium No. 192*), ed. P. Whitelock and R. Cannon, San Francisco: Astron. Soc. Pac., p. 174.

Marconi, G., Tosi, M., Greggio, L. & Focardi, P. 1995, *Astron. J.*, 109, 173.

Marigo, P., Girardi, L. & Chiosi, C. 1996, *Astron. Astrophys.*, 316, L1.

Mark, H., Price, R., Rodriguez, R., Seward, F.D. & Swift, C.D. 1969, *Astrophys. J.*, 155, L143.

Marshall, F.E., Gotthelf, E.V., Zhang, W., Middleditch, J. & Wang, Q.D. 1998, *Astrophys. J.*, 499, L179.

Martin, A.H.M. & Downes, D. 1972, *Astrophys. Lett.*, 11, 219.

Martin, N., Prévot, L., Rebeirot, E. & Rousseau, J. 1976, *Astron. Astrophys.*, 51, 31.

Martínez-Delgado, D. & Aparicio, A. 1997, in *The Nature of Elliptical Galaxies* (*ASP Conference Series Vol. 116*), ed., M. Arnaboldi, G.S. Da Costa, and P. Saha, San Francisco: Astron. Soc. Pac., p. 304.

Martínez-Delgado, D. & Aparicio, A. 1998, *Astron. J.*, 115, 1462.

Martínez-Delgado, D., Gallart, C. & Aparicio, A. 1998, in *Galactic Halos* (*ASP Conf. Series Vol. 136*), San Francisco: Astron. Soc. Pac., p. 81.

Massey, P. 1998a, *Astrophys. J.*, 501, 153.

Massey, P. 1998b, in the Initial Mass Function (*ASP Conf. Series Vol. 142*), San Francisco: Astron. Soc. Pac., p. 17.

Massey, P. & Armandroff, T.E. 1995, *Astron. J.*, 109, 2470.

Massey, P., Armandroff, T.E. & Conti, P.S. 1986, *Astron. J.*, 92, 1303.

Massey, P., Armandroff, T.E., Pyke, R., Patel, K. & Wilson, C.D. 1995, *Astron. J.*, 110, 2715.

Massey, P. & Conti, P.S. 1983, *Astrophys. J.*, 273, 576.

Massey, P. & Hunter, D.A. 1998, *Astrophys. J.*, 493, 180.

Massey, P. & Johnson, O. 1998, *Astrophys. J.*, 505, 793.

Massey, P., Lang, C.C., DeGioia-Eastwood, K. & Garmany, C.D. 1995, *Astrophys. J.*, 438, 188.

Mateo, M. 1987, *Astrophys. J.*, 323, L41.

Mateo, M. 1997, in *The Nature of Elliptical Galaxies* (*ASP Conf. Series Vol. 116*), ed. M. Arnaboldi, G.S. Da Costa, and P. Saha, San Francisco: Astron. Soc. Pac., p. 259.

Mateo, M. 1998, *Annu. Rev. Astron. Astrophys.*, 36, 435.

Mateo, M. 1999, personal communication

Mateo, M., Fischer, P. & Krzeminski, W. 1995, *Astron. J.*, 110, 2166.

Mateo, M., Hodge, P.W. & Schommer, R.A. 1986, *Astrophys. J.*, 311, 113.

Mateo, M., Hurley-Keller, D. & Nemec, J. 1998, *Astrophys. J. Suppl.*, 115, 1856.

Mateo, M., Kubiak, M., Szymański, M., Kałużny, J., Krzemiński, W. & Udalski, A. 1995, *Astron. J.*, 110, 1141.

Mateo, M., Nemec, J., Irwin, M. & McMahon, R. 1991, *Astron. J.*, 101, 892.

Mateo, M., Olszewski, E.W. & Morrison, H.L. 1998, *Astrophys. J.*, 508, L55.

Mateo, M., Olszewski, E.W., Pryor, C., Welch, D.L. & Fischer, P. 1993, *Astron. J.*, 105, 510.

Mateo, M., Olszewski, E.W., Vogt, S.S. & Keane, M.J. 1998, *Astron. J.*, 116, 2315.

Mateo, M., Udalski, A., Szymański, M., Kaluzńy, J. & Kubiak, M. 1995, *Astron. J.*, 109, 588.

Mathewson, D.S. & Clark, J.N. 1972, *Astrophys. J.*, 178, L105.

Mathewson, D.S., Cleary, M.N. & Murray, J.D. 1974, in *The Formation and Dynamics of Galaxies* (*IAU Symposium No. 58*), ed. J.R. Shakeshift, Dordrecht: Reidel, p. 367.

Mathewson, D.S. & Ford, V.L. 1984, in *Structure and Evolution of the Magellanic Clouds* (*IAU Symposium No. 108*), ed. S. van den Bergh and K.S. de Boer, Dordrecht: Reidel, p. 125.

Mathewson, D.S., Ford, V.L. & Visvanathan, N. 1986, *Astrophys. J.*, 301, 664.

Mathewson, D.S., Ford, V.L. & Visvanathan, N. 1988, *Astrophys. J.*, 333, 617.

Mathewson, D.S. & Healey, J.R. 1964 in *The Galaxy and the Magellanic Clouds* (*IAU Symposium No. 20*), ed. F.J. Kerr and A.W. Rodgers, Canberra: Australian Acad. Science, p. 283.

Mathis, J.S. 1990, *Annu. Rev. Astron. Astrophys.*, 28, 37.

Matthews, L.D., Gallagher, J.S. & Littleton, J.E. 1995, *Astron. J.*, 110, 581.

Mayall, N.U. 1935, *Publ. Astron. Soc. Pac.*, 47, 317.

Mayall, N.U. & Eggen, O.J. 1953, *Publ. Astron. Soc. Pac.*, 65, 24.

McCarthy, J.K., Lennon, D.J., Venn, K.A., Kudritzki, R.-P., Puls, J. & Najarro, F. 1995, *Astrophys. J.*, 455, L135.

McCausland, R.J.H., Conlon, E.S., Dufton, P.L., Fitzsimmons, A., Irwin M.J. & Keenan, F.P. 1993, *Astrophys. J.*, 411, 650.

McClure, R.D. 1969, *Astron. J.*, 74, 50.

McClure, R.D. & Racine, R. 1969, *Astron. J.*, 74, 1000.

McClure, R.D. & van den Bergh, S. 1968, *Astron. J.*, 73, 313.

McGee, R.X. & Milton, J.A. 1964, in *The Galaxy and the Magellanic Clouds* (*IAU Symposium No. 20*), ed. F.J. Kerr and A.W. Rodgers, Canberra: Australian Acad. Sci., p. 289.

McGee, R.X. & Milton, J.A. 1966, *Australian J. Phys.*, 19, 343.

McKibben, Nail, V. & Shapley, H. 1953, *Proc. Nat. Acad. Sci. USA*, 39, 358.

McLaughlin, D.B. 1945, *Astron. J.*, 51, 136.

McLean, I.S. & Liu, T. 1996, *Astrophys. J.*, 456, 499.

McNamara, D.H. 1997a, *Publ. Astron. Soc. Pac.*, 109, 857.

McNamara, D.H. 1997b, *Publ. Astron. Soc. Pac.*, 109, 1221.

McNamara, D.H. & Feltz, K.A. 1980, *Publ. Astron. Soc. Pac.*, 92, 587.

McWilliam, A. & Rich, R.M. 1994, *Astrophys. J. Suppl.*, 91, 749.

Meaburn, J., Clayton, C.A. & Whitehead, M.J. 1988, *Mon. Not. R. Astron. Soc.*, 235, 479.

Mebold, U. 1991, in *The Magellanic Clouds* (*IAU Symposium No. 148*), ed. R. Haynes and D. Milne, Dordrecht: Kluwer, p. 463.

Melia, F., Yusef-Zadeh, F. & Fatuzzo, M. 1998, *Astrophys. J.*, 508, 676.

Melotte, P.J. 1926, *Mon. Not. R. Astron. Soc.*, 86, 636.

Merritt, D. 1998, *Comments on Modern Physics* 1, Part E, 39.

Merritt, D. & Sellwood, J.A. 1994, *Astrophys. J.*, 425, 551.

Metzger, P. 1971 in *Dark Nebulae, Globules and Protostars*, ed. B.T. Lynds, Tucson: Univ. Arizona Press, p. 88.

Meuseinger, H., Reimann, H.-G. & Stecklum, B. 1991, *Astron. Astrophys.*, 245, 57.

Meylan, G. & Djorgovski, S. 1987, *Astrophys. J.*, 322, L91.

Michard, R. & Nieto, J.-L. 1991, *Astron. Astrophys.*, 243, L17.

Mighell, K.J. 1997, *Astron. J.*, 114, 1458.

Mighell, K.J. & Rich, R.M. 1995, *Astron. J.*, 110, 1649.

Mighell, K.J. & Rich, R.M. 1996a, *Astron. J.*, 111, 777.

Mighell, K.J. & Rich, R.M. 1996b, in *Formation of the Galactic Halo* (*ASP Conference Series Vol. 92*), ed. H. Morrison and A. Sarajedini, San Francisco: Astron. Soc. Pac., p. 528.

Mighell, K.J., Sarajedini, A. & French, R.S. 1998, *Astron. J.*, 116, 2395.

Mihalas, D. & Binney, J. 1981, *Galactic Astronomy* (2nd Edition), San Francisco: W.H. Freeman and Co.

Miller, B.W., Lotz, J., M. Ferguson, H.C., Stiavelli, M. & Whitmore, B.C. 1998, *Astrophys. J.*, 508, L133.

Minkowski, R. 1965, in *Galactic Structure*, ed. A. Blaauw and M. Schmidt, Chicago: Univ. Chicago Press, p. 321.

Minniti, D. 1995, *Astron. J.*, 109, 1663.

Minniti, D. 1996, *Astrophys. J.*, 459, 599.

Minniti, D., Olszewski, E.W. & Rieke, M. 1993, *Astrophys. J.*, 410, L97.

Minniti, D. & Zijlstra, A.A. 1996, *Astrophys. J.*, 467, L13.

Minniti, D. & Zijlstra, A.A. 1997, *Astron. J.*, 114, 147.

Minniti, D., Zijlstra, A.A. & Alonso, M.V. 1999, *Astron. J.*, 117, 881.

Misselt, K.A., Clayton, G.C. & Gordon, K.D. 1998, *Astrophys. J.*, 515, 128.

Mochejska, B.J., Kaluzny, J., Knockenberger, M., Sassselov, D.D. & Stanek, K.Z. 1998, astro-ph/9806222.

Mochizuki, K., Nakagawa, T., Doi, Y., Yui, Y.Y., Okuda, H., Shibai, H., Yui, M., Nishimura, T. & Low, F.J. 1994, *Astrophys. J.*, 430, L37.

Moffatt, A.F.J., Drissen, L. & Shara, M.M. 1994, *Astrophys. J.*, 436, 183.

Moffat, A.F.J. et al. 1998, *Astrophys. J.*, 497, 896.

Moles, M., Aparicio, A. & Magegosa, J. 1990, *Astron. Astrophys.*, 228, 310.

Montegriffo, P., Bellazzini, M., Ferraro, F.R., Martins, D., Sarajedini, A. & Fusi Pecci, F. 1998, *Mon. Not. R. Astron. Soc.*, 294, 315.

Monteverdi, M.I., Herrero, A., Lennon, D.J. & Kudritzki, R.-P. 1997, in *Wide-Field Spectroscopy*, ed. E. Kontizas et al., Dordrecht: Kluwer, p. 207.

Morgan, D.H. 1994, *Astron. Astrophys. Suppl.*, 103, 235.

Morgan, D.H. & Hatzidimitriou, D. 1995, *Astron. Astrophys. Suppl.*, 113, 539.

Morgan, W.W. 1959, *Astron. J.*, 64, 432.

Morgan, W.W. & Mayall, N.U. 1957, *Publ. Astron. Soc. Pac.*, 69, 291.

Morgan, W.W. & Osterbrock, D.E. 1969, *Astron. J.*, 74, 515.

Morgan, W.W., Whitford, A.E. & Code, A.D. 1953, *Astrophys. J.*, 118, 318.

Morris, M. 1989, *The Center of the Galaxy*, Dordrecht: Kluwer.

Morris, M. 1993, *Astrophys. J.*, 408, 496.

Morris, M. & Serabyn, E. 1996, *Annu. Rev. Astron. Astrophys.*, 34, 645.

Morris, P.W., Reid, I.N., Griffiths, W.K. & Penny, A.J., 1994, *Mon. Not. R. Astron. Soc.*, 271, 852.

Morris, S.L. & van den Bergh, S. 1994, *Astrophys. J.*, 427, 696.

Morse, J.A., Blair, W.P. & Raymond, J.C. 1998, Rev. Mex. *Astron. Astrophys.*, (Conference Series), 7, 21.

Moss, D., Shukurov, A., Sokoloff, D.D., Berkhuijsen, E.M. & Beck, R. 1998, *Astron. Astrophys.*, 335, 500.

Mould, J. 1987, *Publ. Astron. Soc. Pac.*, 99, 1127.

Mould, J. 1997, *Publ. Astron. Soc. Pac.*, 109, 125.

Mould, J. & Aaronson, M. 1983, *Astrophys. J.*, 273, 530.

Mould, J.R., Bothun, G.D., Hall, P.J., Staveley-Smith, L. & Wright, A.E. 1990b, *Astrophys. J.*, 362, L55.

Mould, J., Graham, J.R., Matthews, K., Neugebauer, G. & Elias, J. 1990a, *Astrophys. J.*, 349, 503.

Mould, J., Kristian, J. & Da Costa, G.S. 1983, *Astrophys. J.*, 270, 471.

Mould, J., Kristian, J. & Da Costa, G.S. 1984, *Astrophys. J.*, 278, 575.

Mould, J.R. & Kristian, J. 1986, *Astrophys. J.*, 305, 591.

Mould, J. & Kristian, J. 1990, *Astrophys. J.*, 354, 438.

Murai, T. & Fujimoto, M. 1980, *Publ. Astron. Soc. Jpn.*, 32, 58.

Musella, I., Piotto, G. & Capaccioli, M. 1998, *Astron. J.*, 114, 976.

Nagataki, S., Hashimoto, M. & Sato, K. 1998, *Publ. Astron. Soc. Jpn.*, 50, 75.

Narayanan, V.K. & Gould, A. 1999, astro-ph/9810328.

Neininger, N., Guélin, M., Ungerechts, H., Lucas, R. & Wielebinski, R. 1998, *Nature*, 395, 871.

Nevalainen, J. & Roos, M. 1998, *Astron. Astrophys.*, 339, 7.

Newton, K. 1980, *Mon. Not. R. Astron. Soc.*, 190, 689.

Newton, K. & Emerson, D.T. 1977, *Mon. Not. R. Astron. Soc.*, 181, 573.

Ng, Y.K. 1997, *Astron. Astrophys.*, 328, 211.

Ng, Y.K. 1998, *Astron. Astrophys.*, 338, 435.

Nieto, J.-L. & Aurière, M. 1982, *Astron. Astrophys.*, 108, 334.

Nilson, P. 1974, Uppsala *Astron. Obs. Annu.* Rep. No. 5, p. 1.

Nolthenius, R. & Ford, H.C. 1986, *Astrophys. J.*, 305, 600.

Nolthenius, R. & Ford, H.C. 1987, *Astrophys. J.*, 317, 62.

Nomoto, K., Hashimoto, M., Tsujimoto, T., Thielemann, F.-K., Kishimoto, N., Kubo, Y. & Nakasato, N. 1997, *Nuclear Phys.*, A616, C79.

Norris, J. & Bessell, M.S. 1978, *Astrophys. J.*, 225, L49.

Norris, J.E., Ryan, S.G. & Beers, T.C. 1997, *Astrophys. J.*, 489, L169.

O'Connell, R.W. 1983, *Astrophys. J.*, 267,80.

Oepik, E. 1922, *Astrophys. J.*, 55, 406.

Oestreicher, M.O., Gochermann, J. & Schmidt-Kaler, T. 1995, *Astron. Astrophys. Suppl.*, 112, 495.

Oestreicher, M.O. & Schmidt-Kaler, T. 1996, *Astron. Astrophys. Suppl.*, 117, 303.

Oey, M.S. 1996, *Astrophys. J.*, 465, 231.

Oey, M.S. 1999, in *New Views of the Magellanic Clouds (IAU Symposium No. 190)*, ed. Y.-H. Chu et al., San Francisco: Astron. Soc. Pac., in press.

Oey, M.S. & Massey, P. 1994, *Astrophys. J.*, 425, 635

Olling, R.P. & Merrifield, M.R. 1998, *Mon. Not. R. Astron. Soc.*, 297, 943.

Olsen, K.A.G. 1999, *Astron. J.*, 117, 2244.

Olsen, K.A.G., Hodge, P.W., Mateo, M., Olszewski, E.W., Schommer, R.A., Suntzeff, N.B. & Walker, A.R. 1998, *Mon. Not. R. Astron. Soc.*, 300, 665.

Olszewski, E.W. 1997, astro-ph/9712280.

Olszewski, E.W. & Aaronson, M. 1985, *Astron. J.*, 90, 2221.

Olszewski, E.W., Aaronson, M. & Hill, J.M. 1995, *Astron. J.*, 110, 2120.

Olszewski, E.W., Pryor, C. & Armandroff, T.E. 1996, *Astron. J.*, 111, 750.

Olszewski, E.W., Suntzeff, N.B. & Mateo, M. 1996, *Annu. Rev. Astron. Astrophys.*, 34, 511.

Oort, J.H. 1927, *Bull. Astron. Inst. Netherlands*, 3, 275.

Oort, J.H. 1928, *Bull. Astron. Inst. Netherlands*, 4, 269.

Oort, J.H. 1940, *Astrophys. J.*, 91, 273.

Oort, J.H. 1966, *Bull. Astron. Inst. Netherlands*, 18, 421.

Oort, J.H. 1970, *Astron. Astrophys.*, 7, 381.

Oosterhoff, P.T. 1939, *Observatory*, 62, 104.

Orio, M., Greiner, J. & Della Valle, M. 1998, *Bull. Am. Astron. Soc.*, 30, 1329.

Ortolani, S. & Gratton, R.G. 1988, *Publ. Astron. Soc. Pac.*, 100, 1405.

Ortolani, S., Renzini, A., Gilmozzi, R., Marconi, G., Barbuy, B., Bica, E. & Rich, R.M. 1995, *Nature*, 377, 701.

Ostriker, J.P., Peebles, P.J.E. & Yahil, A. 1974, *Astrophys. J.*, 193, L1.

Oudmaijer, R.D., Groenwegen, M.A.T. & Schrijver, H. 1998, *Mon. Not. R. Astron. Soc.*, 294, L41.

Paczyński, B. 1998, astro-ph/9807173.

Paczyński, B. & Stanek, K.Z. 1998, *Astrophys. J.*, 494, L219.

Pak, S., Jaffe, D.J., van Dishoek, E.F., Johansson, L.E.B. & Booth, R.S. 1998, *Astrophys. J.*, 498, 735.

Palanque-Delabrouille, N. et al. 1998, *Astron. Astrophys.*, 337, 1.

Panagia, N. 1998, in *Views on Distance Indicators*, ed. F. Caputo, Mem. Soc. Astron. Italiana, 69, 225.

Parker, J.W., Clayton, G.C., Winge, C. & Conti, P. 1993, *Astrophys. J.*, 409, 770.

Pascarelle, S.M., Windhorst, R.A. & Keel, W.C. 1998, *Astron. J.*, 116, 2659.

Pascarelle, S.M., Windhorst, R.A., Keel, W.C. & Odewahn, S.C. 1996, *Nature*, 383, 45.

Patel, K. & Wilson, C.D. 1995, *Astrophys. J.*, 453, 162.

Patterson, F.S. 1940, *Harvard Bull. No. 914*, 9.

Payne-Gaposchkin, C. 1957, *The Galactic Novae*, Amsterdam: North Holland Publ. Co., p. 41.

Payne-Gaposchkin, C. 1971a, *Smithsonian Contr. to Astrophys.* No.13.

Payne-Gaposchkin, C. 1971b, in *The Magellanic Clouds*, ed. A.B. Muller, New York: Springer-Verlag, p. 34.

Payne-Gaposchkin, C. 1973, in *Age des Etoiles (IAU Colloquium No. 17)*, ed. G. Cayrel de Stobel and A.M. Delplacc, Paris: Observatoire de Paris-Meudon, p. III-1.

Pease, F.G. 1918, *Proc. Nat. Acad. Sci. USA.*, 4, 21.

Pease, F.G. 1928, *Publ. Astron. Soc. Pac.*, 40, 342.

Peebles, P.J.E. 1994, Astrophys. J., 429, 43.

Peimbert, M. 1990, *Rev. Mex. Astron. Astrophys.*, 20, 119.

Peimbert, M., Bohigas, J. & Torres-Peimbert, S. 1988, *Rev. Mex. Astron. Astrophys.*, 16, 45.

Peimbert, M. & Spinrad, H. 1970, *Astron. Astrophys.*, 7, 311.

Peletier, R.F. 1993, *Astron. Astrophys.*, 271, 51.

Penprase, B.E., Lauer, J., Aufrecht, J. & Walsh, B.Y. 1998, *Astrophys. J.*, 492, 617.

Peres, G., Reale, F., Collura, A. & Fabbiano, G. 1989, *Astrophys. J.*, 336, 140.

Peterson, R.C. & Caldwell, N. 1993, *Astron. J.*, 105, 1411.

Peterson, R.C. & Green, E.M. 1998, *Astrophys. J.*, 502, L39.

Petit, H., Sivan, J.-P. & Karachentsev, I.D. 1988, *Astron. Astrophys. Suppl.*, 74, 475.

Petitpas, G.R. & Wilson, C.D. 1998, *Astrophys. J.*, 496, 226.

Pfenniger, D. 1998, in *Galactic Halos*, (*ASP Conference Series No. 136*), ed. D. Zaritsky, San Francisco: Astron. Soc. Pac.

Pfenniger, D. & Norman, C. 1990, *Astrophys. J.*, 363, 391.

Phelps, R.L. 1997, *Astrophys. J.*, 483, 826.

Phelps, R.L., Janes, K.A. & Montgomery, K.A. 1994, *Astron. J.*, 107, 1079.

Phillips, M.M. & Suntzeff, N.B. 1999, *SN 1987A: Ten Years After (ASP Conference Series)*, San Francisco: Astron. Soc. Pac., in press.

Pietrzyński, G., Udalski, A., Kubiak, M., Szymański, M., Woźniak, P. & Żebruń, K. 1998, astro-ph/9806321.

Pinsonneault, M.H., Stauffer, J., Sonderblom, D.R., King, J.R. & Hanson, R.B. 1998, *Astrophys. J.*, 504, 170.

Piotto, G., Capaccioli, M. & Pellegrini, C. 1994, *Astron. Astrophys.* 287, 371.

Pollard, K.R., Cottrell, P.L. & Lawson, W.A. 1994, *Mon. Not. R. Astron. Soc.*, 268, 544.

Pooley, G.G. 1969, *Mon. Not. R. Astron. Soc.*, 144, 101.

Popowski, P. & Gould, A. 1998, *Astrophys. J.*, 506, 259.

Poretti, E. 1999, *Astron. Astrophys.*, 343, 385.

Preston, G.W., Beers, T.C. & Shectman, S.A. 1994, *Astron. J.*, 108, 539.

Preston, G.W., Shectman, S.A. & Beers, T.C. 1991, *Astrophys. J.*, 375, 121.

Primini, F.A., Forman, W. & Jones, C. 1993, *Astrophys. J.*, 410, 615.

Prinja, R.K. & Crowther, P.A. 1998, *Mon. Not. R. Astron. Soc.*, 300, 828.

Pritchet, C.J. 1988, in *The Extragalactic Distance Scale (ASP Conference Series, Vol. 4)*, ed. S. van den Bergh and C.J. Pritchet, Provo: Astron. Soc. Pac., p. 59.

Pritchet, C. & van den Bergh, S. 1984, *Publ. Astron. Soc. Pac.*, 96, 804.

Pritchet, C. & van den Bergh, S. 1987a, *Astrophys. J.*, 316, 517.

Pritchet, C.J. & van den Bergh, S. 1987b, *Astrophys. J.*, 318, 507.

Pritchet, C. & van den Bergh, S. 1988, *Astrophys. J.*, 331, 135.

Pritchet, C.J. & van den Bergh, S. 1994, *Astron. J.*, 107, 1730.

Pritchet, C.J. & van den Bergh, S. 1999, *Astron. J.*, 118, 833.

Proust, K.M. & Couch, W.J. 1988, *Proc. Astron. Soc. Australia*, 7, 343.

Pryor, C. & Meylan G. 1993, in *Structure and Dynamics of Globular Clusters (ASP Conference Series Vol. 50)*, ed. S.G. Djorgovski and G. Meylan, San Francisco: Astron. Soc. Pac., p. 357.

Putman, M.E. et al. 1998, *Nature*, 394, 752.

Qiao, Q.Y., Qiu, Y.L., Hu, J.Y. & Li, D.W. 1997, *IAU Circular No. 6777*.

Queloz, D., Dubath, P. & Pasquini, L. 1995, *Astron. Astrophys.*, 300, 31.

Raboud, D., Grenon, M., Martinet, L., Fux, R. & Udry, S. 1998, *Astron. Astrophys.*, 335, L61.

Racine, R. 1991, *Astron. J.*, 101, 865.

Ratnatunga, K.U. & van den Bergh, S. 1989, *Astrophys. J.*, 343, 713.

Reagen, M.W. & Wilson, C.D. 1993, *Astron. J.*, 105, 499.

Rebeirot, E., Azzopardi, M. & Westerlund, B.E. 1993, *Astron. Astrophys. Suppl.*, 97, 603.

Reed, L.G., Harris, G.L.H. & Harris, W.E. 1994, *Astron. J.*, 107, 555.

Reid, I.N. 1997, *Astron. J.*, 114, 161.

Reid, I.N. et al. 1991, *Publ. Astron. Soc. Pac.*, 103, 661.

Reid, I.N. & Stugnell, P.R. 1986, *Mon. Not. R. Astron. Soc.*, 221, 887.

Reid, M.J. 1993, *Annu. Rev. Astron. Astrophys.*, 31, 345.

Renzini, A. 1998, *Astron. J.*, 115, 2459.

Renzini, A. & Greggio, L. 1990, in *Bulges of Galaxies (ESO Workshop No. 35)*, ed. B. Jarvis and D. Terndrup, Garching: ESO, p. 47.

Rich, R. 1998, in *Abundance Profiles: Diagnostic Tools for Galaxy History (ASP Conference Series No. 147)*, ed. D. Friedli et al., San Francisco: Astron. Soc. Pac., p. 36.

Rich, R.M. & Mighell, K.J. 1995, *Astrophys. J.*, 439, 145.

Rich, R.M., Mighell, K.J., Freedman, W.L. & Neill, J.D. 1996, *Astron. J.*, 111, 768.

Richer, H.B., Crabtree, D.R. & Pritchet, C.J. 1990, *Astrophys. J.*, 355, 448.

Richer, H.B. et al. 1996, *Astrophys. J.*, 463, 602.

Richer, M.G. & McCall, M.L. 1992, *Astron. J.*, 103, 54.

Richer, M.G. & McCall, M.L. 1995, *Astrophys. J.*, 445, 642.

Richer, M.G., McCall, M.L. & Stasińska, 1998, in *Abundance Profiles: Diagnostic Tools for Galaxy History*, *(ASP Conference Series Vol. 147)*, ed. D. Friedli et al., San Francisco: Astron. Soc. Pac., p. 254.

Richstone, D.O. & Shectman, S.A. 1980, *Astrophys. J.*, 235, 30.

Richter, P. et al. 1998, *Astron. Astrophys.*, 338, L9.

Ritchey, G.W. 1917, *Publ. Astron. Soc. Pac.*, 29, 210.

Roberts, I. 1887, *Mon. Not. R. Astron. Soc.*, 49, 65.

Roberts, M.S. 1962, *Astron. J.*, 67, 431.

Roberts, M.S. 1972, in *External Galaxies and Quasi-stellar Objects* (*IAU Symposium No. 44*), ed. D.S. Evans, Reidel: Dordrecht, p. 12.

Roberts, M.S. & Rots, A.H. 1973, *Astron. Astrophys.*, 26, 483.

Roberts, M.S. & Whitehurst, R.N. 1975, *Astrophys. J.*, 201, 327.

Robin, A.C. 1997 in *The Impact of Large Scale Near-IR Sky Surveys*, ed. F. Garzón et al., Dordrecht: Kluwer, p. 57.

Robin, A.C., Crézé, M. & Mohan, V. 1992, *Astron. Astrophys.*, 265, 32.

Rodgers, A.W. & Roberts, W.H. 1994, *Astron. J.*, 107, 1737.

Rodrigues, C.V., Magalhães, A.M., Coyne, S.J. & Piirola, V. 1997, *Astrophys. J.*, 485, 618.

Rogstad, D.H., Lockhart, I.A. & Wright, M.C.H. 1974, *Astrophys. J.*, 193, 309.

Rogstad, D.H., Wright, M.C.H. & Lockhart, I.A. 1976, *Astrophys. J.*, 204, 703.

Rohlfs, K., Kreitschmann, J., Siegman, B.C. & Feitzinger, J.V. 1984, *Astron. Astrophys.*, 137, 343.

Rolleston, W.R.J. & McKenna, F.C. 1999, in *New Views of the Magellanic Clouds*, (*IAU Symposium No. 190*), ed. Y.-H. Chu et al., San Francisco: Astron. Soc. Pac., in press.

Romaniello, M., Panagia, N. & Scuderi, S. 1998, in *The Large Magellanic Cloud and Other Dwarf Galaxies*, ed. T. Richtler and J.M. Braun, Aachen: Shaker Verlag, p. 197.

Rosado, M., Le Coarer, E. & Georgelin, Y.P. 1994, *Astron. Astrophys.*, 286, 231.

Rosenberg, A., Saviane, I., Piotto, G., Aparicio, A. & Zaggia, S.R. 1998a, *Astron. J.*, 115, 648.

Rosenberg, A., Saviane, I., Piotto, G. & Held, E.V. 1998b, *Astron. Astrophys.*, 339, 61.

Rosino, L. 1964, *Ann. Astrophys.*, 27, 498.

Rosino, L. 1973, *Astron. Astrophys. Suppl.*, 9, 347.

Rosino, L., Capaccioli, M., D'Onofrio, M. & Della Valle, M. 1989, *Astron. J.*, 97, 83.

Rosse, W. Parson, Earl of, 1850, *Phil. Trans. Roy. Soc.*, 140, 499.

Rots, A.H. 1975, *Astron. Astrophys.*, 45, 43.

Rubin, V. 1997, *Observatory*, 117, 129.

Rubin, V.C. & D'Odorico, S. 1969, *Astron. Astrophys.*, 2, 484.

Rubin, V.C. & Ford, W.K. 1970, *Astrophys. J.*, 159, 379.

Rubin, V.C. & Ford, W.K. 1971, *Astrophys. J.*, 170, 25.

Rubin, V.C. & Ford, W.K. 1986, *Astrophys. J.*, 305, L35.

Rubin, V.C., Kumar, C.K. & Ford, W.K. 1972, *Astrophys. J.*, 177, 31.

Rubio, M., Barbá, R.H., Walborn, N.R., Probst, R.G., García, J. & Roth, M.R. 1998, *Astron. J.*, 116, 1708.

Ruiz, M.T., Bergeron, P., Leggett, S.K. & Anguita, C. 1995, *Astrophys. J.*, 455, L159.

Russell, H.N. 1954, *Proc. Nat. Acad. Sci.*, 40, 549.

Russell, S.C. & Dopita, M.A. 1992, *Astrophys. J.*, 384, 508.

Sabalisck, G., Tenorio-Tagle, G., Castañeda, H.O., & Muñoz-Tuñón, C. 1995, *Astrophys. J.*, 444, 200.

Sadler, E.M., Rich, R.M. & Terndrup, D.M. 1996, *Astron. J.*, 112, 171.

Sagar, R. & Panday, A.K. 1989, *Astron. Astrophys. Suppl.*, 79, 407.

Sage, L.J., Welch, G.A. & Mitchell, G.F. 1998, *Astrophys. J.*, 507, 726.

Saha, A., Freedman, W.L., Hoessel, J.G. & Mossman, A.E. 1992, *Astron. J.*, 104, 1072.

Saha, A. & Hoessel, J.G. 1990, *Astron. J.*, 99, 97.

Saha, A., Hoessel, J.G. & Krist, J. 1992, *Astron. J.*, 103, 84.

Saha, A., Hoessel, J.G., Krist, J. & Danielson, G.E. 1996, *Astron. J.*, 111, 197.

Saha, A., Hoessel, J.G. & Mossman, A.E. 1990, *Astron. J.*, 100, 108.

Saha, A., Monet, D.G. & Seitzer, P. 1986, *Astron. J.*, 92, 302.

Sakai, S., Madore, B.F. & Freedman, W.L. 1996, *Astrophys. J.*, 461, 713.

Sakai, S., Madore, B.F. & Freedman, W.L. 1997, *Astrophys. J.*, 480, 589.

Sakai, S., Madore, B.F. & Freedman, W.L. 1999, *Astrophys. J.*, 511, 671.

Salaris, M. & Cassisi, S. 1998, *Mon. Not. R. Astron. Soc.*, 289, 166.

Salpeter, E.E. 1955, *Astrophys. J.*, 121, 161.

Sandage, A. 1961, *The Hubble Atlas of Galaxies*, Washington: Carnegie Institution of Washington.

Sandage, A. 1963a, *Astrophys. J.*, 138, 863.

Sandage, A. 1963b, in *Problems of Extra-Galactic Research*, ed. G.C. McVittie, New York: Macmillan, p. 359.

Sandage, A. 1971, *Astrophys. J.*, 166, 13.

Sandage, A. 1972, *Astrophys. J.*, 176, 21.

Sandage, A. 1983, *Astron. J.*, 88, 1108.

Sandage, A. 1986b, *Astron. Astrophys.*, 161, 89.

Sandage, A. 1986c, *Astron. J.*, 91, 496.

Sandage, A.R., Becklin, E.E. & Neugebauer, G. 1969, *Astrophys. J.*, 157, 55.

Sandage, A. & Bedke, J. 1994, *The Carnegie Atlas of Galaxies*, Washington, DC: Carnegie Institution.

Sandage, A. & Carlson, G. 1983, *Astrophys. J.*, 267, L29.

Sandage, A. & Carlson, G. 1985a, *Astron. J.*, 90, 1019.

Sandage, A. & Carlson, G. 1985b, *Astron. J.*, 90, 1464.

Sandage, A. & Carlson, G. 1988, *Astron. J.*, 96, 1599.

Sandage, A., Freeman, K.C., & Stokes, N.R. 1970, *Astrophys. J.*, 160, 831.

Sandage, A. & Katem, B. 1976, *Astron. J.*, 81, 743.

Sandage, A. & Tammann, G.A. 1968, *Astrophys. J.*, 151, 531.

Sandage, A. & Tammann, G.A. 1981, *A Revised Shapley–Ames Catalog of Bright Galaxies*, Washington: Carnegie Institution.

Santolamazza, P. 1998, *Mem. Soc. Astron. Italiana*, 69, 307.

Sarajedini, A. 1998a, Colloquium given in Victoria, Canada, March 17, 1998.

Sarajedini, A. 1997, *Astron. J.*, 114, 2505.

Sarajedini, A. 1998b, *Astron. J.*, 116, 738.

Sarajedini, A., Claver, C.F. & Ostheimer, J.C. 1997, *Astron. J.*, 114, 2505.

Sarajedini, A., Geisler, D., Harding, P. & Schommer, R. 1998, *Astrophys. J.*, 508, L37.

Sarajedini, A. & Layden, A.C. 1995, *Astron. J.*, 109, 1086.

Sargent, W.L.W., Kowal, C., Hartwick, F.D.A. & van den Bergh, S. 1977, *Astron. J.*, 82, 947.

Sasselov, D.D. et al. 1997, *Astron. Astrophys.*, 324, 471.

Scalo, J. 1998, in *The Stellar Initial Mass Function* (*Proceedings of the 38th Herstmonceaux Conference*), ed. G. Gilmore, I. Parry, and S. Ryan, p. 201.

Saviane, I., Held, E.V. & Piotto, G. 1996, *Astron. Astrophys.*, 315, 40.

Sawa, T. 1999, in *New Views of the Magellanic Clouds* (*IAU Symposium No. 190*), ed. Y.-H. Chu, et al., San Francisco: Astron. Soc. Pac., in press.

Scalo, J. 1998, in *The Stellar Initial Mass Function* (*ASP Conference Series Vol. 142*), ed. G. Gilmore and D. Howell, San Francisco: Astron. Soc. Pac., p. 201.

Schechter, P.L. 1976, *Astrophys. J.*, 203, 297.

Scheiner, J. 1899, *Astron. Nachr.*, 148, 326.

Schild, H., Smith, L.J. & Willis, A.J. 1990, *Astron. Astrophys.*, 237, 169.

Schlegel, D.J. 1998, private communication.

Schlegel, D.J., Finkbein, D.P. & Davis, M. 1998, *Astrophys. J.*, 500, 525.

Schmidt, A.A., Bica, E. & Alloin, D. 1990, *Mon. Not. R. Astron. Soc.*, 243, 620.

Schmidt, M. 1954, *Bull. Astron. Inst. Netherlands.*, 13, 247.

Schmidt-Kaler, T. 1967, *Astron. J.*, 72, 526.

Schmidt-Kaler, T. & Oestreicher, M.O. 1998, *Astron. Nachr.*, 319, 375.

Schmidtke, P.C., Cowley, A.P., Crane, J.D., Taylor, V.A. & McGrath, T.K. 1999, astro-ph/9812217.

Schmidtke, P.C., Cowley, A.P., Frattare, L.M., McGrath, T.K., Hutchings, J.B. & Crampton, D. 1994, *Publ. Astron. Soc. Pac.*, 106, 843.

Schneider, S.E., Spitzak, J.G. & Rosenberg, J.L. 1998, *Astrophys. J.*, 507, L9.

Scholtz, R.-D. & Irwin, M.J. 1994, in *Astronomy from Wide-Field Imaging* (*IAU Symposium No. 161*), ed. H.T. MacGillvray et al., Dordrecht: Kluwer, p. 535.

Schommer, R.A., Christian, C.A., Caldwell, N., Bothun, G.D. & Huchra, J. 1991, *Astron. J.*, 101, 873.

Schommer, R.A., Olszewski, E.W., Suntzeff, N.B. & Harris, H.C. 1992, *Astron. J.*, 103, 447.

Schuster, H.-E. & West, R.M. 1976, *Astron. Astrophys.*, 49, 129.

Schwarzschild, M. 1954, *Astron. J.*, 59, 273.

Schweitzer, A.E., Cudworth, K.M., Majewski, S.R. & Suntzeff, N.B. 1995, *Astron. J.*, 110, 2747.

Schweitzer, A.E., Cudworth, K.M., Majewski, S.R. 1997, in *Proper Motions and Galactic Astronomy* (*ASP Conference Series Vol. 127*), ed. R.M. Humphreys, San Francisco: Astron. Soc. Pac., p. 103.

Scott, J.E., Friel, E.D. & Janes, K.A. 1995, *Astron. J.*, 109, 1706.

Scowen, P.A. et al. 1998, *Astron. J.*, 116, 163.

Searle, L. & Zinn, R. 1978, *Astrophys. J.*, 225, 357.

Secker, J. 1992, *Astron. J.*, 104, 1472.

Sekiguchi, M. & Fukugita, M. 1998, *Observatory*, 118, 73.

Serabyn, E., Schupe, D. & Figer, D.F. 1998, *Nature*, 394, 448.

Sérsic, J.L. & Cerruti, M.A. 1979, *Observatory*, 99, 150.

Seward, F.D., Harnden, F.R. & Helfand, D.J. 1984, *Astrophys. J.*, 287, L19.

Shafter, A.W. 1997, *Astrophys. J.*, 487, 226.

Shapley, H. 1918a, *Astrophys. J.*, 48, 154.

Shapley, H. 1918b, *Publ. Astron. Soc. Pac.*, 30, 42.

Shapley, H. 1938, *Harvard Bull. 908*, 1.

Shapley, H. 1939, *Proc. Nat. Acad. Sci. USA*, 25, 565.

Shapley, H. 1940, *Harvard Bull. 914*, 8.

Shapley, H. 1943, *Galaxies*, Philadelphia: Blakiston.

Shapley, H. 1956, *Am. Scientist*, 44, 73.

Shara, M.M., Fall, S.M., Rich, R.M. & Zurek, D. 1998, *Astrophys. J.*, 508, 570.

Sharina, M.E., Karachentsev, I.D. & Tikhonov, N.A. 1997, *Astron. Lett.*, 23, 373.

Sharov, A.S. 1982, *Tumannost Andromedy (The Andromeda Nebula)*, Moscow: Nauka.

Sharov, A.S. 1990, *Astron. Zh.*, 67, 723.

Sharov, A.S. 1993, *Astron. Lett.*, 19, 7.

Sharov, A.S., Alksnis, A., Nedialkov, P.L., Shokin, Y.A., Kurtev, R.G. & Ivanov, V.D. 1998, *Astron. Lett.*, 24, 445.

Sher, D. 1965, *Mon. Not. R. Astron. Soc.*, 129, 237.

Shetrone, M.D., Bolte, M. & Stetson, P.B. 1998, *Astron. J.*, 115, 1988.

Shetrone, M.D., Briley, M. & Brewer, J.P. 1998, *Astron. Astrophys.*, 335, 919.

Shostak, G.S. & Skillman, E.D. 1989, *Astron. Astrophys.*, 214, 33.

Silbermann, N.A., Harris, H.C. & Smith, H.A. 1996, in *Formation of the Galactic Halo (ASP Conference Series No. 92)*, ed. H. Morrison and A. Sarajedini, San Francisco: Astron. Soc. Pac., p. 536.

Sil'chenko, O.K., Burenkov, A.N. & Vlasyuk, V.V. 1998, *Astron. Astrophys.*, 337, 349.

Simpson, J.A. & Conell, J.J. 1998, *Astrophys. J.*, 497, L85.

Skillman, E.D., Bowmans, D.J. & Kobulnicky, H.A. 1997, *Astrophys. J.*, 474, 205.

Skillman, E.D., Kennicutt, R.C. & Hodge, P.W. 1989, *Astrophys. J.*, 347, 875.

Skillman, E.D., Terlevich, R. & Melnick, J. 1989, *Mon. Not. R. Astron. Soc.*, 240, 563.

Skillman, E.D., Terlevich, R., Teuben, P.J. & van Woerden, H. 1988, *Astron. Astrophys.*, 198, 33.

Sluis, A.P.N. & Arnold, R.A. 1998, *Mon. Not. R. Astron. Soc.*, 297, 732.

Smartt, S.J. & Rolleston, W.R.J. 1997, *Astrophys. J.*, 481, L47.

Smecker-Hane, T.A., Stetson, P.B., Hesser, J.E. & Lehnert, M.D. 1994, *Astron. J.*, 108, 507.

Smith, E.O., Neill, J.D., Mighell, K.J. & Rich, R.M. 1996, *Astron. J.*, 111, 1596.

Smith, E.O., Rich, R.M. & Neill, J.D. 1997, *Astron. J.*, 114, 1471.

Smith, E.O., Rich, R.M. & Neill, J.D. 1998, *Astron. J.*, 115, 2369.

Smith, H.A., Kuhn, J.R. & Hawley, S.L. 1997, in *Proper Motions and Galactic Astronomy (ASP Conference Series Vol. 127)*, ed. R.M. Humphreys, San Francisco: Astron. Soc. Pac., p. 163.

Smith, H.A., Silbermann, N.A., Baird, S.R. & Graham, J.A. 1992, *Astron. J.*, 104, 1430.

Smith, J.A. 1998, *Publ. Astron. Soc. Pac.*, 110, 1251.

Smith, L.F. 1988, *Astron. J.*, 327, 128.

Smith, L.J., Nota, A., Pasquali, A., Leitherer, C., Clampin, M. & Crowther, P.A. 1998, *Astrophys. J.*, 503, 278.

Smith, R.M., Driver, S.P. & Phillipps, S. 1997, *Mon. Not. R. Astron. Soc.*, 287, 415.

Smith, R.C., Kirshner, R.P., Blair, W.P., Long, K.S. & Winkler, P.F. 1993, *Astrophys. J.*, 407, 564.

Smith, S. 1936, *Astrophys. J.*, 83, 23.

Sneden, C., McWilliam, A., Preston, G.W. & Cowan, J.J. 1996, in *Formation of the Galactic Halo...Inside and Out (ASP Conference Series Vol. 92)*, ed. H. Morrison and A. Sarajedini, San Francisco: Astron. Soc. Pac., p. 387.

Sodemann, M. & Thomsen, B. 1998, *Astron. Astrophys. Suppl.*, 127, 327.

Soderblom, D.R., King, J.R., Hanson, R.B., Jones, B.F., Fischer, D., Stauffer, J.R. & Pinsonneault, M.H. 1998, *Astrophys. J.*, 504, 192.

Sofue, Y. 1994, *Astrophys. J.*, 423, 207.

Sommer-Larsen, J., Beers, T.C., Flynn, C., Wilhelm, R. & Christensen, P.R. 1997, *Astrophys. J.*, 481, 775.

Spinrad, H., Taylor, B.J. & van den Bergh, S. 1969, *Astron. J.*, 74, 525.

Spitzer, L. 1969, *Astrophys. J.*, 158, L139.

Sreekumar, P. et al. 1992, *Astrophys. J.*, 400, L67.

Sreekumar, P. et al. 1993, *Phys. Rev. Lett.*, 70, 127.

Sreekumar, P. et al. 1994, *Astrophys. J.*, 426, 105.

Stanek, K.Z. & Garnavich, P.M. 1998, *Astrophys. J.*, 503, L131.

Stanek, K.Z., Mateo, M., Udalski, M., Szymański, M., Kałużny, J., Kubiak, M. & Krzemiński, W. 1996, in *Unsolved Problems of the Milky Way (IAU Symposium No. 169)*, ed. L. Blitz and P. Teuben, Dordrecht: Reidel, p. 103.

Stanek, K.Z., Zaritsky, D. & Harris, J. 1998, *Astrophys. J.*, 500, L141.

Stanghellini, L., Blades, J.C., Osmer, S.J., Barlow, M.J. & Lin, X.-W. 1999, *Astrophys. J.*, 510, 687.

Starrfield, S., Schwarz, G.J., Truran, J.W. & Sparks, W.M. 1998, *Bull. Am. Astron. Soc.*, 30, 1400.

Stasińska, G., Richer, M.G. & McCall, M.L. 1998, *Astron. Astrophys.*, 336, 667.

Staveley-Smith, L., Sault, R.J., Hatzidimitriou, D., Kesteven, M.J. & McConnell, D. 1997, *Mon. Not. R. Astron. Soc.*, 289, 225.

Sterken, C., de Groot, M. & van Genderen, A.M. 1998, *Astron. Astrophys.*, 333, 565.

Stetson, P.B. 1980, *Astron. J.*, 85, 387.

Stetson, P.B. et al. 1999, *Astron. J.*, 117, 247.

Stetson, P.B., Hesser, J.E. & Smecker-Hane, T.A. 1998, *Publ. Astron. Soc. Pac.*, 110, 533.

Stetson, P.B., VandenBerg, D.A., Bolte, M., Hesser, J.E. & Smith, G.H. 1989, *Astron. J.*, 97, 1360.

Stetson, P.B. & West, M.J. 1994, *Publ. Astron. Soc. Pac.*, 106, 726.

Storm, J., Carney, B.W. & Fry, A.M. 1999, astro-ph/9811376.

Strobel, N.V., Hodge, P. & Kennicutt, R.C. 1991, *Astrophys. J.*, 383, 148.

Stryker, L.L. 1983, *Astrophys. J.*, 266, 82.

Suntzeff, N.B. 1992, in *The Stellar Populations of Galaxies (IAU Symposium No. 149)*, ed. B. Barbuy and A. Renzini, Dordrecht: Kluwer, p. 23.

Suntzeff, N.B., Kinman, T.D. & Kraft, R.P. 1991, *Astrophys. J.*, 367, 528.

Suntzeff, N.B., Mateo, M., Terndrup, D.M., Olszewski, E.W., Geisler, D. & Weller, W. 1993, *Astrophys. J.*, 418, 208.

Suntzeff, N.B., Walker, A.R., Smith, V.V., Kraft, R.P., Klemola, A. & Stetson, P.B. 1998, astro-ph/9809358.

Szeifert, T., Stahl, O., Wolf, B., Zickgraf, F.-J., Bouchet, P. & Klare, G. 1993, *Astron. Astrophys.*, 290, 508.

Takano, M., Mitsuda, K., Fukazawa, Y. & Nagase, F. 1994, *Astrophys. J.*, 436, L47.

Tammann, G.A. 1969, *Astron. Astrophys.*, 3, 308.

Tanvir, N.R. 1999, astro-ph/9812356.

Terndrup, D.M. 1988, *Astron. J.*, 96, 884.

Thackeray, A.D. 1973, *IAU Circ. No. 2584*.

Thackeray, A.D. & Wesselink, A.J. 1953, *Nature*, 171, 693.

Thronson, H.A., Hunter, D.A., Casey, S. & Harper, D.A. 1990, *Astrophys. J.*, 355, 94.

Thuan, T.X. & Martin, G.E. 1979, *Astrophys. J.*, 232, L11.

Tikhonov, N. 1996, *Astron. Nachr.*, 317, 175.

Tikhonov, N. 1999, in *Stellar Content of the Local Group (IAU Symposium No. 192)*, ed. P. Whitelock and R.D. Cannon, San Francisco: Astron. Soc. Pac., in press.

Tikhonov, N. & Makarova, L. 1996, *Astron. Nachr.* 317, 179.

Tinsley, B.M. & Spinrad, H. 1971, *Astrophys. Space. Sci.*, 12, 118.

Tolstoy, E. 1996, *Astrophys. J.*, 462, 684.

Tolstoy, E. 1999, in *The Stellar Content of the Local Group (IAU Symposium No. 192)*, ed. P.A. Whitelock and R. Cannon, San Francisco: Astron. Soc. Pac., in press.

Tolstoy, E., Gallagher, J.S., Cole, A.A., Hoessel, J.G., Saha, A., Dohm-Palmer, R.C., Skillman, E.D., Mateo, M. & Hurley-Keller, D. 1998, *Astron. J.*, 116, 1244.

Tolstoy, E., Saha, A., Hoessel, J.G. & Danielson, G.E. 1995, *Astron. J.*, 109, 579.

Tomaney, A.B. & Shafter, A.W. 1993, *Astrophys. J.*, 411, 640.

Toomre, A. 1977, in the *Evolution of Galaxies and Stellar Populations*, ed. B.M. Tinsley and R.B. Larson, New Haven: Yale, p. 401.

Tosi, M., Greggio, L., Marconi, G. & Focardi, P. 1991, *Astron. J.*, 102, 951.

Tosi, M., Pulone, L., Marconi, G. & Bragaglia, A. 1998, *Mon. Not. R. Astron. Soc.*, 299, 834.

Tremaine, S. 1995, *Astron. J.*, 110, 628.

Tremaine, S.D., Ostriker, J.P. & Spitzer, L. 1975, *Astrophys. J.*, 196, 407.

Trentham, N. 1998a, *Mon. Not. R. Astron. Soc.*, 294, 193.

Trentham, N. 1998b, *Mon. Not. R. Astron. Soc.*, 295, 360.

Trinchieri, G. & Fabbiano, G. 1991, *Astrophys. J.*, 382, 82.

Trinchieri, G., Fabbiano, G. & Peres, G. 1988, *Astrophys. J.*, 325, 531.

Tripicco, M.J., Bell, R.A., Dorman, B. & Hufnagel, B. 1995, *Astron. J.*, 109, 1692.

Tsujimoto, T., Nomoto, K., Yoshii, Y., Hashimoto, M., Yanagida, S. & Thielemann, F.-K. 1995, *Mon. Not. R. Astron. Soc.*, 277, 945.

Tufte, S.L., Reynolds, R.J. & Haffner, L.M. 1999, in *Stromlo Workshop on High Velocity Clouds*, (*ASP Conference Series No. 166*), ed. B.K. Gibson and M.E. Putman, San Francisco: Astron. Soc. Pac., p. 231.

Twarog, B.A., Anthony-Twarog, B.J. & Bricker, A.R. 1999, astro-ph/9901251.

Twarog, B.A., Ashman, K.M. & Anthony-Twarog, B.J. 1997, *Astron. J.*, 114, 2556.

Udalski, A., Szymański, M., Kubiak, M., Pietrzyński, G., Woźniak, P. & Żebruń, K. 1998, *Acta Astron.* 48, 1.

Unavene, M., Wyse, R.F.G. & Gilmore, G. 1996, *Mon. Not. R. Astron. Soc.*, 278, 727.

Valtonen, M.J., Byrd, G.G., McCall, M.L. & Innanen, K.A. 1993, *Astron. J.*, 105, 886.

van Agt, S.L.T. 1978, *Publ. David Dunlop Obs.*, 3, 205.

van de Hulst, H.C., Raimoud, E. & van Woerden, H. 1957, *Bull. Astron. Inst. Netherlands*, 14, 1.

van de Rydt, F., Demers, S. & Kunkel, W.E. 1991, *Astron. J.*, 102, 130.

van de Rydt, F., Grebel, E.K., Roberts, W.J. &Geisler, D. 1994, in *Dwarf Galaxies*, ed. G. Meylan and P. Prugniel, Garching: ESO, p. 245.

van den Bergh, S. 1957, *Z. Astrophys.*, 43, 236.

van den Bergh, S. 1958, *Astron. J.*, 63, 492.

van den Bergh, S. 1964, *Astrophys. J. Suppl.*, 9, 65.

van den Bergh, S. 1966a, *Astron. J.*, 71, 922.

van den Bergh, S. 1966b, *Astron. J.*, 71, 990.

van den Bergh, S. 1968a, *The Galaxies of the Local Group* (*David Dunlap Observatory Publication No. 195*), Toronto: David Dunlap Observatory.

van den Bergh, S. 1968b, *Observatory*, 88, 168.

van den Bergh, S. 1968c, *Astrophys. Lett.*, 2, 71.

van den Bergh, S. 1969, *Astrophys. J. Suppl.*, 19, 145.

van den Bergh, S. 1971, *Nature*, 231, 35.

van den Bergh, S. 1972a, *Astrophys. J.*, 171, L31.

van den Bergh, S. 1972b, *Astrophys. J.*, 178, L99.

van den Bergh, S. 1974a, *Astrophys. J.*, 193, 63.

van den Bergh, S. 1974b, *Astrophys. J.*, 191, 271.

van den Bergh, S. 1975, *Annu. Rev. Astron. Astrophys.*, 13, 180.

van den Bergh, S. 1976, *Astrophys. J.*, 203, 764.

van den Bergh, S. 1979, *Astrophys. J.*, 230, 95.

van den Bergh, S. 1981a, *Publ. Astron. Soc. Pac.*, 93, 428.

van den Bergh, S. 1981b, *Astron. Astrophys. Suppl.*, 46, 79.

van den Bergh, S. 1981c, *Astron. J.*, 86, 1464.

van den Bergh, S. 1986, *Astron. J.*, 91, 271.

van den Bergh, S. 1987, *Nature*, 328, 768.

van den Bergh, S. 1988a, in *Supernova Shells and Their Birth Events*, ed. W. Kundt, Berlin: Springer-Verlag, p. 44.

van den Bergh, S. 1988b, *Publ. Astron. Soc. Pac.*, 100, 1486.

van den Bergh, S. 1988c, *Astrophys. J.*, 327, 156.

van den Bergh, S. 1990, *Astron. J.*, 99, 843.

van den Bergh, S. 1991a, *Publ. Astron. Soc. Pac.*, 103, 609.

van den Bergh, S. 1991b, *Publ. Astron. Soc. Pac.*, 103, 1053.

van den Bergh, S. 1991c, *Physics Rep.*, 204, 385.

van den Bergh, S. 1991d, *Astrophys. J.*, 369, 1.

van den Bergh, S. 1993, *Astrophys. J.*, 411, 178.

van den Bergh, S. 1994a, in *The Local Group*, ed. A. Layden, R.C. Smith, and J. Storm, Garching: ESO, p. 3.

van den Bergh, S. 1994b, *Astron. J.*, 107, 1328.

van den Bergh, S. 1995a, *Astrophys. J.*, 446, 39.

van den Bergh, S. 1995b, in *The Local Group: Comparative and Global Properties* (*ESO Workshop No. 51*), ed. A. Layden, R.C. Smith, and J. Strom, Garching: ESO, p. 3.

van den Bergh, S. 1996a, *Observatory*, 115, 103.

van den Bergh, S. 1996b, *Astrophys. J.*, 471, L31.

van den Bergh, S. 1997a, *Astron. J.*, 113, 197.

van den Bergh, S. 1997b, *Astron. J.*, 113, 2054.

van den Bergh, S. 1998a, *Astrophys. J.*, 495, L79.

van den Bergh, S. 1998b, *Astrophys. J.*, 505, L127.

van den Bergh, S. 1998c, *Astron. J.*, 116,1688.

van den Bergh, S. 1998d, *Astrophys. J.*, 507, L39.

van den Bergh, S. 1999, in *Stellar Content of the Local Group (IAU Symposium No. 192)*, ed. P. Whitelock and R.D. Cannon, San Francisco: Astron. Soc. Pac., p. 3.

van den Bergh, S. & de Boer, K.S., ed., 1984, *Structure and Evolution of the Magellanic Clouds (IAU Symposium No. 108)*, Dordrecht: Reidel.

van den Bergh, S. & Dufour, R.J. 1980, *Publ. Astron. Soc. Pac.*, 92, 32.

van den Bergh, S. & Hagen, G.L. 1968, *Astron. J.*, 73, 569.

van den Bergh, S. & Henry, R.C. 1962, *Publ. David Dunlap Obs.*, 2, 281.

van den Bergh, S., Herbst, E. & Kowal, C.T. 1975, *Astrophys. J. Suppl.*, 29, 303.

van den Bergh, S. & Lafontaine, A. 1984, *Astron. J.*, 89, 1822.

van den Bergh, S. & McClure, R.D. 1980, *Astron. Astrophys.*, 80, 360.

van den Bergh, S. & Morbey, C.L. 1984a, *Astrophys. J.*, 283, 598.

van den Bergh, S. & Morbey, C.L. 1984b, *Astron. Expr.*, 1, 1.

van den Bergh, S. & Pritchet, C.J. 1992, in *The Stellar Populations of Galaxies (IAU Symposium No. 149)*, ed. B. Barbuy and A. Renzini, Dordrecht: Reidel, p. 161.

van den Bergh, S. & Racine, R. 1981, *Publ. Astron. Soc. Pac.*, 93, 35.

van den Bergh, S. & Tammann, G.A. 1991, *Annu. Rev. Astron. Astrophys.*, 29, 363.

van der Hulst, J.M. 1997, in *High-Sensitivity Radio Astronomy*, ed. N. Jackson and R.J. Davis, Cambridge: Cambridge University Press, p. 106.

van der Kruit, P.C. 1986, *Astron. Astrophys.*, 157, 230.

van der Kruit, P.C. 1989, in *The Milky Way as a Galaxy*, ed. G. Gilmore, I.R. King, and P.C. van der Kruit, Sauverny: Geneva Observatory, p. 331.

van der Marel, R.P., Cretton, N., de Zeeuw, P.T. & Rix, H.-W. 1998, *Astrophys. J.*, 493, 613.

van der Marel, R.P., de Zeeuw, P.T. & Rix, H.-W. 1997, *Astrophys. J.*, 488, 119.

van Dokkum, P.G. & Franx, M. 1995, *Astron. J.*, 110, 2027.

van Driel, W., Kraan-Korteweg, R.C., Binggeli, B. & Huchtmeier, W.K. 1998, *Astron. Astrophys. Suppl.*, 127, 397.

van Dyk, S.D., Puche, D. & Wong, T. 1998, *Astron. J.*, 116, 2341.

van Leeuwen, F., Feast, M.W., Whitelock, P.A. & Yudin, B. 1997, *Mon. Not. R. Astron. Soc.*, 287, 955.

Velázquez, H. & White, S.D.M. 1995, *Mon. Not. R. Astron. Soc.*, 275, L23.

Venn, K.A. 1999, astro-ph/9901306.

Venn, K.A., McCarthy, J.K., Lennon, D.J. & Kudritzki, R.P. 1998, in *Abundance Profiles: Diagnostic Tools for Galaxy History (ASP Conference Series Vol. 147)*, ed. D. Friedli et al., San Francisco: Astron. Soc. Pac., p. 54.

Vesperini, E. 1998, *Mon. Not. R. Astron. Soc.*, 299, 1019.

Vetešnik, M. 1962, *Bull. Astron. Inst. Czechoslovakia*, 13, 180.

Vílchez, J.M., Pagel, B.E.J., Díaz, A.I., Terlevich, E. & Edmunds, M.G. 1988, *Mon. Not. R. Astron. Soc.*, 235, 633.

Visvanathan, N. 1989, *Astrophys. J.*, 346, 629.

Vogt, S.S., Mateo, M., Olszewski, E.W. & Keane, M.J. 1995, *Astron. J.*, 109, 151.

Volders, L. & Högbom, J.A. 1961, *Bull. Astron. Inst. Netherlands*, 15, 307.

von Hippel, T. 1998, *Astron. J.*, 115, 1536.

Voors, R.H.M., Walters, L.B.F.M., Morris, P.W., Trams, N.R., de Koter, A. & Bouwmans, J. 1999, *Astron. Astrophys.*, 341, L67.

Wakker, B., Howk, J.C., Chu, Y.-H., Bomans, D. & Points, S.D. 1998, *Astrophys. J.*, 499, L87.

Wakker, B.P. & van Woerden, H. 1997, *Annu. Rev. Astron. Astrophys.*, 35, 217.

Wakker, B.P., van Woerden, H. & Gibson, B.K. 1999, in *Stromlo Workshop on High-Velocity Clouds (ASP Conference Series Vol. 165)*, ed. B.K. Gibson and M.E. Putman, San Francisco: Astron. Soc. Pac., in press.

Walborn, N.R., Barbá, R.H., Brandner, W. & Rubio, M. 1999, *Astron. J.*, 117, 225.

Walborn, N.R. & Blades, J.C. 1997, *Astron. J.*, 112, 457.

Walborn, N.R., Prévot, L., Wamsteker, W., González, R., Gilmozzi, R. & Fitzpatrick, E.L. 1989, *Astron. Astrophys.*, 219, 229.

Walker, A.R. 1992, *Astrophys. J.*, 390, L81.

Walker, A.R. 1999, in *Post-Hipparcos Cosmic Candles*, ed. F. Caputo and A. Heck, Dordrecht: Kluwer, p. 125.

Walker, A.R. & Mack, P. 1988, *Astron. J.*, 96, 871.

Walker, M.F. 1962, *Astrophys. J.*, 136, 695.

Walker, M.F. 1964, *Astron. J.*, 69, 744.

Walker, M.F., Blanco, V.M. & Kunkel, W.E. 1969, *Astron. J.*, 74, 44.

Walsh, J.-R., Dudziak, G., Minniti, D. & Zijlstra, A.A. 1997, *Astrophys. J.*, 487, 651.

Walterbos, R.A.M. & Braun, R. 1994, *Astrophys. J.*, 431, 156.

Walterbos, R.A.M. & Kennicutt, R.C. 1987, *Astron. Astrophys. Suppl.*, 69, 311.

Walterbos, R.A.M. & Kennicutt, R.C. 1988, *Astron. Astrophys.*, 198, 61.

Wang, L., Höflich, P. & Wheeler, J.C. 1997, *Astrophys. J.*, 483, L29.

Wang, Q., Hamilton, T., Helfand, D.J. & Wu, X. 1991, *Astrophys. J.*, 374, 475.

Wang, Q.D. 1999, in *New Views of the Magellanic Clouds (IAU Symposium No. 190)*, ed. Y.-H. Chu et al., San Francisco: Astron. Soc. Pac., in press.

Wang, Q.D. & Gotthelf, E.V. 1998, *Astrophys. J.*, 494, 623.

Welch, D.L. 1998, *IAU Circular No. 6802.*

Welch, D.L. et al. 1996, *IAU Circular No. 6434.*

Welch, D.L. et al. 1997, in *Proceedings of the 12th IAP Colloquium "Variable Stars and the Astrophysical Returns of Microlensing Surveys,"* ed. R. Ferlet, J.-P. Milliard, and B. Raban, Paris: Editions Frontières, p. 205.

Welch, D.L., McLaren, R.A., Madore, B.F. & McAlary, C.W. 1987, *Astrophys. J.*, 321, 162.

Welch, G.A., Sage, L.J. & Mitchell, G.F. 1998, *Astrophys. J.*, 499, L209.

Welty, D.E., Frisch, P.S., Sonneborn, G. & York, D.G. 1999, *Astrophys. J.*, 512, 636.

Westerhout, G. 1954, *Bull. Astron. Inst. Netherlands*, 13, 201.

Westerlund, B.E. 1997, *The Magellanic Clouds*, Cambridge: Cambridge Univ. Press.

Westerlund, B.E., Azzopardi, M., Breysacher, J. & Rebeirot, E. 1991, *Astron. Astrophys. Suppl.*, 91, 425.

Whiting, A.B., Irwin, M.J. & Hau, G.K.T. 1997, *Astron. J.*, 114, 996.

Whitelock, P.A. & Cannon, R.D., ed. 1999, *Stellar Content of the Local Group (IAU Symposium No. 192)*, San Francisco: Astron. Soc. Pac.

Whitelock, P.A., Irwin, M. & Catchpole, R.M. 1996, *New Astronomy*, 1, 57.

Widmann, H. et al. 1998, *Astron. Astrophys.*, 338, L1.

Wiklind, T. & Rydbeck, G. 1986, *Astron. Astrophys.*, 164, L22.

Wilcots, E.M. & Miller, B.W. 1998, *Astron. J.*, 116, 2363.

Wilhelm, R. 1996, *Publ. Astron. Soc. Pac.*, 108, 726.

Williams, R.M., Chu, Y.-H., Dickel, J.R., Beyer, R., Pete, R., Smith, R.C. & Milne, D.K. 1997, *Astrophys. J.*, 480, 618.

Wilson, A.G. 1955, *Publ. Astron. Soc. Pac.*, 67, 27.

Wilson, C.D. 1990, Cal. Tech. Ph.D. thesis.

Wilson, C.D. 1991, *Astron. J.*, 104, 1374.

Wilson, C.D. 1994, *Astrophys. J.*, 434, L11.

Wilson, C.D., Freedman, W.L. & Madore, B.F. 1990, *Astron. J.*, 99, 149.

Wilson, C.D. & Matthews, B.C. 1995, *Astrophys. J.*, 455, 125.

Wilson, C.D. & Rudolph, A.L. 1993, *Astrophys. J.*, 406, 477.

Wilson, C.D. & Scoville, N. 1989, *Astrophys. J.*, 347, 743.

Wilson, C.D., Scoville, N., Freedman, W.L., Madore, B.F. & Sanders, D.B. 1988, *Astrophys. J.*, 333, 611.

Wilson, C.D., Scoville, N. & Rice, W. 1991, *Astron. J.*, 101, 1293.

Wilson, C.D., Welch, D.L., Reid, I.N., Saha, A. & Hoessel, J. 1996, *Astron. J.*, 111, 1106.

Wirth, A. & Gallagher, J.S. 1984, *Astrophys. J.*, 282, 85.

Wolf, M. 1906, *Mon. Not. R. Astron. Soc.*, 67, 91.

Wolf, M. 1923, *Astron. Nachr.*, 217, 476.

Wood, P.R., Arnold, A.S. & Sebo, K.M. 1997, *Astrophys. J.*, 485, L25.

Woosley, S.E. 1988, in *Supernova 1987A in The Large Magellanic Cloud*, ed. M. Kafatos and A. Michalitsianos, Cambridge: Cambridge Univ. Press, p. 289.

Worthey, G. 1998, *Publ. Astron. Soc. Pac.*, 110, 888.

Wright, M.C.H. 1974, *Astron. Astrophys.*, 31, 317.
Wright, M.C.H. 1979, *Astrophys. J.*, 233, 35.
Wright, M.C.H., Warner, P.J. & Baldwin, J.E. 1972, *Mon. Not. R. Astron. Soc.*, 155, 337.
Wyder, T.K., Hodge, P.W. & Skelton, B.P. 1997, *Publ. Astron. Soc. Pac.*, 109, 927.
Wyse, R.F.G. & Gilmore, G. 1992, *Astron. J.*, 104, 144.
Xu, C. & Helon, G. 1994, *Astrophys. J.*, 426, 109.
Xu, J. & Crotts, A.P.S. 1998, astro-ph/9808274.
Xu, J., Crotts, A.P.S. & Kunkel, W.E. 1995, *Astrophys. J.*, 451, 806.
Yahil, A., Tammann, G.A. & Sandage, A. 1977, *Astrophys. J.*, 217, 903.
Yang, H., Chu, Y.-H., Skillman, E.D. & Terlevich, R. 1996, *Astron. J.*, 112, 146.
Yang, H. & Skillman, E.D. 1993, *Astron. J.*, 106, 1448.
Ye, T, 1998, *Mon. Not. R. Astron. Soc.*, 294, 422.
Ye, T., Amy, S.W., Wang, Q.D., Ball, L. & Dickel, J. 1995, *Mon. Not. R. Astron. Soc.*, 275, 1218.
Young, L.M. 1998, in *Star Formation in Early-type Galaxies*, ed. p. Carral and J. Cepa, in press.
Young, L.M. 1999, *Astron. J.*, 117, 1758.
Young, L.M. & Lo, K.Y. 1996a, *Astrophys. J.*, 462, 203.
Young, L.M. & Lo, K.Y. 1996b, *Astrophys. J.*, 464, L59.
Young, L.M. & Lo, K.Y. 1997a, *Astrophys. J.*, 476, 127.
Young, L.M. & Lo, K.Y. 1997b, *Astrophys. J.*, 490, 710.
Yungelson, L., Livio, M. & Tutukov, A. 1997, *Astrophys. J.*, 481, 127.
Yusef-Zadeh, F., Cotton, W.D. & Reynolds, SP. 1998, *Astrophys. J.*, 498, L55.
Zaritsky, D. 1999, in *the Galactic Halo* (*ASP Conf. Series No. 165*), ed. B.K. Gibson, T.S. Axelrod and M.E. Putman, San Francisco: Astron. Soc. Pac., p. 34.
Zaritsky, D., Elston, R. & Hill, J.M. 1989, *Astron. J.*, 97, 97.
Zaritsky, D., Harris, J. & Thompson, I., 1997, *Astron. J.*, 114, 2545.
Zaritsky, D. & Lin, D.N.C. 1997, *Astron. J.*, 114, 2545.
Zaritsky, D., Olszewski, E.W., Schommer, R.A., Peterson, R.C. & Aaronson, M. 1989, *Astrophys. J.*, 345, 759.
Zhao, H.-S. 1998, *Astrophys. J.*, 500, L149.
Ziegler, B.L. & Bender, R. 1998, *Astron. Astrophys.*, 330, 819.
Zijlstra, A.A., Dudziak, G. & Walsh, J.R. 1997, in *Planetary Nebulae* (*IAU Symposium No. 180*), ed. H.J. Habing and H.J.G.L.M. Lamers, Dordrecht: Kluwer, p. 479.
Zijlstra, A.A. & Minniti, D. 1999, *Astron. J.*, 117, 1743.
Zinn, R. 1978, *Astrophys. J.*, 225, 790.
Zinn, R. 1985, *Astrophys. J.*, 293, 424.
Zinn, R. 1993, in *The Globular Cluster–Galaxy Connection* (*ASP Conf. Series Vol. 48*), ed. G.H. Smith and J.P. Brodie, San Francisco: Astron. Soc. Pac., p. 302.
Zinnecker, H. 1994, in *Dwarf Galaxies*, ed. G. Meylan and P. Prugniel, Garching: ESO, p. 231.
Zubko, V.G. 1999, *Astrophys. J.*, 513, L29.
Zwaan, M.A., Biggs, F.H., Sprayberry, D. & Sorar, E. 1997, *Astrophys. J.*, 490, 173.
Zwicky, F. 1933, *Helv. Phys. Acta.*, 6, 110.
Zwicky, F. 1942, *Phys. Rev. II*, 61, 489.
Zwicky, F. 1957, *Morphological Astronomy*, Berlin: Springer-Verlag.

Object Index